LEARNING FROM
FUKUSHIMA

NUCLEAR POWER IN EAST ASIA

LEARNING FROM
FUKUSHIMA

NUCLEAR POWER IN EAST ASIA

EDITED BY PETER VAN NESS
AND MEL GURTOV

WITH CONTRIBUTIONS FROM ANDREW BLAKERS,
MELY CABALLERO-ANTHONY, GLORIA KUANG-JUNG HSU,
AMY KING, DOUG KOPLOW, ANDERS P. MØLLER,
TIMOTHY A. MOUSSEAU, M. V. RAMANA, LAUREN RICHARDSON,
KALMAN A. ROBERTSON, TILMAN A. RUFF, CHRISTINA STUART,
TATSUJIRO SUZUKI, AND JULIUS CESAR I. TRAJANO

Australian
National
University

PRESS

ANU PRESS

Published by ANU Press
The Australian National University
Acton ACT 2601, Australia
Email: anupress@anu.edu.au
This title is also available online at press.anu.edu.au

National Library of Australia Cataloguing-in-Publication entry

Title: Learning from Fukushima : nuclear power in East Asia /
             Peter Van Ness, Mel Gurtov, editors.

ISBN: 9781760461393 (paperback) 9781760461409 (ebook)

Subjects: Nuclear power plants--East Asia.
             Nuclear power plants--Risk assessment--East Asia.
             Nuclear power plants--Health aspects--East Asia.
             Nuclear power plants--East Asia--Evaluation.

Other Creators/Contributors:
             Van Ness, Peter, editor.
             Gurtov, Melvin, editor.

Cover design and layout by ANU Press.

Cover image: 'Fukushima apple tree' by Kristian Laemmle-Ruff. Near Fukushima City, 60 km from the Fukushima Daiichi Nuclear Power Plant, February 2014.

The number in the artwork is the radioactivity level measured in the orchard—2.166 microsieverts per hour, around 20 times normal background radiation.

Contents

Part I The state of the nuclear industry

Part II Country studies

Part III The real costs of going nuclear

Part IV A post-nuclear future

Figures

Tables

Acronyms and abbreviations

3S	safety, security, and safeguards
ABWR	advanced boiling water reactor
ADEME	Agence de l'environnement et de la maîtrise de l'énergie
AEC	Atomic Energy Council
AELB	Atomic Energy Licensing Board
ANU	The Australian National University
APP	atomic power plant
Areva NC	Areva Nuclear Cycle
Areva NP	Areva Nuclear Power
ASEAN	Association of Southeast Asian Nations
ASEANTOM	ASEAN Network of Regulatory Bodies on Atomic Energy
ASN	Autorité de sûreté nucléaire
AUM	assets under management
BAPETEN	Nuclear Energy Regulatory Agency of Indonesia
BATAN	National Nuclear Energy Agency of Indonesia
BEIR	Biological Effects of Ionizing Radiation
Bq	becquerel
Bq/m^3	becquerel per cubic metre
BWR	boiling water reactor
CAC40	Commissariat aux Comptes 40
CCS	carbon capture and storage
CEA	Commissariat à l'énergie atomique et aux énergies alternatives

CGNPC	China General Nuclear Power Corporation
CI	confidence intervals
CIA	Central Intelligence Agency
Cigéo	Centre industriel de stockage géologique
CNEA	China National Energy Administration
CNNC	China National Nuclear Corporation
CNRS	Centre National de la Recherche Scientifique
CO_2	carbon dioxide
COP	Conference of the Parties
CP	Contract Programme
CPV	concentrating photovoltaics
CSSI	Chung Shan Science Institute
CT	computed tomography
CWIP	construction work-in-progress
DPP	Democratic Progressive Party
EDF	Électricité de France
EIA	Environmental Impact Assessment
EPR	European pressurised reactor
ESBWR	economic simplified boiling water reactor
EU	European Union
FBR	fast breeder reactor
FEPC	Federation of Electric Power Companies
GCR	gas-cooled, graphite-moderated reactor
GDP	gross domestic product
GE	General Electric
GW	gigawatt
GWe	gigawatt electric
GWh	gigawatt hour
Gy	gray
HLW	high-level radioactive waste
HTGR	high-temperature gas-cooled reactor
HVDC	high voltage direct current

HWGCR	heavy water–moderated, gas-cooled reactor
IAEA	International Atomic Energy Agency
ICRP	International Commission on Radiological Protection
IEA	International Energy Agency
IMF	International Monetary Fund
INER	Institute of Nuclear Energy Research
INES	International Nuclear and Radiological Event Scale
IU	Indiana University
JAEA	Japan Atomic Energy Agency
JAEC	Japan Atomic Energy Commission
kBq/m^2	kilobecquerel per square metre
KEPCO	Korea Electric Power Company
KFEM	Korean Federation for Environmental Movements
KHNP	Korea Hydro & Nuclear Power
KMT	Kuomintang
kV	kilovolt
kW	kilowatt
kWh	kilowatt hour
kW/m^2	kilowatt per square metre
LCOE	levelised cost of energy
LNT	linear no-threshold
LWR	light-water reactor
METI	Ministry of Economy, Trade and Industry
mGy	milliGray
MNPC	Malaysia Nuclear Power Cooperation
mSv	milliSievert
MW	megawatt
MWe	megawatt electric
MWh	megawatt hour
NATO	North Atlantic Treaty Organization
NDRC	National Development and Reform Commission

NEA	Nuclear Energy Agency
NEC-SSN	Nuclear Energy Cooperation Sub-Sector Network
NEPIO	Nuclear Energy Program Implementing Organisation
NGO	non-governmental organisation
NPP	nuclear power plant
NRC	Nuclear Regulatory Commission
NSSC	Nuclear Safety and Security Commission
NTS	Non-Traditional Security
OAP	Office of Atoms for Peace
OECD	Organisation for Economic Co-operation and Development
PHES	pumped hydro energy storage
PHWR	pressurised heavy-water reactor
PRIS	Power Reactor Information System
Pu	plutonium
PV	photovoltaics
PWR	pressurised water reactor
RECNA	Research Center for Nuclear Weapons Abolition
RSIS	S. Rajaratnam School of International Studies
SCGR	sodium-cooled, graphite-moderated reactor
SNPTC	State Nuclear Power Technology Corporation
SOE	state-owned enterprise
Sv	sievert
TECRO	Taipei Economic and Cultural Representative Office
TEPCO	Tokyo Electric Power Company
TFU	Tohoku Fukushi University
tHM	tonnes of heavy metal
TSE	Total Support Estimate
TWh	terawatt hour
UAE	United Arab Emirates
UK	United Kingdom

UN	United Nations
UNGG	Uranium Naturel Graphite Gaz
UNSCEAR	United Nations Scientific Committee on the Effects of Atomic Radiation
US	United States
VARANS	Vietnam Agency for Radiation and Nuclear Safety
VINATOM	Vietnam Atomic Energy Institute
VVER	water–water energetic reactor
W	watts
WHO	World Health Organization
WIPP	Waste Isolation Pilot Plant
μSv/h	microSievert per hour

Preface

Work for this book began shortly after the earthquake, tsunami, and nuclear meltdown in March 2011 in Fukushima, when we asked Japanese friends and colleagues in the Tohoku region what we could do to help the victims. Many replied that they simply did not believe the information that the government, the company (Tokyo Electric Power Company, TEPCO), much of the press, or even some of the academic specialists were telling them about the implications of the disaster.

We decided to put together an international workshop, in collaboration with Tohoku Fukushi University (TFU) in Sendai, on the topic of 'Nuclear Disaster Response: The Need to Know', and we searched for the most knowledgeable people we could find, in Japan and abroad, to try to answer the questions raised. Koki Hagino, President of TFU, generously hosted the workshop, and Professor Norifumi Namatame worked with us to organise two days of meetings, and then a public presentation of some of the results, in Japanese, to some 400 people. Two of our colleagues, Richard Tanter and Rikki Kersten, made presentations in Japanese, and other workshop participants spoke via translated filmed interviews. In the end, we published a special issue of *Asian Perspective* 37(4) 2013, with papers from that meeting edited by Professor Namatame.

Yet the deeper we got into the global debate about nuclear power, the more concerned we became about the quality of the discussion. We were surprised to find the amount of misinformation and even disinformation that sometimes characterised debates about nuclear power. At best, in an organised debate, proponents and opponents typically would talk past each other, one side focusing on some aspects, and the other side emphasising different aspects. Moreover, while neither Australia nor any of the 10 member-countries of the Association of Southeast Asian

Nations (ASEAN), who are our closest neighbours, had built nuclear power plants, several were very interested, and some had already made plans for nuclear power.

This prompted us to organise a second international workshop in 2014, this time at The Australian National University (ANU), on the topic 'Nuclear Power in East Asia: The Costs and Benefits'. We identified nine key aspects of any decision to build a nuclear power plant (e.g. costs of construction, regulation, liability in the event of accident, decommissioning, disposal of nuclear waste, and the relationship of nuclear power to climate change), and we looked for experts on these particular aspects, no matter whether they were publicly committed to supporting or opposing nuclear power.

Participants came from the United States, Japan, Singapore, Taiwan, and Australia. We met for two days at the China in the World building at ANU, and then made a presentation of results for the public on the third day. We also published a second special issue of *Asian Perspective* 39(4) 2015, with papers from that workshop this time edited by Tilman Ruff.

This book presents the results of our work since Fukushima, the findings from our workshops, and the insights of additional contributions from other colleagues who have joined us in the project. It is our best effort to assess the role of nuclear power in East Asia.

Our thanks go first and foremost to the authors of the 11 chapters, and to Mary-Louise Hickey, who has so carefully copy-edited their work. Thanks also to the other scholars and students who participated in the workshops and helped out in the arrangements. Koki Hagino and Norifumi Namatame were wonderful hosts in Sendai at TFU, and at ANU we owe thanks to the Department of International Relations, the Japan Institute, the ANU–IU (Indiana University) Pan Asia Institute, and to China in the World for the use of their excellent facilities. The Global Nuclear Power Database, from *The World Nuclear Industry Status Report* and the *Bulletin of the Atomic Scientists*, has provided a comprehensive empirical foundation for our study, and we are especially grateful to Julie Hazemann and Mycle Schneider, who have been so helpful for our work.

Thanks to all.
Canberra
September 2017

Contributors

Andrew Blakers is Professor of Engineering at The Australian National University (ANU), Canberra. He was a Humboldt Fellow and has held Australian Research Council Queen Elizabeth II and Senior Research Fellowships. He is a Fellow of the Academy of Technological Sciences and Engineering, the Institute of Energy, and the Institute of Physics. He has published approximately 300 papers and patents. His research interests are in the areas of photovoltaic and solar energy systems, particularly advanced thin film silicon solar cell technology and solar concentrator solar cells, components, and systems. He also has an interest in sustainable energy policy, and is engaged in detailed analysis of energy systems with high (50–100 per cent) penetration by wind and photovoltaics.
Email: andrew.blakers@anu.edu.au

Mely Caballero-Anthony is Associate Professor and Head of the Centre for Non-Traditional Security (NTS) Studies at the S. Rajaratnam School of International Studies (RSIS), Nanyang Technological University, Singapore. She previously served as the Director of External Relations at the Association of Southeast Asian Nations (ASEAN) Secretariat, and currently serves on the UN Secretary-General's Advisory Board on Disarmament Matters and Security. She is also Secretary-General of the Consortium of Non-Traditional Security Studies in Asia (NTS-Asia) and is a member of the World Economic Forum Global Agenda Council on Conflict Prevention. Dr Caballero-Anthony's research interests include regionalism and regional security in the Asia-Pacific, multilateral security cooperation, politics and international relations in ASEAN, conflict prevention and management, as well as human security. She has published extensively in peer-reviewed journals on a broad range of security issues in the Asia-Pacific.
Email: ISMCAnthony@ntu.edu.sg

Mel Gurtov is Professor Emeritus of Political Science at Portland State University, Oregon, and Senior Editor of *Asian Perspective*. He previously served on the staff of the RAND Corporation in Santa Monica, California (1966–71), and at the University of California, Riverside (1971–86). He has published over 20 books and numerous articles on East Asian affairs, US foreign policy, and global politics from a human interest perspective. His most recent books are *Will This Be China's Century? A Skeptic's View*, Lynne Rienner Publishers, 2013; *Global Politics in the Human Interest*, Lynne Rienner Publishers, 2007; and *Superpower on Crusade: The Bush Doctrine in US Foreign Policy*, Lynne Rienner Publishers, 2006. His blog on foreign affairs, 'In the Human Interest', can be found at melgurtov.com. Email: mgurtov@aol.com

Gloria Kuang-Jung Hsu is Professor in the Department of Atmospheric Sciences, National Taiwan University. She received a Bachelor of Science from National Taiwan University, and a PhD in Chemistry from the University of Pittsburgh in the US. She also holds a Master's in Public Administration from the Kennedy School of Government at Harvard University. Her interests include ozone chemistry and air pollution, and environment and energy policies. She has published articles in *Science of the Total Environment*, *Journal of Atmospheric Chemistry*, *Tellus*, *Atmospheric Environment*, and *Carbon Economy Monthly* (in Chinese). Over the years, she has served as Vice Chair and Chair of the Taiwan Environmental Protection Union, and she is now serving as Chair of the newly founded Mom Loves Taiwan Association. She has been a Commissioner of the Environmental Impact Assessment Commission at both the national and local levels, and a Commissioner of the National Sustainable Development Committee. She is currently an advisor to the Executive Yuan. Email: kjhsu@ntu.edu.tw

Amy King is Senior Lecturer in the Strategic and Defence Studies Centre at The Australian National University, specialising in Chinese foreign and security policy, China–Japan relations, and the international relations and security of the Asia-Pacific region. She is concurrently an Australian Research Council Discovery Early Career Researcher Award (DECRA) Fellow and a Westpac Research Fellow, and is engaged in a three-year research project examining China's role in shaping the international economic order. Amy is the author of *China–Japan Relations after World War Two: Empire, Industry and War, 1949–1971*, Cambridge University

Press, 2016. She received her DPhil in International Relations and MPhil in Modern Chinese Studies from the University of Oxford, where she studied as a Rhodes Scholar.

Email: amy.king@anu.edu.au

Doug Koplow is the founder of Earth Track in Cambridge, MA (www.earthtrack.net), focused on making the scope and cost of environmentally harmful subsidies more visible, and identifying reform strategies. His work on natural resource subsidies spans more than 25 years, and has included detailed reviews of commonly employed subsidy valuation approaches. His most recent focus has been on the distortionary effects of subsidies to fossil fuels and nuclear power. Doug has advised a variety of governmental agencies, environmental groups, foundations, and trade associations on resource subsidy measurement and reform, and his work is regularly cited across the political spectrum. He holds an MBA from the Harvard Business School and a BA in economics from Wesleyan University.

Email: dkoplow@earthtrack.net

Anders P. Møller is an evolutionary biologist employed by the Centre National de la Recherche Scientifique (CNRS), France, working since 1991 on the ecological and evolutionary implications of low-dose radiation. He is a highly cited biologist and has been involved in research concerning the consequences of the Chernobyl disaster since 1991 and the Fukushima disaster since July 2011, with more than 800 peer-reviewed scientific papers (more than 100 related to radiation effects) and more than 60,000 career citations. He co-directs the Chernobyl + Fukushima Research Initiative along with his research partner, Timothy A. Mousseau. His research has been featured in the *New York Times*, *The Economist*, CBS TV's *60 Minutes*, *Scientific American*, the BBC, CNN, PBS, and many other media outlets. He holds a PhD from the University of Århus, Denmark.

Email: anders.moller@u-psud.fr

Timothy A. Mousseau is a Professor of Biological Sciences at the University of South Carolina and the Co-Director of the Chernobyl + Fukushima Research Initiative. Past positions include Dean of the Graduate School and Associate Vice President for Research at the University of South Carolina, and Program Officer for Population Biology at the US National Science Foundation. He is a leading authority concerning the impacts of radioactive fallout from the Chernobyl and Fukushima

disasters on natural populations of animals, plants, and microbes. He has written or edited 11 books and published more than 200 scientific papers. His research has been featured in the *New York Times*, *The Economist*, CBS TV's *60 Minutes*, *Scientific American*, the BBC, CNN, PBS, and many other media outlets. He is a member of the International Union of Radioecology, the American Nuclear Society, and the New York Academy of Sciences. He is a Fellow of the American Association for the Advancement of Science, the American Council of Learned Societies, and the Explorers Club, and he has served on US National Academy of Sciences review panels on the risks and hazards of radioactive emissions for nuclear power plants. He holds a PhD from McGill University.
Email: MOUSSEAU@mailbox.sc.edu

M. V. Ramana is the Simons Chair in Disarmament, Global and Human Security with the Liu Institute for Global Issues at the University of British Columbia, Vancouver, Canada. The research for his chapter was conducted when he was affiliated with the Program on Science and Global Security at Princeton University, where he worked until 2016. He is the author of *The Power of Promise: Examining Nuclear Energy in India*, Penguin Books, 2012, and co-editor of *Prisoners of the Nuclear Dream*, Orient Longman, 2003. He is a member of the International Panel on Fissile Materials, the Global Council of Abolition 2000, and the National Coordinating Committee of India's Coalition for Nuclear Disarmament and Peace. He is the recipient of a Guggenheim Fellowship and a Leo Szilard Award from the American Physical Society, and has been selected as a Distinguished Lecturer by Sigma Xi for 2016–17.
Email: m.v.ramana@ubc.ca

Lauren Richardson is Teaching Fellow in Japanese–Korean Relations and Politics at the University of Edinburgh, and incoming Lecturer and Director of Studies at the Asia-Pacific College of Diplomacy at The Australian National University (ANU). Her research is focused on the role of non-state actors in shaping policy and diplomatic transitions in Northeast Asia. Lauren is currently completing a book manuscript entitled, 'Reshaping Japan–Korea Relations: Transnational Advocacy Networks and the Politics of Redress'. She holds a PhD from ANU and Master's degrees from Keio and Monash Universities.
Email: lauren.richardson@ed.ac.uk

Kalman A. Robertson is a Stanton Nuclear Security Postdoctoral Fellow in the International Security Program and the Project on Managing the Atom at the Belfer Center for Science and International Affairs of Harvard University. His research focuses on nuclear safeguards and nuclear cooperation agreements. He holds a PhD in International, Political, and Strategic Studies from the Strategic and Defence Studies Centre of The Australian National University (ANU), where he also received the University Medal for Physics and First Class Honours in Law. Prior to his arrival in Cambridge, he worked as a lecturer in the School of Politics and International Relations at the ANU, where he convened the course 'Politics of Nuclear Weapons'. He was Australia's representative to the Nuclear Energy Experts Group of the Council for Security Cooperation in the Asia Pacific from 2012 to 2016.

Email: kalman.robertson@anu.edu.au

Tilman A. Ruff, AM, is a public health and infectious diseases physician, Associate Professor in the Nossal Institute for Global Health, University of Melbourne, and medical advisor to the International Program of Australian Red Cross. Since 2012, he has served as the first Australian Co-President of International Physicians for the Prevention of Nuclear War (Nobel Peace Prize 1985). He was founding Australian and international Chair of the International Campaign to Abolish Nuclear Weapons. He teaches on the public health dimensions of nuclear technology in the University of Melbourne's graduate medicine program, five Master's subjects and two undergraduate breadth subjects. He is past national president of the Medical Association for Prevention of War (Australia). In 2008, he was the first civil society representative on an Australian government Nuclear Non-Proliferation Treaty conference delegation, and one of two civil society advisors to the International Commission on Nuclear Non-proliferation and Disarmament. In 2012, Tilman was appointed a Member of the Order of Australia 'for service to the promotion of peace as an advocate for the abolition of nuclear weapons, and to public health through the promotion of immunisation programs in the South-East Asia-Pacific region'.

Email: tar@unimelb.edu.au

Christina Stuart is an energy-climate expert at Carbone4, a leading energy consulting firm specialising in low-carbon strategies, based in Paris. She is a Physics graduate from the Sorbonne University Pierre-et-Marie-Curie in Paris and holds a Bachelor's degree in Political Science from the Paris School of International Affairs at Sciences Po Paris. Her Master's degree from both institutions is in Environmental Sciences and Policy, specialising in sustainable energy. Christina's academic and research focus resides in the energy–climate nexus. Her latest publication is 'Energy and Climate Adaptation in Developing Countries', German Corporation for International Cooperation, 2017. Since 2014, she has been a member of the European Union's Climate–Knowledge Innovation Community. Previously, Christina worked in environmental consulting at Ernst & Young and in renewable energy development at the Australian Renewable Energy Agency.

Email: christina.stuart@sciencespo.fr

Tatsujiro Suzuki is Director and Professor of the Research Center for Nuclear Weapons Abolition (RECNA), Nagasaki University, Japan. Before joining RECNA, he was a Vice Chairman of the Japan Atomic Energy Commission of the Cabinet office from January 2010 to April 2014. Before that he was an Associate Vice President of the Central Research Institute of Electric Power Industry in Japan (1996–2009); a Visiting Professor at the Graduate School of Public Policy, University of Tokyo (2005–09); an Associate Director of MIT's International Program on Enhanced Nuclear Power Safety (1988–93); and a Research Associate at MIT's Center for International Studies (1993–95), where he co-authored a report on Japan's plutonium program. He is also a Council Member of Pugwash Conferences on Science and World Affairs (2007–09, and 2014–). Dr Suzuki has a PhD in nuclear engineering from Tokyo University (1988).

Email: suzukitatsu@nagasaki-u.ac.jp

Julius Cesar I. Trajano is Associate Research Fellow with the Centre for Non-Traditional Security (NTS) Studies at the S. Rajaratnam School of International Studies (RSIS), Nanyang Technological University, Singapore. He is a member-participant of the Nuclear Energy Experts Group and the Energy Security Study Group of the Council for Security Cooperation in the Asia-Pacific (CSCAP). He has published journal articles, reports, and op-ed commentaries on nuclear energy governance in the Asia-Pacific, regional energy security, South China Sea disputes, and humanitarian assistance and disaster relief. His current research focuses on regional cooperation on nuclear safety and security in Asia.

Email: isjtrajano@ntu.edu.sg

Peter Van Ness is a Visiting Fellow in the Department of International Relations at The Australian National University (ANU), and has served as convener for this project on nuclear power in East Asia. He is a specialist on Chinese foreign policy and the international relations of the Asia-Pacific. He has published books on Chinese support for revolution during the Maoist period, market reforms in socialist societies, the human rights debate in Asia, and Asian responses to the Bush Doctrine. His articles have been published in the *New York Times*, *Washington Post,* and *The Nation*, as well as in academic journals. For many years a member of the faculty at the Graduate School of International Studies, University of Denver, he has won grants from the Social Science Research Council and the American Council of Learned Societies, and two Fulbright fellowships to Japan. He has taught at four Japanese universities, including Keio University and the University of Tokyo, and has been a research fellow at ANU, the Center for Chinese Studies at the University of Michigan, the Woodrow Wilson International Center for Scholars in Washington, DC, and the Inter-University Program for Chinese Language Studies in Taipei. He holds a PhD from the University of California, Berkeley.

Email: peter.van-ness@anu.edu.au

Introduction: Nuclear energy in Asia

Mel Gurtov

The Fukushima nuclear disaster of March 2011 has raised serious questions about nuclear power. In our work since Fukushima, we have tried to answer two questions: What is the current status of nuclear energy in Asia? Does nuclear power have a future in East Asia? By answering those questions, we hope to contribute to the global debate about nuclear energy. To be sure, questions of such magnitude can rarely be answered with a simple 'yes' or 'no'. Decisions on energy are made at the national level, on the basis of both objective factors such as cost-effectiveness and notions of the national interest, and less objective ones, such as influence peddled by power plant operators, corruption, and bureaucratic self-interest. Nevertheless, by closely examining the status and probable future of nuclear power plants in specific countries, the authors come up with answers, albeit mostly of a negative nature.

At the start of 2017, 450 nuclear power reactors were operating in 30 countries, with 60 more under construction in 15 countries (Nuclear Energy Institute 2016).[1] Thirty-four reactors are under construction in Asia, including 21 in China (*Bulletin of the Atomic Scientists* 2017; see Figure I.1). The 'Fukushima effect' has clearly had an impact in Asia, however. In China, no new construction took place between 2011 and 2014, although since then there has been a slow increase of licences (*Bulletin of the Atomic Scientists* 2017). Nevertheless, the full story of China's embrace of nuclear power, as told here by M. V. Ramana and Amy King, is that the onset of a 'new normal' in economic growth aims

---

1   *Bulletin of the Atomic Scientists* (2017) reports 55 nuclear reactors under construction in 13 countries as of 1 January 2017.

and structural changes in the economy have led to a declining demand for electricity and the likelihood of far less interest in nuclear power than had once been predicted. On the other hand, in South Korea, which relies on nuclear power for about 31 per cent of its electricity, Lauren Richardson's chapter shows that the Fukushima disaster and strong civil society opposition have not deflected official support of nuclear power, not only for electricity but also for export.

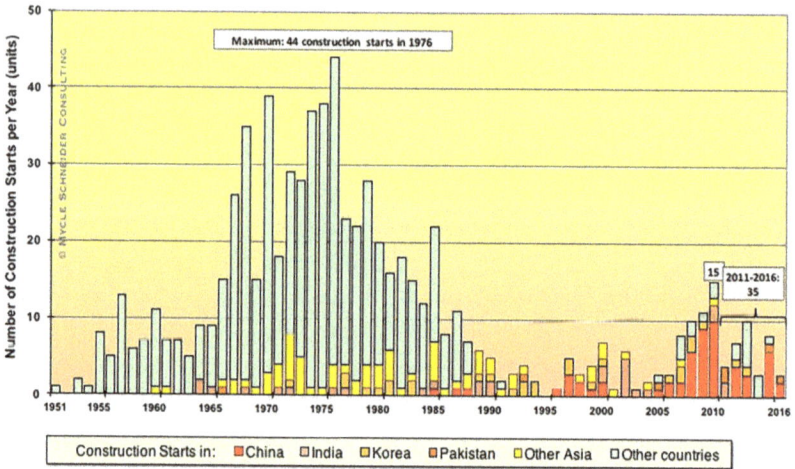

Figure I.1 Construction starts of nuclear reactors in the world by year, 1951–2016 (in units)

Source: Schneider et al. (2017). Reprinted with permission of the authors.

Meanwhile, the 10 countries that comprise the Association of Southeast Asian Nations (ASEAN) are divided about pursuing the nuclear-energy option, with Vietnam deciding to opt out in 2016, and Cambodia, Indonesia, Malaysia, and the Philippines at various stages of evaluation. Even so, the chapter by Mely Caballero-Anthony and Julius Cesar I. Trajano shows that only about 1 per cent of ASEAN's electricity will derive from nuclear power in 2035, whereas renewables will account for 22 per cent.

Factors in the declining attractiveness of the nuclear option

How viable nuclear power is finally judged to be will depend primarily on the decisions of governments, but increasingly also on civil society. ASEAN has established a normative framework that emphasises safety, waste disposal, and non-proliferation; and civil society everywhere is increasingly alert to the dangers and costs, above-board and hidden, of nuclear power plants. As Doug Koplow's chapter shows, for example, the nuclear industry, like fossil fuels, benefits from many kinds of government subsidies that distort the energy market against renewable energy sources. Costs are politically as well as environmentally consequential: even if construction begins on a nuclear power plant, it will be cancelled and construction abandoned in 12 per cent of all cases. It is important to note that of the 754 reactors constructed since 1951, 90 have been abandoned and 143 plants permanently shut down. When construction does proceed, it takes between five to 10 years on average for completion (338 of 609), with some 15 per cent taking more than 10 years (*Bulletin of the Atomic Scientists* 2017). And, in the end, old and abandoned reactors will have to be decommissioned, as Kalman A. Robertson discusses, with costs that may double over the next 15–20 years. As Robertson points out, the problem of safe disposal of radioactive waste and the health risk posed by radiation released during decommissioning should be factored into the total price that cleanup crews and taxpayers will eventually pay. On top of all that, there isn't much experience worldwide in decommissioning.

Then there is the issue of trust in those who make decisions. Tatsujiro Suzuki's chapter shows that in Japan, the chief legacy of Fukushima is public loss of trust in Japanese decision-makers and in the nuclear industry itself. Several years after the accident, costs continue to mount, a fact that pro-nuclear advocates elsewhere in Asia might want to consider.[2] They also need to consider the issue of transparency for, as Suzuki shows, the nuclear industry has consistently dodged the fairly obvious lessons of Fukushima with regard to costs, nuclear energy's future, and communication with the public. Similarly, in Taiwan, as Gloria Kuang-Jung Hsu's study

---

2   Six years after the Fukushima Daiichi accident, radiation readings at one of the three reactors being decommissioned were at their highest level. Estimated costs for decommissioning, decontamination, compensation to victims, and storage of radioactive waste now run over US$180 billion. See McCurry (2017).

shows, transparency about safety issues has been notoriously lacking, and a history of efforts to obfuscate nuclear weapon ambitions means that constant vigilance over nuclear regulators is necessary. Of course, if public opinion does not count in a country—say, in China and Vietnam—the issue of trust is muted. But we know that, even there, people are uneasy about having a nuclear power plant in their backyard.

Issues of hidden cost and public trust are also embedded in the biological and health threat posed by nuclear energy. Tilman A. Ruff, a long-time student of radiation effects on human health, demonstrates how these effects have been underestimated. He offers a detailed explanation of what exposure to different doses of radiation, such as from the Fukushima accident, means for cancer rates and effects on DNA. Timothy A. Mousseau and Anders P. Møller, who have undertaken field research for many years on the genetic effects of the Chernobyl accident, look at how nuclear plant accidents affect the health of humans and other species. Combined, these two chapters offer a potent, often overlooked, argument against the nuclear option.

A sustainable future?

In all, these chapters put to rest many misconceptions about costs and investment risks of nuclear energy. The fact is, the economics of energy point to a declining future for nuclear power. Even in France, where (as Christina Stuart's chapter points out) nuclear energy accounts for 77 per cent of electrical output, the highest in the world, the 'French exception' is undergoing new scrutiny. Cost factors may finally neutralise the traditional argument that nuclear power is cheap, efficient, and *the* answer to global warming concerns. Andrew Blakers underscores that idea by closely examining sustainable energy options: namely, wind power, photovoltaics (PV), and hydro. He finds that wind and PV are already price-competitive with fossil fuels and increasingly outpacing them in rates of installation.[3]

---

3   In support of Blakers' view, the World Economic Forum (2016: 6) reports: 'the two major sources of non-hydro renewable energy have reached grid parity in a number of countries. In an increasingly larger number of countries, it has become more economical to install solar and wind capacity than coal capacity. It is estimated that more than 30 countries have already reached grid parity without subsidies, and around two thirds of the world should reach grid parity in the next couple of years'.

According to the World Economic Forum (2016: 4), '[r]enewable infrastructure has reached sufficient maturity to constitute a sound investment proposition and the best chance to reverse global warming'. The bad news is that worldwide investment in renewables is far below what it would take to arrest global warming.

Conclusions from the United Nations Framework Convention on Climate Change Conference of the Parties (COP) in December 2015 (COP21) highlight the need for an additional US$1 trillion in annual renewable infrastructure investment by 2030 to meet the goal of limiting global warming to 2 degrees Celsius. This need compares to a current annual average capacity investment of around US$200 billion. Furthermore, among the top 500 asset owners, including foundations, pensions, and endowments, only 0.4 per cent of total assets under management (AUM) have been identified as low-carbon investments (US$138 billion versus US$38 trillion AUM).

The overriding energy challenge in Asia, and elsewhere, is how to wean decision-makers away from reliance on nuclear power and fossil fuels, and into deep investments in wind, solar, and water power. The solution rests above all in politics much more than in science or economics, for otherwise the rational choice would be to abandon nuclear power, oil, and natural gas, whose short- and long-term costs are beyond excessive from a planetary point of view. Whether or not such a dramatic shift in understanding of the energy picture is possible at a time when all countries demand more energy for higher growth must be doubted.

References

Bulletin of the Atomic Scientists, 2017. Global nuclear power database: World nuclear power reactor construction, 1951–2017. thebulletin.org/global-nuclear-power-database (accessed 9 February 2017).

McCurry, Justin, 2017. Fukushima nuclear reactor radiation at highest level since 2011 meltdown. *Guardian*, 3 February.

Nuclear Energy Institute, 2016. World statistics: Nuclear energy around the world. www.nei.org/Knowledge-Center/Nuclear-Statistics/World-Statistics (accessed 9 February 2017).

Schneider, Mycle, and Antony Froggatt, with Julie Hazemann, Tadahiro Katsuta, M. V. Ramana, Juan C. Rodriguez, and Andreas Rüdinger, 2017. *The World Nuclear Industry Status Report 2017*. Paris: Mycle Schneider Consulting Project.

World Economic Forum, 2016. Renewable infrastructure investment handbook: A guide for institutional investors. Geneva: World Economic Forum, December. www3.weforum.org/docs/WEF_Renewable_Infrastructure_Investment_Handbook.pdf (accessed 9 February 2017).

Part I
The state of the nuclear industry

1

Nuclear energy policy issues in Japan after the Fukushima nuclear accident

Tatsujiro Suzuki

Abstract

The Fukushima nuclear accident of 11 March 2011 was a turning point for Japan's nuclear energy and overall energy policy. The biggest impact was the loss of public trust, not only in relation to nuclear safety, but also overall energy policy. More than five years after the catastrophe, this is still the case and more than 80 per cent of the public want to phase out nuclear power eventually. In short, the effects of the accident are not over yet. On 11 April 2014, the Japanese government adopted a new national Strategic Energy Plan declaring its intention to reduce dependence on nuclear energy while considering it as one of the important base-load electricity sources. Regardless of the future of nuclear energy, there are five key policy issues that Japan needs to face: spent fuel management, plutonium stockpile management, radioactive waste disposal, human resources management, and restoring public trust. This chapter discusses those five critical issues and possible policy alternatives that Japan should pursue.

Introduction

The Pacific Ocean earthquake and resulting tsunamis that struck the Tohoku District and Fukushima Daiichi and Daini nuclear power stations at 14.46 on 11 March 2011 (3/11) were followed by a nuclear accident unprecedented in both scale and time frame. Since then, 3/11 has become a historic day for all nuclear experts to remember not only in Japan but also in the rest of the world. Although the earthquake occurred in 2011, the effects of the accident continue. About 100,000 evacuated residents in Fukushima still live in temporary housing and are uncertain as to when they will be able to return to their original hometowns. Although conditions at the Fukushima power stations have improved, it will take more than 40 years to remove melted fuel debris from the site and decommission the plant. We need to draw lessons based on the knowledge and information available to ensure the safety of existing nuclear facilities as much as possible, and to understand potential implications for future nuclear energy policy.

This chapter summarises the current status, both on-site and off-site, of the Fukushima Daiichi nuclear power plant, and reviews possible impacts on Japan's energy policy as well as on global nuclear power development. The chapter identifies key policy issues that are important regardless of the future direction of nuclear power in Japan.

Current status and future prospects of the Fukushima Daiichi nuclear power plant and the environment

On 12 June 2015, the Inter-Ministerial Council for Contaminated Water and Decommissioning Issues (2015) published an updated 'Mid-and-long-term roadmap towards the decommissioning of TEPCO's Fukushima Daiichi nuclear power station'. The report emphasised 'risk reduction', implying that the decommissioning process still poses significant risk to workers and the public. It also delayed the first phase (removing spent fuel from the storage pools of Units 1–3) by more than three years.

The Tokyo Electric Power Company (TEPCO), the owner and operator of the Fukushima nuclear power plant, is responsible for decommissioning the plant. It has been struggling with the storage of a huge and increasing amount of contaminated water (roughly 400 tonnes per day), some of which, it is suspected, has leaked into the sea. In order to contain the contaminated water, TEPCO and the Ministry of Economy, Trade and Industry (METI) decided to install a so-called 'frozen wall' to stop water flowing in and out of the site. The wall is almost complete, but the Nuclear Regulatory Authority concluded that its effectiveness is limited and that alternative methods (such as pumping out underground and contaminated water) need to be continued (*Asahi Shimbun* 2016).

Contaminated water is just one of the unprecedented challenges that TEPCO and METI face. The roadmap for decommissioning Fukushima Daiichi estimates that it will take at least 30 to 40 years to complete decommissioning. The first stage involves removal of the spent fuel from the pools in all four units (in two to three years), the second stage involves removal of the melted core debris from Units 1–3 (in at least 10 years), and the third stage encompasses decontamination of the whole plant (in 30 to 40 years). Removal of spent fuel (1,331 spent fuel assemblies and 202 unirradiated fuel assemblies) from Unit 4's storage pool was successfully completed on 22 December 2014. Operations to remove spent fuel from Units 1–3 are now underway. For removal of the melted cores, the information available on melted debris is very limited and no one is sure where they are or what form they now take. It is not possible to get close to the reactor buildings of Units 1–3 due to high radiation, and it is necessary to develop remote-control equipment or sophisticated, radiation-resistant robots.

On 1 April 2014, TEPCO established a new company, the Fukushima Daiichi Decontamination and Decommissioning Engineering Company, as a dedicated institution to manage this huge, complex, and challenging operation. An International Research Institute for Nuclear Decommissioning was also established in August 2013 by METI, TEPCO, and other interested parties, including nuclear vendors and the Japan Atomic Energy Agency (JAEA). The institute's purpose is to promote necessary research and development efforts for decommissioning in general, but especially for the Fukushima Daiichi nuclear reactors. However, there are still concerns about a lack of transparency and independent oversight in regard to the whole decommissioning process. The Japan Atomic Energy Commission (JAEC) recommended that the government should establish an independent (third-party) organisation with overseas experts as members to assess and audit the entire measures in order to maximise transparency (JAEC 2012b). However, such an independent organisation has not been established by the government.

Decontamination and reconstruction of evacuated zones

There are three different levels of evacuated zones designated by the government, depending on the level of monitored radiation levels: a 'non-return' zone (above 50 milliSieverts (mSv) per year); a 'preparation for return' zone (below 50 mSv per year and above 20 mSv per year); and a 'possible to return' zone (below 20 mSv per year). Due to natural radiation decay and decontamination efforts, more areas are now designated as 'possible to return' zones. On 31 August 2016, the government announced that some of the 'non-return' zones would be designated as 'recovering centres' and that life infrastructure would be re-established so that people could return soon (Recovery Council, Nuclear Accident Emergency Response Headquarters 2016). However, the criteria of 20 mSv per year has been a source of public debate as it is much higher than the 5 mSv per year level that was the evacuation criteria for the Chernobyl accident five years after that accident.

The issue of returning to the hometown is connected to the compensation issue. Under current rules, once the town is no longer considered as an evacuated zone, citizens are no longer eligible for compensation. More importantly, there is not enough public participation in the decision-making process, which will lead to a loss of public trust, as discussed below.

Loss of public trust

On 24 February 2015, TEPCO (2015) issued a press release stating that the source of high radiation levels in one of its drains originated from a puddle of rainwater that had accumulated on the rooftop of Unit 2 at the Fukushima Daiichi nuclear power station. The drain leads to open seawater. It was thus suspected that contaminated water may have leaked into the sea, although TEPCO found 'no increase in radioactivity' in the seawater in the area.

This illustrates just one episode in a series of many adverse events in Japan's nuclear industry that have been reported over the past four years. However, this particular incident was worse than usual because TEPCO was aware of the high level of radioactivity in the drain but failed to notify the Nuclear Regulation Authority or the local government. It was also very bad timing. After long negotiations with the local fishing industry, TEPCO was about to release some of the accumulated radioactive groundwater, which had been cleaned through a water treatment process, into the Pacific Ocean. On 25 February 2015, the local fishing industry association criticised TEPCO, with Hiroyuki Sato, the chairman of the Soma-Futaba Fisheries Cooperative Association, stating that 'trust has been lost'.

Lack of trust is a fundamental problem that underlies the challenges facing Japan's nuclear industry since the Fukushima disaster. The public has lost faith in nuclear safety regulation. Faith has not been fully restored even though a newly independent Nuclear Regulation Authority was established in 2012, and much tougher regulatory standards were introduced. According to poll results, the proportion of the public that want to shut down all nuclear power plants immediately increased from 13.3 per cent in June 2011 to 30.7 per cent in March 2013. The same polling data also suggested that about 80 per cent of the public still believed that serious nuclear accidents would occur again in Japan (Hirose 2013).

In polling undertaken in August 2014, the proportion of the public who oppose restarting the existing reactors rose to 56 per cent, an increase of 4 percentage points from previous polling on this question. The same poll indicated that 61 per cent of the public were willing to accept higher electricity prices if existing nuclear power plants remained closed (*Nihon Keizai Shimbun* 2014). Hirose's (2013) polling also suggested that government agencies were considered to be the 'most untrustworthy' organisations of those that were listed. This loss of trust is the most serious challenge that nuclear policymakers and the nuclear industry now face in Japan. Six years after the accident, it has not been addressed adequately.

Two recent important policy developments have occurred that have further eroded public trust. First, on 20 December 2016, the TEPCO Reform Committee (2016) published a new report concerning TEPCO reform, in which it outlined new estimates for total accident-related costs and its financing scheme. The total estimated cost of the accident is now about ¥22 trillion (US$200 billion), which is two times higher than the previous estimate. The estimated costs of each item are as follows: Fukushima Daiichi decommissioning (¥8 trillion), compensation costs (¥8 trillion), and the decommissioning of contaminated land (¥6 trillion). The report also announced that TEPCO should bear about ¥16 trillion of the cost, but that the rest should be financed by other electricity companies—¥4 trillion from new and conventional utilities—and ¥2 trillion from the government.

Second, on 21 December 2016, the Cabinet Ministers' Meeting on Nuclear Energy Policy released two policy documents: 'Basic policy of fast reactor development' and 'Government policy on the fast reactor prototype reactor "Monju"' (Cabinet Ministers' Meeting on Nuclear Energy 2016a, 2016b). These policy documents emerged in response to the 'Recommendation by the Nuclear Regulatory Authority on Monju' issued in November 2015, which recommended that the government must find an alternative operating/managing institution to the JAEA as the JAEA was judged to be incapable of operating Monju (NRA 2015). The documents stated that the government had decided to decommission Monju from 2017, while fast reactor development would continue without Monju. The government also renewed its commitment to build a 'demonstration fast reactor' and to achieve 'future commercialisation of fast reactor'. But this policy decision was based on a statement made by a series of closed meetings, called 'Fast reactor development meeting', consisting of the JAEA, METI, the Ministry of Education, Culture, Sports, Science and Technology, the Federation of Electric Power Companies (FEPC), and Mitsubishi Heavy Industry. Without open debate and a thorough review of the Monju project, the credibility and feasibility of a fast reactor program is now in serious doubt.

Possible impacts on Japan's energy policy

The economic impact of shutting down nuclear power plants is also significant. According to a study carried out by the Institute of Energy Economics, Japan (2013), about ¥3.6 trillion (US$36 billion) of extra payments were made because of the shutdown of nuclear power plants during Fiscal Year 2011 and Fiscal Year 2012 (the Japanese fiscal year starts in April and ends in March), while energy demand decline contributed to about ¥1.2 trillion (US$12 billion) of savings during the same period. In addition, carbon dioxide emissions in 2012 increased by about 70 million tonnes, that is, an increase of about 5.8 per cent from 2011 levels, which was roughly equal to the emission increase in the Middle Eastern region or India alone in 2012 (IEA 2013).

On 11 April 2014, the new Strategic Energy Plan was adopted by the Japanese Cabinet (METI 2014a). The plan stated that the government would not only decrease its dependence on nuclear energy as much as possible, but also that nuclear power should be used as an important base-load energy source and thus the necessary level of nuclear energy should be maintained.

The METI Advisory Council set up one working group to determine the future energy mix targeted for 2030, and another working group to re-examine the generation cost of nuclear power compared with other power sources. On 5 April 2015, it was reported that METI's new cost estimate for newly built nuclear power would be about ¥1 per kilowatt hour (kWh) more expensive than the ¥8.9 per kWh previously estimated by the government in 2012, but still believed to be less expensive than newly built fossil fuel power plants (*Nihon Keizai Shimbun* 2015). On 7 April 2015, METI stated that so-called 'base-load' electricity should supply about 60 per cent of total power generation, with nuclear power, coal, and geothermal power being part of such base-load power sources. In July 2015, METI published its new long-term energy outlook based on its Strategic Energy Plan of 2014 (METI 2015b). According to the outlook, the share of nuclear energy in total power generation will be around 20–22 per cent, which is a slight decline from 2010 (26 per cent), and the share of renewable energy will be around 22–24 per cent. Maintaining the nuclear share of 20–22 per cent is likely to require extension of the 40-year lifetime operating period of current nuclear power plants, or the building of new nuclear power plants. This policy has been criticised as being inconsistent with the goal of 'reducing the dependency on nuclear power as much as possible' (*Asahi Shimbun* 2015a). The Ministry of Environment also published its future energy mix plan, suggesting that the share of renewable energy could be increased to 24–35 per cent by 2030 (*Asahi Shimbun* 2015b).

Policy issues and challenges regardless of future directions of nuclear power in Japan

Although Japan's future energy policy is still under discussion, certain important issues need to be overcome: spent fuel management, plutonium stockpile management, high-level waste disposal, securing human resources, and restoring public trust.

Spent fuel management

Even before the Fukushima accident, the question of the management of accumulating spent fuel on-site at nuclear power plants was a major policy issue for nuclear utilities and the government. By the end of 2011, about 17,000 tonnes of spent fuel were in storage, out of which about 14,000 tonnes were at nuclear power plant sites and 2,900 tonnes were at the Rokkasho reprocessing plant. The total spent fuel pool storage capacity at nuclear power plant sites is about 20,630 tonnes, and this is roughly 70 per cent full (Takubo and von Hippel 2013). For some reactor sites, the pool will be full within a few years if reactors restart operations. The Rokkasho reprocessing plant, with planned capacity to reprocess 800 tonnes of spent fuel per year, has only one storage pool with a 3,000 tonne capacity. The plant is currently shutdown after a period of hot testing and the repair of vitrification equipment, and it is not clear when it will start commercial operation, due to new regulatory standards. Since the storage pool is almost full, unless the plant starts commercial operation, it may not be able to accept further spent fuel.

Another option is an 'away-from-reactor' centralised storage facility at Mutsu city, which is also under construction. Its capacity is 5,000 tonnes but it is not yet fully operational and will accept only spent fuel from TEPCO and the Japan Atomic Power Company. Safe and secure dry cask storage on-site is technically possible, as proven at the Fukushima Daiichi site, where dry casks loaded with spent reactor fuel withstood the earthquake and tsunami without significant damage, and at the Tokai Daini nuclear power plant site. But local communities at nuclear power plant sites are not in favour of accepting further spent fuel storage on-site.

In short, finding additional storage capacity (possibly dry cask storage) is a top priority issue for nuclear utilities and the government, in order to increase the flexibility of spent fuel management, as uncertainty regarding reprocessing still remains.

Plutonium stockpile management

The basic policy for spent fuel management in Japan has been (and still is) 'reprocessing and recycling plutonium' for energy use. Since plutonium can also be used to manufacture nuclear bombs, the JAEC introduced a 'no plutonium surplus' policy from 1991, and strengthened its policy in 2003 by introducing new guidelines to improve its transparency when the Rokkasho commercial reprocessing plant was expected to start operations. According to the guidelines, utilities are expected to submit a 'plutonium usage plan' annually before they reprocess and recover plutonium. In short, this is intended to ensure that Japan will not possess plutonium without plans for its use. However, in reality, the plutonium usage program (recycling as mixed-oxide fuel into existing reactors and fast breeder reactors in the future) has been delayed significantly. As a result, by the end of 2015, Japan possessed about 48 tonnes of separated plutonium (10.8 tonnes in Japan, and 37.1 tonnes in France and the UK where Japan has commercial reprocessing contracts; see Table 1.1) (JAEC 2016). This is the largest stockpile among non–nuclear weapon states and could increase further if the Rokkasho reprocessing plant starts operation, and if its recycling program into 15–18 reactors as currently planned does not smoothly move ahead. As a result, if the Rokkasho plant starts operating, Japan's plutonium stockpile is likely to grow (Takubo and von Hippel 2013).

Table 1.1 Japan's stockpile of separated plutonium

	Stockpile at the end of 2014 (kg)	Stockpile at the end of 2015 (kg)
Stock in Japan (Pu total)		
Reprocessing plants	4,322	4,126
Mixed-oxide fuel plant	3,404	3,596
Stored at reactors	3,109	3,109
Sub-total (fissile plutonium)*	10,835 (7,310)	10,832 (7,307)
Stocks in Europe (Pu total)		
United Kingdom	20,696	20,868
France	16,278	16,248
Sub-total: Pu total (fissile plutonium)	36,974 (24,511)	37,115 (24,574)
Total (fissile plutonium)	47,809 (31,821)	47,947 (31,881)

* Fissile plutonium (Pu 239 and Pu 241) is typically about 60 per cent of total plutonium, which includes non-fissile isotope of plutonium (Pu 240 and Pu 242).

Pu = plutonium

Source: JAEC (2016).

Meanwhile, due to heightened concern over nuclear proliferation and nuclear security, international attention on Japan's plutonium stockpile is also increasing. For example, the US–Japan Nuclear Working Group of the Maureen and Mike Mansfield Foundation published its recommendations on nuclear energy policy for Japan:

> The disposition of Japan's sizeable plutonium stockpile is an outstanding issue that must be addressed regardless of whether or not Japan decides to move forward with nuclear power ... Absent a credible strategy for reducing Japan's plutonium stockpile, nonproliferation and security concerns will grow over time, undermining Japan's international leadership on nuclear nonproliferation (US–Japan Nuclear Working Group 2014: 4).

In order to reduce such concern and to minimise proliferation and nuclear security risks, Japan may need to produce a new plutonium management plan. I propose three new principles for plutonium management in Japan:

1. Demand comes first: Reprocessing should take place only when plutonium demand (use) is specified.
2. Stockpile reduction: Matching demand/supply is not good enough. The existing stockpile should be reduced before further reprocessing.
3. Flexible plan: The current plutonium use plan (mixed-oxide recycling in 16–18 units) is no longer certain. Other options (plutonium ownership transfer, disposition as waste, and so on) need to be pursued. Such options should minimise cost, transportation, and time required for disposal (Suzuki 2013).

In addition, a multilateral approach to managing nuclear fuel cycle facilities can be a good way to improve international confidence in Japan's nuclear fuel cycle program. One such idea is to put both enrichment and reprocessing facilities under international control (Diesendorf 2014). In the future, this approach could even be applied to facilities in other countries in the region, including China and North Korea.

High-level radioactive waste disposal

Like many other countries, Japan has not found a final repository site for high-level radioactive waste (HLW). Since 2000, when the Law on Final Disposal of Specified Radioactive Waste (i.e. vitrified HLW) was passed and the Nuclear Waste Management Organisation was established as the principal implementation institution for final disposal, all efforts to find even a single candidate for possible investigation did not succeed. Japan's approach was to wait for local communities to volunteer as candidates; only one town (Toyo-town) volunteered, but later cancelled the request due to strong public opposition. In 2010, the JAEC asked the Science Council of Japan for their advice on how to improve public communication on HLW, and the Science Council published its response in 2012 (Science Council of Japan 2012). The report recommended a fundamental reform of Japan's HLW disposal policy. In particular, it was recommended that '(long term) temporary storage' be used instead of 'geological disposal', for which it argued that scientific knowledge is still too uncertain to commit to geological disposal in Japan.

The JAEC responded with its own policy statement in December 2012 (JAEC 2012d). The JAEC agreed with the Science Council that the current HLW disposal program needed to be reviewed, but maintained the basic conclusion of its advisory committee report that was published in 1998, which recommended 'geological disposal' as the most appropriate policy option under current circumstances. Still, the JAEC also agreed with the Science Council that constant review of the program is necessary and 'retrievability' and 'reversibility' should be clearly integrated into the disposal program. Further, the JAEC also recommended that the government 'establish an independent and functionally effective third party organization to provide suitable advice to the government and related parties in time'.

METI set up two working groups to review the HLW disposal program. One was to examine the whole process and programs including public participation, and the other was to review scientific knowledge on HLW disposal in Japan especially after 3/11. Based on its findings (METI 2014b), a 'Basic plan for final disposal of specified radioactive waste' was adopted by the Cabinet on 22 May 2015 (METI 2015a). The new plan now places stronger responsibility with the government and introduced some flexibility, including the concept of 'retrievability' and 'reversibility'. Still, the future of the HLW disposal program is very uncertain.

The Science Council of Japan published a report to follow up its 2012 report, re-emphasising the importance of a 'consensus building process' for HLW disposal and proposing the creation of a 'national people's conference on radioactive waste' (Science Council of Japan 2015). The Science Council proposed to use a period established by the 'temporal storage' (not 'interim storage', which assumes that the final decision on HLW disposal has been made) for gaining national consensus. Whether such a proposal will be accepted by the government remains to be seen.

Securing human resources and research and development

Since the future prospects of nuclear power have become uncertain, it is likely that attracting young and capable talent to nuclear energy fields may become difficult. Further, there is an emerging demand for new tasks such as the decommissioning of Fukushima reactors. Therefore, it is important to secure human resources to meet such new and challenging tasks in the coming decades. Research and development programs also need to be re-examined to meet new challenges and to provide future human resources. In order to meet such challenges, the JAEC published policy statements on human resources and on research and development in 2012 (JAEC 2012c, 2012e).

For human resource management, the JAEC recommended, among other things, to draw a 'human resource demand/supply map'—'the related government agencies and demand side, including the nuclear industry, [should] clarify when, in what areas and how much manpower is required based on operational plans' (JAEC 2012c). This cannot be done by the government agencies, but should be undertaken by related industry organisations as they have better knowledge and data. Other important recommendations included education based on lessons learned from the Fukushima accident, providing new education opportunities for mid-career experts, enhancing human resource development for nuclear safety, security, and safeguards, providing incentives for nuclear businesses to maintain human resources, securing human resources for maintaining the operation of domestic nuclear power plants, and human resource development for international deployment of nuclear energy and technology.

Restoring public trust

Last, but not least, public trust must be restored. As noted above, the loss of public trust in the government's handling of nuclear energy policy is one of the biggest consequences of the Fukushima accident. The JAEC issued a policy statement on this issue in 2012 (JAEC 2012a), and listed four basic principles for restoring public confidence.

Accountability

First, it is important that the individuals/organisations tackling such challenges explain their mission to the public—what they do, and why and how they do it. Such individuals/organisations should be aware of their primary responsibility to seek solutions to challenges and manage risks in the public interest, and be accountable for their plans and the results of their actions. They have an obligation to continuously explain to the public how their actions fulfill their responsibilities and their commitment to public well-being and safety.

Correct information disclosure

Second, it is important to remember that these explanations should be provided based on sufficient and correct information to the public on a timely basis. For example, in discussing a plant operator's actions for nuclear power safety, we should carefully explain the nature of the threat facing a facility, the operator's target, and how it intends to reach the target. In doing so, explaining by using comparisons with other facilities is acceptable but must be done carefully. Evaluations should be made including all relevant factors, including costs, environmental impacts, and stability, and comparison based on one point alone may be inappropriate, even if accurate. However, we should also note that speed is sometimes more important than accuracy. In that case, details should immediately be provided about what has happened and why, and what can be expected to happen in the future, while explaining uncertainties in such information and the range of possible outcomes.

Transparency/fairness and public involvement in decision processes

Third, it is important to design fair decision-making processes, as the basis for administrative decisions, and, in making the process open, to provide opportunities for public participation in the process. In this case, the parties concerned should deeply appreciate that securing transparency means the public can view the decision-making process, access information, and provide input into these processes. Based on this acknowledgment, the greater the public interest in a decision, the more carefully the public should be involved at the earliest possible stage before decisions are made. Organisations involved should strive to give the public opportunities to express their views.

Further, administrative bodies should establish verifiable decision-making processes, with full and accessible documentation: from the creation of administrative documents, and hearings from experts, interested parties and the public, to final decision-making.

Easy-to-understand explanations

Fourth, public explanations should be clear and plain, with accuracy a prerequisite. It is often noted that if the public cannot understand information released, it cannot be considered transparent, even if it is believed that transparency is attained in doing so. It is not easy to ensure material is both accurate and comprehensible, but court decisions have long been written in normal Japanese. Administrative bodies must not forget to check the processes of creating documents and preparing explanations from this perspective, continuously educating and training themselves in this area.

Conclusion

Nuclear energy policy after 3/11 needs to be changed to reflect lessons learned from the Fukushima accident and the different priorities and tasks required after the Fukushima accident, such as the decommissioning of the Fukushima site and restoring lives and livelihoods for people in Fukushima and other affected areas; enhancing safety and security, spent fuel management, plutonium stockpile management, waste disposal, and human resource development; and, most of all, restoring public trust. The Japanese government should also initiate a national debate to re-examine the risks and benefits of nuclear energy involving various stakeholders and civil society. Establishing an independent commission to conduct a comprehensive, non-biased assessment of nuclear energy policy would be desirable. These are necessary changes regardless of the future directions of nuclear energy in Japan.

References

Asahi Shimbun, 2015a. Reduction target for greenhouse gases set at 25% at 2030: Government's draft plan submitted. 24 April.

Asahi Shimbun, 2015b. Share of renewable energy will be 'around mid 20%' for energy mix in 2030, METI says. 8 April.

Asahi Shimbun, 2016. 'Effectiveness of the wall is limited', the Nuclear Regulatory Authority concluded. 27 December.

Cabinet Ministers' Meeting on Nuclear Energy, 2016a. Kosokuro Kaihatsu no Hoshin [Basic policy of fast reactor development]. 21 December. www.cas.go.jp/jp/seisaku/genshiryoku_kakuryo_kaigi/pdf/h281221_siryou1.pdf (accessed 23 January 2017).

Cabinet Ministers' Meeting on Nuclear Energy, 2016b. Monju no Toriatsukai ni Kansuru Seifu Houshin [Government policy on the fast reactor prototype reactor 'Monju']. 21 December. www.cas.go.jp/jp/seisaku/genshiryoku_kakuryo_kaigi/pdf/h281221_siryou2.pdf (accessed 23 January 2017).

Diesendorf, Mark, 2014. *Sustainable Energy Solutions for Climate Change*. Sydney: UNSW Press.

Hirose, Hirotada, 2013. Genshiryoku Hatsuden wo meguru yoron no henka [Changes of public opinion regarding nuclear power]. Presentation to the Japan Atomic Energy Commission regular meeting, 17 July. www.aec.go.jp/jicst/NC/iinkai/teirei/siryo2013/siryo27/siryo2.pdf (accessed 23 January 2017).

IEA (International Energy Agency), 2013. Redrawing the energy-climate map. 10 June.

Institute of Energy Economics, Japan, 2013. Dengen Betsu Kosuto Jisseki Hyoka to Denki Jigyo Zaimu he no eikyo [Assessment of power generation cost based on actual data and its impacts on financial performance of electric utility companies]. Presentation to the Japan Atomic Energy Commission, 20 August. www.aec.go.jp/jicst/NC/iinkai/teirei/siryo2013/siryo31/siryo3.pdf (accessed 27 February 2017).

Inter-Ministerial Council for Contaminated Water and Decommissioning Issues, 2015. Mid-and-long-term roadmap towards the decommissioning of TEPCO's Fukushima Daiichi nuclear power station. 12 June. www.meti.go.jp/english/earthquake/nuclear/decommissioning/pdf/20150725_01b.pdf (accessed 23 January 2017).

JAEC (Japan Atomic Energy Commission), 2012a. Efforts to build public confidence. 25 December. www.aec.go.jp/jicst/NC/about/kettei/121225-2_e.pdf (accessed 23 January 2017).

JAEC (Japan Atomic Energy Commission), 2012b. Progress of medium- and long-term efforts to decommission Fukushima Dai-ichi NPP of TEPCO. 27 November. www.aec.go.jp/jicst/NC/about/kettei/121127-1_e.pdf (accessed 27 February 2017).

JAEC (Japan Atomic Energy Commission), 2012c. Promotion of measures to secure and develop human resources for nuclear energy. 27 November. www.aec.go.jp/jicst/NC/about/kettei/121127-2_e.pdf (accessed 23 January 2017).

JAEC (Japan Atomic Energy Commission), 2012d. Renewing approaches to geological disposal of high-level radioactive waste (HLW). 18 December. www.aec.go.jp/jicst/NC/about/kettei/121218_e.pdf (accessed 23 January 2017).

JAEC (Japan Atomic Energy Commission), 2012e. Research and development on nuclear power in the future should be. 25 December. www.aec.go.jp/jicst/NC/about/kettei/121225-1_e.pdf (accessed 23 January 2017).

JAEC (Japan Atomic Energy Commission), 2016. The status report of plutonium management in Japan – 2015. 27 July. www.aec.go.jp/jicst/NC/iinkai/teirei/siryo2016/siryo24/siryo1_e.pdf (accessed 23 January 2017).

METI (Ministry of Economy, Trade and Industry), 2014a. Strategic energy plan. April. www.enecho.meti.go.jp/en/category/others/basic_plan/pdf/4th_strategic_energy_plan.pdf (accessed 23 January 2017).

METI (Ministry of Economy, Trade and Industry), 2014b. Sogo Shigen Enerugi Chosakai, Denryoku/Gasu jigyo bunnkakai, Gennsiryoku Shoiinnkai, Houshasei Haikibutsu Waakingu Guruupu, 'Houshasei Haikibutsu WG Chukan Torimatome' [Working group on high-level radioactive waste, interim report]. May. www.meti.go.jp/committee/sougouenergy/denryoku_gas/genshiryoku/houshasei_haikibutsu_wg/report_001.pdf (accessed 23 January 2017).

METI (Ministry of Economy, Trade and Industry), 2015a. Basic plan for final disposal of specified radioactive waste. 22 May.

METI (Ministry of Economy, Trade and Industry), 2015b. Choki Enerugi Jyukyu Mitoshi [Long term energy supply demand outlook]. 7 July. www.meti.go.jp/press/2015/07/20150716004/20150716004_2.pdf (accessed 23 January 2017).

Nihon Keizai Shimbun, 2014. Genpatsu Saikado 'Susumete' 32%, Honsha Yoron Chosa [Public polling result shows 32% for restart up of existing nuclear power plants]. 24 August.

Nihon Keizai Shimbun, 2015. Genpatsu no Hatsuden Kosuto 1 wari zo, Keisansho Shisan, Kakaku Yuisei wa Iji [Nuclear power plant cost will be 10% more expensive than the previous cost estimate, METI says, while its cost competitiveness will remain]. 5 April.

NRA (Nuclear Regulation Authority), 2015. Monbu Kagaku Daijin he no Kankoku [Recommendation to Minister of Education, Sports, Culture and Technology (MEXT)]. 13 November. www.nsr.go.jp/data/000129633.pdf (accessed 7 September 2017).

Recovery Council, Nuclear Accident Emergency Response Headquarters, 2016. Kikan Konnan Kuiki no Toriatsukai ni Kansuru Kangaekata [Basic thinking on non-return zone]. 31 August. www.meti.go.jp/earthquake/nuclear/kinkyu/pdf/2016/0831_01.pdf (accessed 23 January 2017).

Science Council of Japan, 2012. Ko Reberu Hoshasei Haikibutsu no Shobun Ni Tsuite [Regarding final disposal of high level radioactive waste: Answers to the Japan Atomic Energy Commission]. 11 September. www.scj.go.jp/ja/info/kohyo/pdf/kohyo-22-k159-1.pdf (accessed 23 January 2017).

Science Council of Japan, 2015. Ko Reberu Hoshasei Haikibutsu no Shobun Ni Kansuru Seisaku Teigen – Kokumin Teki Goi Keisei ni Muketa Zantei Hokan [Policy proposal for final disposal of high-level radioactive waste—temporal storage for gaining national consensus]. 24 April. www.scj.go.jp/ja/info/kohyo/pdf/kohyo-23-t212-1.pdf (accessed 23 January 2017).

Suzuki, Tatsujiro, 2013. Purutoniumu Riyo Keikaku Eno 3tsu no Teian [Japan Atomic Energy Commission magazine 123]. 29 March. www.aec.go.jp/jicst/NC/melmaga/2013-0123.html (accessed 23 January 2017).

Takubo, Masafumi, and Frank N. von Hippel, 2013. Ending reprocessing in Japan: An alternative approach to managing Japan's spent nuclear fuel and separated plutonium. Research Report No. 12. Princeton, NJ: International Panel on Fissile Material, Program on Science and Global Security, Princeton University.

TEPCO (Tokyo Electric Power Company), 2015. Unit 2 reactor building and large carry-in entrance rooftop accumulated water quality results. 24 February. www4.tepco.co.jp/en/nu/fukushima-np/handouts/2015/images/handouts_150224_01-e.pdf (accessed 7 September 2017).

TEPCO (Tokyo Electric Power Company) Reform Committee, 2016. Toden Kaikaku Teigen [A recommendation for TEPCO reform]. 20 December. www.meti.go.jp/committee/kenkyukai/energy_environment/touden_1f/pdf/161220_teigen.pdf (accessed 23 January 2017).

US–Japan Nuclear Working Group, 2014. Statement on shared strategic priorities in the aftermath of the Fukushima nuclear accident. New York: Maureen and Mike Mansfield Foundation.

2

The French exception: The French nuclear power industry and its influence on political plans to transition to a new energy system

Christina Stuart

Abstract

The Fukushima accident was a turning point for French energy policy, as it prompted the country to put forth legislation to reduce the share of nuclear power and accelerate renewable energy growth. Even so, France's nuclear fleet remains intact and dominates the French energy sector with powerful political momentum. An explanation and outlook for the evident tension between the two realities is the goal of this research. By first examining how the French nuclear industry came to be so exceptionally powerful, the industry's response to environmental, safety, and economic concerns is analysed. An outlook for the French nuclear industry within a new political framework is ultimately proposed. Based on a sociopolitical analysis, the nuclear industry currently overrides environmental and climate concerns by putting forward its low-carbon technology. The industry also controls safety concerns at the cost of extensive safety investments. However, this research shows that the finances that are currently keeping the industry afloat are ultimately creating an unstable economic situation. The balance of power between energy transition policy and nuclear industry growth is a question of economics over politics.

Introduction

In August 2015, former President François Hollande of France passed the Energy Transition for Green Growth Law (Energy Transition Law), which aims to reduce the share of nuclear generation in the national electricity mix. This unprecedented energy policy change commits the country to a suite of ambitious environmentally friendly targets, including reducing the share of nuclear electricity generation from 76.3 per cent (RTE 2016) to 50 per cent by 2025, and increasing the share of renewable electricity generation up to 40 per cent by 2030 (by increasing the total renewable energy share to 32 per cent). Whilst the objectives pertaining to greenhouse gas emission reductions and renewable energy generation are generally accepted, the legislated rapid decrease in the nuclear share remains controversial. How exactly these concrete energy transition targets will be met is unclear. According to the French Court of Audit, the nuclear industry would have to shut down seven to 20 reactors by 2025, if electricity consumption and export levels remain stable. However, two years since the law was passed, Électricité de France (EDF), the electricity utility monopoly, has still not shut down its oldest, most unstable, and internationally contested nuclear plant in Fessenheim, at the border with Germany.

There is evident conflict and tension between, on the one hand, France's plans to transition to a new energy system that is less reliant on nuclear power and, on the other hand, the ease with which the French industry is refusing to implement these plans. To understand this disconnect, one has to recognise France's exceptional nuclear industry and powerful nuclear lobby. French nuclear power is exceptional as it has an industrial configuration inseparable from political power, it is the nervous system of the current centralised state, and it is so ingrained and established in French culture that this is the first time that it is being truly put into question. Due to this exceptional configuration, France's nuclear fleet is the second largest in the world after the United States' fleet, and the leading country in terms of the greatest share of nuclear generation in the national electricity mix. With 58 reactors accounting for 77 per cent of France's electricity generation, France relies on nuclear power more than any other country in the world.

The international literature on nuclear energy generation exposes a recent trend for nuclear decline (Schneider et al. 2016). Even the projections by the International Atomic Energy Agency (IAEA) on nuclear expansion are being updated each year to consider both the consequences of ageing reactors and their corresponding closures, as well as a reduction in the rate of new nuclear reactor builds (IAEA 2014/2015). Even though the reasons for this global decline are complex and vary in nature between countries, there is clearly a global transition to a new energy paradigm. This new paradigm goes beyond an increase in renewable energy production and the slowing down of fossil fuel and nuclear energy growth. Energy sustainability policy is now being more directly driven by security of supply and the environment (including climate change). In contrast to the international literature, the French literature on nuclear energy generation is much less forward-looking and predominantly focuses on analysing the nuclear industry's historic decisions (Topçu 2013). As a consequence, there is no consensus on how French nuclear power is developing. In fact, whereas the international literature points to a struggling French nuclear industry, the French projections are optimistic and view the industry as an essential energy asset for low-carbon growth. A knowledge gap exists around how to reconcile how France is reacting to conflicting views on nuclear power.

In this chapter, I articulate why there is this conflict between the current state of affairs in France and international and national policies, as well as expose how this situation is developing within the context of a global energy system transition. More completely, the aims of this chapter are to explain how the French nuclear industry is an exception compared to other nuclear industries, and to understand to what extent the French nuclear industry's exceptional status might influence the outlook of France's energy system. Could the French nuclear establishment have enough power to resist current political objectives to reduce the share of nuclear and favour renewables? In an initial section, I discuss how the French nuclear industry has historically dictated energy policy and created a strong industrial configuration resistant to nuclear decline trends. In a second section, I address the role that nuclear has to play in the global environment and climate policy setting. A third section focuses on the French nuclear industry's strategy to mitigate nuclear safety risk aversion. Finally, the chapter concludes with an economic analysis of the French nuclear industry, questioning the extent to which it may remain politically influential.

Development of the French nuclear industry

Before beginning to analyse the balance of power between the nuclear industry and French policies, one has to understand what makes the French nuclear industry so exceptional. For this, one has to go back to the explosive birth and introduction of nuclear technology into France. There are many elements that contribute to the success story of France's nuclear industry. I develop three of these, each of which has led to the political strength of the nuclear establishment. The first essential contributing factor to this success story is the nuclear industry's military origin. Second, the main actors that make up this industry are closely interlinked with each other and the government. Third, the French economy has grown to rely on its nuclear export capacity.

The 'dissuasive weapon' (*l'arme de dissuasion*) was the name given to nuclear fission technology by French President Charles De Gaulle and father of France's *grandes programmes*. After the Second World War, and in response to the successful tests of the first Soviet atomic bombs, Europe both feared and respected the power of nuclear fission technology. In 1952, the European Defence Community Project designed a treaty to protect member states from nuclear threats during the Cold War. Six member states including France agreed upon this treaty, which notably included a clause prohibiting all signatories of being in possession of an atomic bomb. However, after the defeat of the French Union by the Viet Minh during the first Indochina war in 1954, General De Gaulle made the decision to begin producing atomic weapons, in the name of national defence. Thus, France officially rejected the treaty.

In 1958, when De Gaulle became president of the fifth Constitution, he was appointed, by definition, head of the army and therefore of nuclear arms control. Complete control over such a powerful and potentially dangerous technology by one person was, however, difficult for the public to accept. To legitimise possessing such power, De Gaulle announced a referendum in 1962 resulting in a modification of the presidential election procedure, which replaced indirect suffrage with a direct universal suffrage system. By making his presidency a democratic statement, De Gaulle gave legitimacy to his nuclear arms control (Chantebout 1986).[1] De Gaulle viewed nuclear as much more than a military strategy; it was at the heart of his national independence policy strategy. For him, military nuclear was not enough and civil nuclear power became his solution to ensuring conformity with Article 5 of the French Constitution, which states that the president is the 'guarantor of national independence'. The paradigm shift during the 1960s from nuclear as a military strategy to a civil commodity marked a historic turning point in French history.

National energy independence was translated as independence from fossil fuel imports, a political strategy used to push civil nuclear power generation. Independence especially from North American oil imports was paramount. In fact, France left the North Atlantic Treaty Organization (NATO) in 1966, symbolising its independence. Ironically, as soon as De Gaulle was no longer president (1969), the American pressurised water reactor (PWR) design was imported and used for new builds from then on as they were cheaper than the French Uranium Naturel Graphite Gaz (UNGG) reactor design (Reuss 2007: 68). Of course, complete 'energy independence' was an exaggeration, as nuclear only covers electricity generation. Oil and gasoline imports for transport were hardly affected and unavoidable.

EDF commenced construction of pilot nuclear reactor projects in the 1960s, whilst a gradual shift towards electrifying all possible energy flows began. Military funding conveniently supported the investment in civil nuclear and the continuation of these pilot projects. Most often, the reasons given to explain France's transition to civil nuclear power and to justify colossal nuclear investments made by the French government are linked to the oil crisis of 1973. However, to remain chronologically coherent, the civil nuclear investments have to be considered in conjunction with military ambitions during the 1970s Arab–Israeli conflicts. In May 1973, before the oil crisis, Prime Minister Pierre Messmer, during an inter-ministerial committee, had already announced that France would accelerate his initial plan for 8,000 megawatts (MW), to be built between 1972 and 1977, to 13,000 MW of nuclear capacity (INA 1975).

---

1   'It is therefore impossible not to draw a comparison between the development of the first French atomic weapons and the constitutional reform of 1962 on the election of the president by direct universal suffrage. It is certain that only a legal personality who enjoys the full legitimacy conferred to himself/herself by direct election by the people can find in himself/herself the moral strength necessary to decide upon the use of such a lethal device' (Chantebout 1986).

Only a few months later, the oil crisis shocked economies worldwide, giving civil nuclear power generation the perfect reason to exist and flourish. Offering energy independence and military defence, nuclear reactors took on a construction speed not seen in any other country. Although the oil crisis happened almost simultaneously with the decision to accelerate the nuclear electrification of France, it is important to note that before the crisis the civil nuclear plan had already been formulated. This culminated in Prime Minister Messmer's 1974 famous plan: 'all nuclear' (*le tout-nucléaire*). The first phase of the Messmer Plan was to construct 13 nuclear reactors, each of 900 MW in capacity, in two years. Ten years later, 50 Westinghouse PWRs were under construction all over the country.[2] As for the French population, the justification for the civil nuclear investment came down to promising cheap national electricity.[3]

Today's nuclear fleet was constructed in three main phases: CP0 (Contract Programme 0), CP1, and CP2 were constructed between 1971 and 1982, P4 and P'4 between 1977 and 1986, and the N4 reactor series was constructed between 1984 and 1993 (see Table 2.1).

Table 2.1 The three phases of construction of the current French nuclear reactor fleet

Reactor series	Construction dates	Number of reactors		Class in MWe
CP0	1971–74	6	34	900
CP1	1974–81	18		900
CP2	1976–82	10		900
P4	1977–80	8	20	1300
P'4	1980–86	12		1300
N4	1984–93	4	4	1450
Total	1971–93		58	Total net capacity in MWe: 63,130

MWe = megawatt electric

Source: Based on Brottes and Baupin (2014).

2   Although six nuclear reactors of French UNGG design had been built (at the Chinon, Saint Laurent des Eaux, and Bugey sites), the decision was made to switch to the cheaper American PWR model. The Alternative Energies and Atomic Energy Commission (Commissariat à l'énergie atomique et aux énergies alternatives, CEA) was strongly supportive of the UNGG model in contrast to EDF, which supported the Westinghouse model.
3   At the time, France's electricity was mainly derived from imported oil power plants. Hence, electricity was very expensive considering the oil crisis.

Although the civil 'all nuclear' plan was not fully realised, almost 80 per cent of France's electricity production is nuclear-based and even today not entirely disconnected from military interest. In addition to the electronuclear reactors, uranium enrichment facilities such as at the Tricastin site provide joint military and civil services. Indeed, as well as being part of France's fuel recycling capacity, these facilities enable plutonium extraction. Although the military origin is part of the fleet's grand history, the military connection lingers.

The same actors that enabled the development of the current nuclear fleet still exist today—more than exist, they are still at the heart of the nuclear industry in France. Three main actors constitute this industrial configuration. First, EDF is France's monopoly electricity utility. In 1946, it was decided that this company be a nationalised integrated monopoly to avoid competition and benefit from state support. Unlike in the United States or the United Kingdom where many electricity utilities coexisted, the French National Innovation System did not encourage market competition. Even though EDF has been a limited-liability corporation under private law and no longer state-owned since 2004, 84.5 per cent of its shares are still retained by the French government (EDF 2014). Accordingly, EDF still holds its historic title of a 'state within a state' (*etat dans un etat*).

Second, supporting EDF, Areva is the main nuclear engineering firm and second major actor in the nuclear industry. Areva is divided into Areva Nuclear Power (Areva NP) and Areva Nuclear Cycle (Areva NC). Areva NP constructs the nuclear reactors, whereas Areva NC is responsible for fuel cycle expertise, including reprocessing spent fuel and waste management. Areva NC has had experience in dealing with nuclear waste and spent fuel since the construction of the recycling facility in La Hague in 1966. The third actor at the core of the industry is the Commissariat à l'énergie atomique et aux énergies alternatives (CEA), which was institutionalised in 1945 to undertake nuclear research for national defence interests. Today, it still provides a dual military and civil research service.

Coordination between and within institutions was guaranteed by the traditional French technocracy. After the Second World War, the workforce within an engineering firm in France was almost always entirely composed of members of the 'grand corps of the state' (*corps d'etat*).[4] This meant that, during the nuclear acceleration, not only the Minister for Industry but also the chief executive officers and senior executive staff in the nuclear industry at the time had all studied at the same prestigious university, Ecole des Mines. Due to the homogeneity of their education, a natural efficiency arose between these three institutions and their executive leaders. As government officials often had previously worked within or with these prestigious institutions, the efficiency was extended to the political sphere. The proximity these historic institutions have with one another and with the government means that political *grandes programmes* could easily reach consensus from both right- and left-wing Members of Parliament, without much public debate (Gerbault 2011). This efficiency was moreover facilitated by the vertical integration of the nuclear value chain within the single institutions. The grand result of this centralised and integrated configuration was an easy and undisputed nuclear acceleration.[5]

The last defining element responsible for the French nuclear industry's political strength is France's reliance on nuclear exports. First, France is the largest electricity exporter in the world and part of the European Union's (EU) interconnected electricity grid. In 2015, France generated 546 terawatt hours (TWh) compared to 476 TWh consumed (RTE 2016). This tells us that the electricity capacity in France is on average much higher than it needs to be in order to sustain domestic consumption. For technical reasons, it is costly to shut down a nuclear power plant and restart it. Therefore, as electricity demand fluctuates, exports are essential to keep nuclear power plants operating at a roughly constant rate. France enjoys nuclear dominance thanks to the European interconnected grid, which gives the industry the opportunity to generate constantly at a surplus.

Also in 2015, France exported 91.3 TWh of electricity and imported 29.9 TWh (RTE 2016). The quantity that is being imported reflects the electricity price volatility and that nuclear is not always the cheapest option in the EU. On the one hand, Switzerland imports most of the French exported electricity, followed by Italy, Germany, Belgium, the UK, and Spain. On the other hand, France imports mainly from Germany, when renewable electricity is produced more cheaply than nuclear power,

4   The *corps d'etat* historically includes graduates from prestigious engineering schools (such as les Mines and Des Ponts) as well as the National Institute of Statistics and Economic Studies.
5   Other actors such as the Autorité de sûreté nucléaire (ASN) are discussed later.

as well as from Switzerland and Spain. As heating-cooling systems are primarily electric in France, it is the most thermo-sensitive state in the western European region. Even with an average surplus of nuclear power, during winter cold spells France does not have enough dispatchable capacity and is forced to import electricity from neighbouring countries.

Second, France exports nuclear reactors. Areva NP constructs the nuclear reactors and has exported 102 light-water reactors (LWR) worldwide. Most recently, Areva is exporting a new reactor design, the European pressurised reactor (EPR). There are four examples of current EPR builds: one in Finland, two in China, and one in France. However, this new technology is proving problematic, symbolised by the controversy related to the Hinkley Point C investment for the construction of two EPRs in the UK. The details concerning the underlying issues about the EPR design and its export opportunity will be discussed later.

Third, France exports reactor safety expertise.[6] Reactor safety standards have become a major export opportunity since the international accidents of Chernobyl and Fukushima. In fact, France has been very active in developing security norms at a European level since the aforementioned disasters made additional precautions necessary.

Fourth, as part of France's safety expertise, Areva NC has been able to export a fuel reprocessing plant, similar to the one in La Hague, in Rokkasho, Japan. More on the topic of radioactive waste management is discussed in the third section of this chapter.

Together, if all direct and indirect value streams of nuclear exports are considered, according to a study by PricewaterhouseCoopers Advisory (2011), close to €6 billion worth of exported nuclear electricity and other nuclear goods and services are gained annually, making nuclear power represent a potential €45 billion turnover and 2 per cent of French gross domestic product (GDP). Nuclear exports represent a significant revenue stream that France currently heavily relies on.

---

6   According to French Law No. 2006-686 on transparency and security in nuclear matters, nuclear security includes both nuclear safety (referring to the construction, functioning, and decommissioning of nuclear reactors) as well as radioprotection, prevention against malevolent acts, and civil security in the event of an accident.

In the 1970s, France made decisions that locked the energy sector into a nuclear technology pathway for electricity generation. Its military origin enabled nuclear investment, the main industry actors are close to government, and the economy depends on nuclear exports. This explains how the industry historically became so exceptional. For the industry to actually be able to trump political plans to reduce the share of nuclear power, the configuration has to remain stable and exports successful. However, a critical technicality threatens these two elements: the end of the current nuclear fleet's lifetime. No matter the policy, the fleet will have to be replaced this decade. This gives additional room and strength to the French policy, which fits into a broader global energy transition, to step in and propose an alternative to replacing the fleet with new nuclear reactors. Pressure for a new global energy paradigm coincides with the necessity to have a new power fleet, and potentially to transition to a more renewable system if the nuclear industry cannot prove its new reactor design and how nuclear will fit into this new energy paradigm.

A new energy paradigm

Globally, environmental and climate concerns are creating the foundation for a new energy paradigm; the divorce from fossil fuel–based growth and the increase of renewable energy generation. France indeed has a low-carbon economy based on nuclear power; however, as the reactor fleet in France has an average age of 31 years in 2016, it will soon reach the end of its official lifetime.[7] France's political energy plans and the global transition align with the need for the current fleet to be replaced, creating a unique opportunity for a new energy system. Hollande's Energy Transition Law is a first indication of what the political overarching plans and objectives are for the new system.

This section discusses the role that the nuclear industry in France has had in the conversation about the new energy paradigm. First, I analyse the details of the Energy Transition Law for the nuclear industry and how it is being enforced in France. Next, there is a growing trend for increased renewable electricity capacity, which is accompanied by a downward market pressure on nuclear electricity, as renewable energy becomes competitive. This section concludes with an analysis of how nuclear

---

7   The initial lifetime of a nuclear reactor in France is 40 years.

energy was perceived during the United Nations Framework Convention on Climate Change Conference of the Parties' (COP) negotiations in Paris (COP21) and whether it will be part of the new energy paradigm.

The Energy Transition Law stipulates a series of objectives for France to transition to a more sustainable economy. The motivations for this law originated with the Fukushima accident. During the presidential campaign of 2012, a debate around the future of energy began between Nicolas Sarkozy supporting nuclear power and Hollande supporting renewable energy. The final law text is based on the conclusions that were made during the 2012–13 national public energy debate on how to achieve a sustainable energy system. Although the law suffered a fierce political battle with more than 1,000 amendments,[8] in 2015 it was finally presented as the most ambitious energy law to date by the former Minister for the Environment, Ségolène Royal. The 66 articles contain most notably:

- a reduction in the share of nuclear energy production in the electricity mix: down to 50 per cent by 2025;
- a cap on nuclear power capacity: at the current level of 63.2 gigawatt electric (GWe);
- an increase in the share of renewable energy in the electricity mix: up to 40 per cent of final electricity production by 2030[9] (by increasing the share of renewable energy up to 32 per cent of final energy consumption by 2030);
- a reduction of final fossil fuel energy consumption: 30 per cent reduction compared to 2012 figures by 2030;
- a reduction of final energy consumption: 50 per cent reduction compared to 2012 figures by 2050; and
- a reduction of greenhouse gas emissions: 40 per cent reduction compared to 1990 figures by 2030.

---

8   The main topics that were disputed within the text of the law and that formed the amendments related to the date and even inclusion or not of a date for the reduction down to 50 per cent of nuclear, the nuclear cap exact figure, figures around the reduction in final energy consumption, and more ambitious targets for sustainable and low-energy intensive housing (Energiewende Team 2015).
9   This would be up to 32 per cent of the total final energy share, which was at 18.7 per cent in 2014 (RTE 2016).

The Energy Transition Law, although ambitious and concrete, does not give any indication as to how the objectives will be met. This is the role of the Multi-Annual Energy Programme, which aims to give a detailed plan of action for how to reduce the nuclear share. On 1 July 2016, Minister Royal published a 275-page document on the ministerial website. The Programme was reviewed by the Energy Transition Expert Committee, the Autorité de sûreté nucléaire (ASN), and public consultation was undertaken, before it was officially adopted on 27 October 2016. There is not yet any mention in this document of how many reactors will be shut down. Decisions on closures and lifetime extensions beyond 40 years are stated to commence in 2019. The only quantitative detail concerning nuclear electricity generation is that annual nuclear production will be reduced by 10 to 65 TWh by 2023, meaning a reduction of only 2.5 to 15.6 per cent of current production. For reference, a 10 TWh reduction only just corresponds to the shutting down of the two oldest reactors in Fessenheim.[10] On average, 65 TWh corresponds to 10 reactors being shut down.

Both the lower and upper limits are very different from the Court of Audit's estimations. In their report, 17 to 20 reactors will need to be shut down before 2025 in order to conform with the Energy Transition Law. Of course, these numbers are subject to gross electricity consumption and production increases; however, it still seems that there is a gap between the law and the document pertaining to how it should be enforced. Practically, this gap means that the nuclear lobby has material influential inertia and played a role in determining the Multi-Annual Energy Programme. Not a single reactor has been shut down in France since the Energy Transition Law was passed. The oldest still-operational nuclear plant in Fessenheim is at the centre of the controversy.

The pressures on Fessenheim to shut down are threefold: its official licence expiry date is 2018, it lies in a seismic risk zone at the border with Germany, thus receiving international pressure for closure, and the Energy Transition Law itself. Concerning the end of its lifetime, former President Hollande had made it a campaign promise to shut down the Fessenheim reactor before the end of his mandate (2017). He was, however, unable to meet this promise, the responsibility of which has now been passed on to Presdient Emmanuel Macron. The international concerns regarding the

---

10   In 2015, the two reactors at the Fessenheim site generated 13 TWh.

Fessenheim reactors and their proximity to the border are also legitimate, as the reactors actually have been stopped by the emergency systems on multiple occasions in the past because of technical issues.

The third legal deadline is due to a combination of the cap on nuclear capacity stated in the Energy Transition Law and the new EPR being built in Flamanville 3. Although the build of the new 1,650 MW EPR, the first of its kind in France, has been significantly delayed and has encountered numerous problems, it is expected that it will be commissioned in 2018. As a result of the cap in the law, the two reactors at Fessenheim will have to be shut down before the Flamanville 3 reactor can come online. Despite immense pressure internationally and by the French government, EDF has only recently accepted to commence the legal requirements that would lead to the anticipated closure of the two reactors on the basis of previous unsatisfactory compensation. EDF stated that it would refuse the plant closure unless it received €2 billion to €3 billion, which is nothing close to the €80 million to €100 million that was proposed by former Minister Royal in May 2016 (*Le Monde* 2016). On 24 January 2017, €490 million were proposed by the government and seem to be the adequate compromise for EDF. Even so, the Fessenheim reactors remain fully operational today.

Since the election of President Macron, there has been renewed confidence that the energy transition will indeed accelerate. The nomination of environmental activist Nicolas Hulot as Minister of Ecological and Solidarity Transition is a sign that the energy transition will move forward. An illustration of this was the loss in EDF share value by 6.57 per cent the day of his nomination (Stothard 2017). President Macron has underlined that the reduction of the share of nuclear electricity production to 50 per cent is a priority, that he stands by the closure of the Fessenheim plant, and that the renewable energy objectives described by the Energy Transition Law will be taken seriously (Macron 2017). He remains, however, unclear and doubtful about the 2025 objective. While he has presented himself as active in the fight against climate change, nuclear is still a controversial topic. After five years of promised nuclear reduction with no results during Hollande's mandate, Macron remains realistic and not idealistic in his approach to nuclear politics. Indeed, he has cautiously said that the future of the nuclear industry will depend on two factors: the results from the ASN real cost of nuclear electricity evaluation to be held in 2018, and the possibility to reform the governance structure at EDF considering the difficulty that policy has to drive nuclear decisions.

Even with activist Minister Hulot, the prospects for nuclear power under President Macron will be determined by external factors as he has recognised what little political power he has over the industry.

The nuclear industry is currently on track to scale-down the level of ambition of French energy policy plans; however, there is also downward pressure on the industry at a market level. Global market trends are favouring alternative renewable energy over nuclear power. Since the Kyoto Protocol was signed in 1997, countries worldwide are slowly implementing new renewable energy capacity. Between 2000 and 2015, wind, solar, and nuclear grid connections represented net cumulated capacities of 417 GWe, 229 GWe, and 27 GWe[11] respectively (Schneider et al. 2016). In the EU, the shift from nuclear to renewables is even more noticeable as electricity generated from nuclear plants has decreased by 65 TWh a year since 1997, whereas wind and solar generation have both increased by 303 TWh and 109 TWh respectively. The consequence of this renewables trend is an increase in the competitiveness of renewable energy compared to nuclear-generated electricity.

In order to compare competitiveness of different energy generation resources, the levelised cost of energy (LCOE) indicator is used.[12] In the Organisation for Economic Co-operation and Development (OECD) countries, wind's LCOE fell by 50 per cent between 2009 and 2014, reaching in 2015 a median LCOE of just over US$60 per megawatt hour (MWh) (ranging from US$33 to US$135 per MWh) (IEA 2015). In 2015, nuclear power's median LCOE was just over US$52 per MWh (ranging from US$29 to US$64 per MWh). The LCOE estimates represent a very large range, making it arguable that in certain contexts renewable power today is already cheaper than nuclear electricity.

Contrary to popular belief and certainly to many French politicians' views, reaching 100 per cent renewable electricity is not necessarily more expensive than nuclear power. The French Environment and Energy Management Agency (Agence de l'environnement et de la maîtrise de l'énergie, ADEME) published a pilot study in 2015 concluding that France could solely depend on renewable electricity by 2050, and estimated that this scenario would be at a comparable cost to a situation including

---

11   This figure includes long-term operation reactors.
12   The LCOE includes installation and operational costs, taxes, maintenance, and revenue requirements of the system over its lifetime.

nuclear power. The study provides two major conclusions in favour of renewable energy. First, renewable electricity production potential[13] is estimated at 1,268 TWh per year, which is three times the estimated 2050 demand for electricity in France.[14] The renewable electricity mix proposed by ADEME is 63 per cent wind (offshore and onshore), 17 per cent solar (photovoltaics and thermal), 13 per cent hydro, and 7 per cent renewable thermal (biomass and geothermal). The second conclusion to be drawn from the study is that the price of 100 per cent renewable electricity is in the same order of magnitude as the current political proposition of 40 per cent renewable and 50 per cent nuclear. With either the 100 per cent or 40 per cent renewables scenario, the cost of electricity has been estimated to be around €120 per MWh. In 2016, electricity costs were estimated at €90 per MWh. Ultimately, the price of electricity for end consumers would increase proportionally to the cost by 30 per cent compared to today's price, regardless of whether the government decides to transition to a 100 per cent renewable electricity mix or remain at 40 per cent for 2050 (ADEME 2015). These underlying conclusions, however, are very controversial. In short, even in France, there is growing market pressure favouring renewable electricity.

As the global energy paradigm is being primarily dictated by environmental concerns, to determine what role nuclear will have in this new paradigm, this last segment focuses on past COP21 negotiations. Whereas French policy and market trends are favouring renewables to the detriment of nuclear power, the international sphere is more ambiguous about what role nuclear will have in the global energy system. Although not a *renewable* energy source, nuclear is still a *low-carbon* energy source in comparison with fossil fuel plants. In 2014, the Intergovernmental Panel on Climate Change published its fifth assessment report and concluded that nuclear power has the second lowest life-cycle emissions analysis after wind generation sources (Schlömer et al. 2014: 1335). As nuclear power's carbon footprint is low, a controversy exists around its role in the climate change negotiations. Since COP21, held in Paris over a two-week period in December 2015, all 195 states have signed a binding agreement aiming to not exceed 2 degrees Celsius of global warming and, if possible, only 1.5 degrees Celsius by the end of the century. Furthermore, states plan to

---

13   Including solar, wind, biomass, geothermal, hydro, and marine.
14   This estimate takes into account technology-related energy efficiencies.

review their Intended Nationally Determined Contributions pertaining to greenhouse gas mitigation and adaptation strategies every five years, the aim being that each new contribution will be more ambitious than the last.

As COP21 took place in the heart of France, it was expected that nuclear energy would be debated in the context of climate change, especially as EDF was one of its major sponsors. Surprisingly, it seemed nuclear was the elephant in the room. Although organisations with strong public opinions were present and had representative booths and side-events, direct attention was hardly put on the subject of the future for nuclear energy. Prior to the event, EDF had been tweeting so evidently in favour of nuclear that it was being accused of greenwashing. However, during the event itself, no discussions on this topic were initiated. The IAEA refused to explicitly advocate for nuclear power.

On the other side of the spectrum, Greenpeace, a strong environmentalist and anti-nuclear association, did not address the nuclear issue at all. The OECD organised the only official event referring to nuclear energy: 'Why the climate needs nuclear energy'. The latter was a small side-event that took place one day before the final text was agreed upon; enough to say it did not have an impact on the agreement. Nuclear power, although represented, remained passive. Even though no decision was made to explicitly include nuclear in the new energy paradigm, no decision was made to exclude it either. Concerning nuclear energy in the context of climate change, ambiguity is all that can be taken away from COP21. For now, climate change negotiations are shaping the new energy paradigm to be low-carbon but not necessarily only renewable.

It is understandable that the literature on the French energy projections does not come to a consensus; the position of the present French nuclear industry seems to contradict both long-term policy goals and market trends. Confirming this ambiguity, during historical events, such as COP21, where the decisions are being made about how future energy will be produced, there is no common global nuclear outlook. Environmental concerns are strengthening the renewables sector but not directly modifying nuclear power production. In fact, the only historical events that have ever led to dramatic changes in global nuclear production have been nuclear accidents. It was, in fact, the Fukushima accident that led to the political compromise between the Ecologist Party, which supports a complete nuclear phase-out, and former President Hollande, to include the nuclear reduction share in the Energy Transition Law.

Nuclear safety

Both the Chernobyl and Fukushima accidents sent global messages on nuclear safety risks and were consequently followed by nuclear reactor closures. France has, however, historically been somewhat disconnected to this risky reality, as the industry has tried to dissimulate the danger of nuclear. This next section discusses how the French nuclear industry has resisted and is resisting the nuclear decline trend in response to safety risks. First, the French nuclear industry built a strong communication strategy to keep the reactors safe from public opinion concerns. This strategy was historically successful, especially after the Chernobyl accident. Second, however, after Fukushima, despite the industry's investments in nuclear safety reinforcements to maintain trust with the domestic population, French nuclear became unpopular. The last part of this section focuses specifically on waste and how the French industry is dealing with this particular aspect of nuclear safety.

The French nuclear industry has always paid great attention to nuclear risks and the importance of nuclear safety. Realising that the sheer concept of nuclear risk could be detrimental to the industry, it felt forced to dissimulate the danger of nuclear accidents. Perhaps the biggest nuclear propaganda in French history was the media release related to the Chernobyl accident: 'the Chernobyl radioactive cloud stopped at France's borders' (Morice 2011). Simplistically, the reason that was given to explain how this could be possible was that the wind was headed north and not west.

The nuclear industry focused on the positive side of nuclear energy, by referencing the independence from Russian gas and the modernity of nuclear technology. 'In France we may not have oil, but we have brains' was the national slogan used by Valéry Giscard d'Estaing during the 1974 political campaign. The nuclear industry's great efforts to fight nuclear scepticism related to nuclear accidents paid off, as the common belief that French nuclear technology is safe and that no accident has occurred is widespread and is even expressed in public declarations. In 2012, former French President Sarkozy explained: 'For as long as nuclear power has been generated in France, we have never experienced a major accident' (Vie Publique 2012). Pierre Tanguy, the Inspector General for

EDF security, also asserted that there has never been a nuclear accident in France. There have, however, been numerous nuclear *incidents*, as is easy to believe considering the number of reactors.

Moreover, two nuclear *accidents* have also occurred but were not fully communicated through the media, giving the impression that the French reactors are of exceptional quality. One accident in particular, at the Saint-Laurent-des-Eaux nuclear plant in 1980, was classified a level 4 accident on the International Nuclear and Radiological Event Scale (INES; see IAEA and OECD/NEA 2013).[15] To gain perspective, both the Fukushima and the Chernobyl accidents were ranked at the maximal level of 7. The Saint-Laurent-des-Eaux nuclear accident resulted in a partial fuel meltdown and an automatic shutdown of the reactor. A level 4 nuclear accident indicates that there is 'limited off-site risk' (detailed definitions are provided in IAEA and OECD/NEA 2013) and minor radioactive release or 'public exposure of the order of prescribed limits'. However, it was publicised in 2015 that there was also radioactive release into the river Loire due to the 1980 Saint-Laurent-des-Eaux accident.

The investigators who provided the evidence for the documentary *Nucléaire, la politique du mensonge*, revealed that there had been an ongoing release of plutonium into the river for a duration of five years since the accident occurred (Canal+ 2015). By comparison, the Three Mile Island accident in Dauphin County, Pennsylvania, in 1979, which included a nuclear meltdown and release of radioactive material into the Susquehana River, was ranked at level 5.[16] It is surprising how the accident of Saint-Laurent-des-Eaux did not have a similar media effect on the French public's perception of nuclear safety as the Three Mile Island accident did in the US. Vague communication has been a useful tool to reassure the French public after such accidents. Neither the Saint-Laurent-des-Eaux, the Three Mile Island, nor the Chernobyl accidents affected nuclear energy growth in France owing to the industry's danger dissimulation efforts.

In contrast to the accidents from the twentieth century, the accident in 2011 in Fukushima occurred in a new globalised world. In other words, since the nuclear industry is a global market, the accident had more wide-

---

15   The lower levels refer to incidents, whereas the higher levels refer to more significant accidents.
16   Level 5: 'Limited release: Likely to require partial implementation of planned counter-measures' (see IAEA and OECD/NEA 2013).

ranging impacts, such as the announced nuclear phase-outs. Still, France did not choose at the time to change energy policy. Sarkozy reaffirmed on 24 March, just after the accident, that the nuclear energy policy was not to be questioned (INA 2011). However, the French industry's global position in the nuclear market was undeniably affected.

It is worth studying how France reacted to the Fukushima accident as a means to test the strength of the industry against trends for nuclear decline. At a European level, France had to prove that it still had the most reliable fleet by exporting safety expertise. France boosted public trust by being extremely active at a European level and providing maintenance and security measures to reinforce security thresholds, which were consequently integrated into the European Nuclear Safety Framework.[17] For example, nuclear plants in the EU since the Fukushima accident are obliged to withstand an incoming aeroplane crash.

At a national level, the French reaction to the Fukushima accident was to reaffirm the safety of the French reactors. To do so, the independent nuclear safety authority ASN was given permission for the first time to carry out audits of all 58 reactors.[18] The conclusions of these audits noted that even though no immediate shutdowns would take place, safety reinforcements had to be made as soon as possible. The suggestions made by the ASN included increasing back-up cooling capacity, crisis control centres, and the obligation that each plant have a 'rapid reaction force', meaning staff that are trained specifically to control nuclear emergencies. At the time, these safety reinforcements were estimated to cost €13 billion (Crumley 2012).

Despite the industry's efforts to regain trust through safety investments, these were not enough to avoid a serious dip in public opinion polls regarding the popularity of nuclear energy.

In June 2011, the Ipsos Mori poll was conducted and collected data suggesting that 67 per cent of the French population either somewhat or strongly opposed nuclear power (Ipsos 2011). Another poll showed that

---

17   As the initiator of the Western European Nuclear Regulators Association, the French nuclear industry was very active in informing how to undertake the 'stress tests', drawing from its own experiences and expertise. The 'stress test' refers to the risk and safety assessments to be carried out on all EU nuclear plants after the Fukushima accident (Ministère du Developpement Durable 2012; Dehousse with Verhoeven 2014).

18   French Law No. 2006-686 on transparency and security in nuclear matters mandated the founding of the ASN, an independent administrative authority.

57 per cent of French respondents were in favour of phasing out nuclear power (Buffery 2011). At this stage, there was evident conflict between public opinion of nuclear and the nuclear industry, the latter dismissing the safety concerns. Despite continued nuclear power generation, public opinion in 2011 showed signs of the precarious situation that the French nuclear industry is now in. The risk of nuclear accidents, although not strong enough to shut down any reactors in France, is undermining the 'safe' image that the industry had worked so hard on and thus is weakening the industry's political influence.

A case study for controversy between public opinion and the French industry related to nuclear safety is nuclear waste management. The difference with waste, as opposed to nuclear accidents, is that closing down reactors will not eliminate the problem. Half a century's worth of waste has already been produced and is waiting to be disposed of properly. Although for many countries the issue of nuclear radioactive waste alone has been enough to discourage them from engaging in nuclear power activities, France has once again taken a strategic route to dissimulating the danger involved. The promise was made that nuclear energy produces little waste, in fact no more than the equivalent volume of 'an Olympic swimming pool', according to Anne Lauvergeon during her time as president at Areva between 2001 and 2011.

In fact, the total volume of radioactive waste in France at the end of 2013 was estimated at 1,460,000 m^3 (ANDRA 2013), 60 per cent of which was derived from nuclear totalling 880,000 m^3. Although this is much less than any fossil fuel power industry has created, it is certainly more than an Olympic swimming pool's volume (close to 3,000 m^3). Not only political euphemisms originating from the strong nuclear lobbies are used, but also direct influence on what is officially considered as nuclear waste shows the stronghold that the industry has on policy. The articles related to radioactive waste management in French Law No. 2006-686, for example, state that if a material can be potentially reused it is not considered as waste.[19] As materials containing uranium and plutonium can both be potentially reprocessed, any radioactive component containing these elements has been excluded from the official quantity of waste.

---

19 'A radioactive material shall include any radioactive substance that is intended for future use, after treatment, if need be. Radioactive waste shall include any radioactive substance for which no further use is prescribed or considered'.

France's partial solution to the waste issue so far has been its fuel cycle expertise, led by Areva NC. The site in La Hague was built by Areva NC to 'recycle' nuclear waste by reprocessing spent uranium fuel, and separating the plutonium. The separated plutonium can then be used as mixed-oxide fuel for France's LWRs. There is a difference, however, between what has been recycled and what is recyclable. The La Hague site stores its recyclable material before reprocessing it and, if not used, the recyclable or 'recoverable material'[20] ends up in the waste category. Not all the 'recoverable material' is French. For example, Germany, Belgium, the Netherlands, Switzerland, and Japan have used the facilities in La Hague to process their spent fuel and to avoid conflict with their publics over waste storage; these orders, however, are declining (Schneider and Marignac 2008; International Panel on Fissile Materials 2015).

Nuclear fuel reprocessing and enriching poses several issues. First, plutonium extraction capacities raise the controversy of nuclear proliferation. In fact, to foster transparency, as Mark Diesendorf suggests, 'all civil uranium enrichment and reprocessing facilities [should] be placed under international control' (Diesendorf 2014). Second, reprocessing fuel is costly and generates large volumes of low-level waste.[21] The La Hague facility can only process up to 1,700 tonnes of heavy metal (tHM) of spent fuel a year. Knowing that LWRs use on average 21 tonnes of fuel and discharge 20 tonnes of spent fuel annually, France produces roughly 1,000 tonnes of spent fuel each year (Feiveson et al. 2011). Since the beginning of reprocessing, about 30,000 tHM of fuel have been reprocessed. Currently, roughly 1,200 tHM of fuel is being processed each year.[22] By the end of 2007, 13,500 tHM of nuclear spent fuel were still awaiting reprocessing in cooling ponds and dry cask storage areas at La Hague as well as on nuclear plant sites (Feiveson et al. 2011).

It has been known since its creation that the facility at La Hague produced radioactive waste that had to be managed in the long-term and the evident accumulation of waste finally encouraged the development of a geological underground radioactive waste disposal solution. Geological disposal of nuclear waste has been undergoing research for over 20 years in France, and has been actively controversial since the first mention of this possible

---

20 Recoverable material includes natural uranium, enriched uranium, plutonium, thorium, and spent fuel.
21 Low radioactivity level and long half-life.
22 Of which only 8.5 tHM of plutonium can be reused, the rest would end up in storage.

solution in the 2006 law and development of the Cigéo (Centre industriel de stockage géologique) project. This project was developed to design a solution for high-level and long half-life radioactive long-term waste management. If the project is accepted, the underground geological storage site would be situated at 500 m depth in the commune of la Bure in the department of Meuse in eastern France, and would be able to store up to 80,000 m³ of waste for millennia. The only other approved site of its kind is in Olkiluto, Finland. According to French law, it must be assured that the solution is also reversible in the event that a better waste management solution is discovered (Cigéo.com 2013).

The Cigéo project is currently the only viable solution for managing this type of waste; it is, however, extremely unpopular as residents near Meuse consider environmental and health impacts of living near a radioactive garbage dump (*Le Monde* 2015). Costs also contribute to the controversy; the Court of Audit estimates that the cost of the Cigéo project could reach €43.6 billion (taking into account increases in waste quantities) (Collet 2016). If the industry cannot manage waste correctly, its political influence will be further weakened.

After the Chernobyl accident, media and communication dissimulation was enough to keep public opinion stable. However, since the Fukushima accident, public opinion has been clearly disturbed. Still, the industry has been influential enough to keep its number one placeholder in the electricity market. The waste issue is more complex as it is an existing and growing long-term safety concern, exposing added costs to an already expensive safety investment. Instead of responding to nuclear risks by shutting down reactors, France has invested in nuclear safety, which will be an investment with never-ending, escalating costs.

The economics of nuclear energy in France

This last section focuses on the financial risks that are threatening the viability of France's nuclear industry. The extent to which the industry will be able to influence or resist government plans to reduce the share of nuclear power will ultimately be determined by whether it can financially continue as a commercial business. Each previous section has hinted at the economic concerns related to renewing the fleet, competitive renewable energy, and safety standards.

This section analyses the industry's financial situation. First, the current state of financial affairs in France shows signs of a weakened industry. Second, based on the factors that got the industry into this financial position, an analysis of the 'nuclear bill' sheds light on the industry's prospects for recovery. Third, the Hinkley Point C reactor controversy sums up the uncertainty on the economics of future nuclear builds.

In recent years, the world has witnessed a French nuclear industry financial crisis. Since 2011, Areva has been facing debt and bankruptcy challenges. In November 2014, Areva was rated by Standards and Poor's and downgraded to BB+ ('junk'). The monopolistic reactor engineering company accumulated €6.3 billion worth of debt in 2015 (Areva 2016), by which time its rating was once again downgraded to BB–.[23] Areva faces not only the problem of capital debt, but also of a skills gap. The last nuclear plant to be built in France was in 1993. The 1993 workforce lived through the 1970s nuclear expansion and had sound knowledge and experience of the nuclear field. However, as the construction of the Flamanville EPR began in 2007, 14 years since the last build, the active French nuclear workforce is either old and retiring or young and inexperienced. The skill gap and increasing operational costs are contributing to Areva's debt. Therefore, encouraged by the French government, a rescue plan for Areva's financial issues was decided upon.

EDF bought the majority of Areva NP for €2.7 billion in 2015. EDF now bears the entirety of the nuclear investment risk for France, and so its own fate is also heading towards an increasing debt, which by the end of 2015 was at €37.4 billion (EDF 2016). On 21 December 2015, EDF's risk profile got the company eliminated from the Commissariat aux Comptes 40 (CAC40).[24] Since entering the CAC40 10 years ago with shares worth €32, EDF's share value has decreased by 70 per cent, reaching €9.75 in January 2017 (Boursier.com 2017).

In addition, EDF faced deregulated electricity prices for the first time in 2016. Even though France passed legislation in 2007 to accommodate the European directive 92/96 for a liberalised European Electricity Market, in France the electricity market had hardly changed. In order to fully liberalise the market to competition, the Nouvelle Organisation

---

23 Areva's turnover was €4.2 billion in 2015 (Areva 2016).
24 The CAC40 is a benchmark in the French stock exchange market index. The 40 highest performing largest equities are part of the CAC40 and typically include France's most successful businesses.

du Marché d'Electricité Law of 2010 states that in 2016 the prices shall no longer be regulated for residential and professional consumers that have high electricity power requirements (at least 36 kilo-volt-amperes).[25] Instead of the state determining the electricity price based on EDF's financial needs, electricity prices shall be determined each year by the Commission de régulation de l'énergie.

EDF is losing clients and electricity price stability benefits. Especially as operational costs for EDF are increasing, a decrease in the electricity wholesale market price because of competition is affecting EDF's revenue stream significantly. Slowly but surely a wedge between political support and the nuclear industry, practically represented by EDF, is undermining the industry's political authority.[26]

For EDF to have any hope of regaining its monopolistic position in the electricity market, it will have to return to a financially stable position. To understand whether this development could eventuate, an analysis of the future economic risks associated with EDF's current financial position are discussed in this section. The 'nuclear bill' that symbolises the magnitude of these economic risks will consider the major costs that the industry will have to pay to continue doing business. For this analysis, we will only consider certain major costs, including fleet reparations (in line with safety measures), decommissioning, and, most importantly, the costs associated with the new EPR design.

Fleet reparations

As the fleet is reaching its life's end and it seems that in the short-term France is still on its way to continue nuclear production and not a phase-out, reactor's lives will have to be prolonged. Prolonging the lifetime includes reparations and maintenance in accordance with new safety standards for the whole fleet. EDF has estimated that it would cost €55 billion to prolong the lifetime of the entire current fleet by 10 years. If this is not already an enormous investment, the Court of Audit has estimated that this figure could well turn out to be double that.

---

25   The deregulation of electricity prices for residential customers is due to start in 2019.
26   Other significant costs, such as those pertaining to safety investments as mentioned earlier, will also contribute greatly to the 'nuclear bill'.

Decommissioning

EDF's debt could also increase due to decommissioning costs. France has little experience with even the beginning phases of decommissioning. Brennilis is the first French plant to have begun a decommissioning process. Since 1985, when it was first shut down, decommissioning of this plant has cost almost €500 million over a duration of 35 years. As the definition of a 'decommissioned power plant' has not been clearly articulated, EDF refers to a decommissioned site as having a status equivalent to 'unrestricted public reuse'. EDF has estimated that it could cost €30 billion (€500 million per plant) to decommission the entire current fleet. However, according to the Court of Audit, this is clearly an underestimate. The main reason given by the Court of Audit is that cost estimates of decommissioning in other countries have been much higher. For example, in the UK, decommissioning of 35 reactors is likely to cost €103 billion.[27]

A new EPR design

The final major cost for the industry relates to the renewal of the nuclear fleet in order to sustain current nuclear capacity after plants are decommissioned. Areva and EDF's solution to this issue is the EPR. Examples of the building time frames and costs will enable this analysis to determine how future investments in this technology might turn out. Areva boasts that the EPR is 'among the most powerful reactors in the world'. It is considered a Generation III model because of its high performance (1,650 MW capacity as opposed to 1,450 MW for Generation II reactors) and safety structures. Areva promised that electricity production costs would be 10 per cent less than with the Generation II plants. Other advantages of this new technology include a lifetime increase of 20 years, easier maintenance, and shorter construction time (only 57 months) (Areva 2017). The last promise made by the industry was that the EPR design would also be cheaper to construct compared to previous reactor designs.

Worldwide, there are four EPR plants being constructed by Areva. The unit 3 reactor in Flamanville has been under construction since 2007 and was originally set to come online in 2012. The original cost for construction

27   Each UK reactor has an average capacity of 900 MW (Goldberg 2010).

was estimated by EDF at €3.3 billion. Since 2007, construction delays and cost overruns have occurred after security assessments made by the ASN. In May 2008, construction was suspended due to 'anomalies' of the iron framework. Repairs resulted in an increase in costs by 20 per cent and an extra year of construction. After the Fukushima accident in 2011, along with new safety measures, new tests had to be completed, materials reinforced, and construction times increased, pushing back the operational date by another two years. After material shipping delays and a recent anomaly on the reactor vessel discovered in April 2015 by the ASN (2016), the plant should be operational by 2018 at a cost of €10.5 billion (see Figure 2.1).

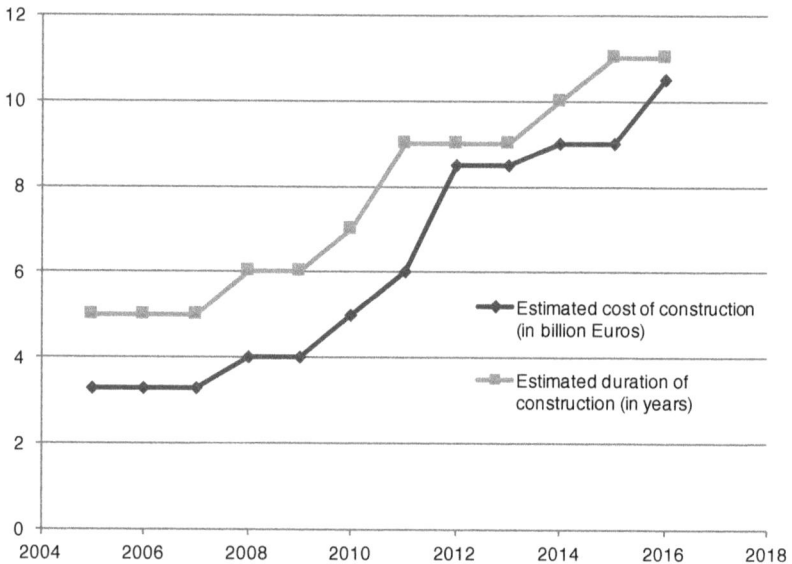

Figure 2.1 Evolution of costs and duration of construction estimates for the EPR in Flamanville

Source: Based on Soleymieux (2015) and reprinted with permission of the author.

In 2010, former EDF president François Roussely highlighted the reasons behind the Flamanville EPR failures by compiling a series of much contested radical recommendations to improve the outlook of the industry (Roussely 2010). Roussely provided insight into French-specific challenges. During the early stages of the EPR construction, EDF and Areva were distanced for business-related reasons. A lack of strategic coordination between both bodies impaired communication and thus commercial inefficiencies arose. Roussely also pointed to the complexity of the design of the EPR itself to be one of the reasons why it is encountering so many construction problems. According to his analysis, the fact that the French nuclear industry has not been able to attract any private investment and has always relied on state backing is another sign pointing to why the EPR projects are so economically challenging—they are simply not economically competitive. Although EDF and Areva are closer than they have ever been before, the issues that the EPR is facing clearly expose the fragility of this design.

So far, it seems that to invest in such costly and technically risky technology may not be feasible. Only if the new EPR model can be standardised would it be able to experience a decreasing learning curve thanks to economies of scale. There are EPRs under construction in Flamanville, Olkiluoto, and Taishan; as the same technology is being used in all cases, one could expect that the learning curve should already be decreasing or at least that the construction duration should decrease as lessons are being learnt. In reality, this is not the case. Instead, each project has taken on the characteristics of a 'first of a kind' technology. Mauro Mancini (2015) explains that intrinsic nuclear build characteristics disable a decreasing learning curve. These include location-specific characteristics, such as experience of the available workforce and the local political situation. If we compare, for example, the Flamanville nuclear power plant with the other EPR projects, high cost overruns and delay problems have occurred even though the same technology has been used.

The EPR at the Olkiluoto site had an original delivery price of €3 billion and was supposed to be commercially operational by 2005. In 2012, Areva re-evaluated this cost at €8.5 billion and, in January 2016, the operational date was pushed back to 2018. The two Taishan reactors began construction in 2009 and 2010 and were planned to go online in 2013. The current predicted date for commercial operation is the fourth trimester of 2018 (EDF 2017). At a potential cost of €8 billion for both reactors and a four-year delay, these nuclear EPR constructions are a relative success story compared to the French and Finnish experiences.

There are, however, lingering doubts about the quality of certain components and the security of the entire plant design (Radio Free Asia 2016). These concerns have been evident since April 2016, when the ASN discovered excessive carbon in the EPR design's main vessel (ASN 2016), which could lead to serious safety issues if cracks were to develop in consequence to the sub-optimal mechanical properties of the reactors. These concerns would eventually increase costs and push back the operation dates again. The EPR has been unsuccessful in establishing a reliable new design and there is no clear path for new construction in France. The EPR experience therefore puts serious doubt on the potential revenue streams that the industry would be able to derive from both EPR-generated electricity as well as EPR exports.

To highlight the industry's precarious financial situation, this very last section exposes the Hinkley Point C controversy. On the one hand, EDF has its reputation and possible growth at stake; on the other hand, this investment is so costly that the risks involved in going through with the project could lead to the downfall of the entire industry. Owned primarily by EDF and one-third by a Chinese contractor (China General Nuclear Power Corporation, CGNPC), the Hinkley Point C project includes two EPRs being built at Somerset. The two reactors are assumed to eventually cover 7 per cent of the UK's electricity demand. The cost for the two reactors in Somerset was originally estimated at £16 billion in 2013. This estimate was increased by £2 billion in October 2015 and the completion date was shifted from 2023 to 2025.

The cost-related part of the controversy contributed the most to delaying the final investment decision. As Simon Taylor (2016: 3) points out, Hinkley Point C will be 'the most expensive power station in history'. Indeed, it is setting out to be the most expensive object on earth. In March 2016, the chief finance officer of EDF, Thomas Piquemal, resigned because of the financial risks involved with accepting the Hinkley Point C investment. The Workers Council also protested against this investment actively, even though the state supports the project (Schneider and Froggatt 2015). EDF argues that although the cost of the plant was estimated in January 2016 at €25 billion, the payoff for France will be rich, as the UK agreed to pay £92.5 per MWh for 35 years for the generated nuclear electricity. When the first contract was drafted in 2012, this price was double the electricity wholesale price; it is now three times the current electricity wholesale rate (*Economist* 2016), making Hinkley Point C a very attractive deal for EDF—if it can deliver. Piquemal did not think it could and resigned.

In addition, a few hours before the investment decision was made on 28 July 2016, Gérard Mongin, one of the 18 board members who had the responsibility of making the decision at EDF, resigned as well (Cosnard 2016). Two spectacular resignations have occurred to symbolise the risk. Yet, after Mongin left, the decision was finally made to go ahead with Hinkley Point C (Bernard and Cosnard 2016). Bernard Lévy, chief executive officer at EDF, expected that all obstacles had now been overcome and that the project could begin. Another surprise event, however, took place the following day when Greg Clark announced '[t]he government will now consider carefully all the component parts of this project and make its decision in the early autumn' (*Financial Times* 2016). On 15 September 2016, British Prime Minister Theresa May officially gave the 'green light' to the project. The Hinkley Point C reactor clearly shows that the economics of the potential plant do not add up. It is in the hands of governments to support the investment if there is any chance of EDF not collapsing. It is fair to say that to regain financial stability, the industry needs support from the state, both financial and political.

It is obvious that the main actors of the nuclear industry, Areva and EDF, are in a financial crisis. The reasons behind their financial situation are linked to previous costs and the risk of future costs escalating beyond already high estimates. Even though the government is against supporting EDF financially for internal costs related to continuing nuclear power generation, as this would be against national policy, it is in favour of a strong nuclear export capacity. Economics will ultimately determine the future of national nuclear growth. Regarding external growth, however, this will depend on the success of the EPR design. As the design is in some cases economically unfeasible, its success will be determined by politics.

Conclusion

It is fair to say that the nuclear industry in France will not remain as it is today. The exceptional status it once had may give way to a more compromised position. Although the nuclear industry represents enormous political strength and stability, the consequences of the Energy Transition Law and efforts to reduce safety risks are rendering nuclear uncompetitive. The elevated costs of keeping the industry afloat reflect the high level of risk that the entire nuclear industry is facing. Nuclear investments without government support are nearly impossible. In the end, the question is whether the economics of nuclear will trump the political power of the nuclear establishment. Within France's borders, the electricity market is becoming so open to competition that the nuclear industry will cease to dominate electricity generation, as renewables will undercut nuclear prices. This will, however, not be a hastened decline, for as long as the current fleet's reactors are still generating, their electricity will be sold. The industry can only influence the speed of the transition, while it is heading towards an economically hostile environment. Outside of French borders the prospects are also quite bleak, despite the state backing the nuclear export industry. France could continue to extract revenue from safety expertise, waste management processes, and perhaps from the new EPR design. For this last case, it will not be a matter of economics but a matter of political support and, more importantly, proof to foreign governments that the EPR design can indeed deliver, which will determine sales of this Generation III reactor.

Could the French nuclear establishment have enough power to resist current political objectives to reduce the share of nuclear and favour renewables? In the long term, no. In the short term, however, the industry will influence the magnitude and speed of the decline in the context of the new low-carbon energy paradigm. The nuclear industry is going through its own transition, focusing more on exports such as the EPR to secure revenue, as the economics set pessimistic prospects for nuclear to continue dominating the national energy system. The new French energy paradigm does not contradict the growth and strength of the nuclear industry; both will transition together to accommodate a more sustainable system favouring renewables and nuclear exports.

References

ADEME (Agence de l'environnement et de la maîtrise de l'énergie), 2015. Un mix electrique 100% renouvelable? Analyses et optimisations. October. Angers: ADEME. www.ademe.fr/sites/default/files/assets/documents/rapport_final.pdf (accessed 1 February 2017).

ANDRA (Agence nationale pour la gestion des déchets radioactifs), 2013. Les volumes de déchets. www.andra.fr/pages/fr/menu1/les-dechets-radioactifs/les-volumes-de-dechets-11.html (accessed 1 February 2017).

Areva, 2016. 2015 annual results. Press release, 26 February. www.areva.com/EN/news-10717/2015-annual-results.html (accessed 1 February 2017).

Areva, 2017. EPR™ reactor: The very high power reactor (1,650 MWe). www.areva.com/EN/global-offer-419/epr-reactor-one-of-the-most-powerful-in-the-world.html (accessed 9 February 2017).

ASN (Autorité de sûreté nucléaire), 2016. Note d'information: L'ASN met à la disposition du public plusieurs courriers envoyés à AREVA depuis 2006 sur la fabrication de la cuve de l'EPR de Flamanville. 20 April. www.asn.fr/Controler/Actualites-du-controle/Controle-du-reacteur-EPR-en-construction/Anomalies-de-la-cuve-de-l-EPR/Courriers-relatifs-a-la-fabrication-de-la-cuve-de-l-EPR-de-Flamanville (accessed 1 February 2017).

Bernard, Philippe, and Denis Cosnard, 2016. EDF décide de lancer Hinkley Point, Londres tergiverse. *Le Monde*, 28 July.

Boursier.com, 2017. EDF. www.boursier.com/actions/cours/edf-FR00 10242511,FR.html (accessed 9 February 2017).

Brottes, François, and Denis Baupin, 2014. Rapport fait au nom de la commission d'enquête relative aux coûts passés, présents et futurs de la filière nucléaire, à la durée d'exploitation des réacteurs et à divers aspects économiques et financiers de la production et de la commercialisation de l'électricité nucléaire, dans le périmètre du mix électrique français et européen, ainsi qu'aux conséquences de la fermeture et du démantèlement de réacteurs nucléaires, notamment de la centrale de Fessenheim – Tome I. Assemblée nationale, July.

Buffery, Vicky, 2011. Majority of French want to drop nuclear energy-poll. Reuters, 13 April.

Canal+, 2015. *Nucléaire, la politique du mensonge*. Film documentary. May.

Chantebout, Bernard, 1986. La dissuasion nucléaire et le pouvoir presidential. *Pouvoirs 38, L'armée* 21–32.

Cigéo.com, 2013. Operations at Cigéo. 30 August. www.cigéo.com/le-fonctionnement-du-centre (accessed 1 February 2017).

Collet, Philippe, 2016. Cigeo: Segolene Royal fixe l'objectif de cout a 25 milliards d'euros pour le stockage des dechets radioactifs. Actu-environnement.com, 18 January. www.actu-environnement.com/ae/news/dechets-radioactifs-cout-objectif-cigeo-arrete-26065.php4 (accessed 1 February 2017).

Cosnard, Denis, 2016. EDF: hostile au projet Hinkley Point, un administrateur du groupe claque la porte. *Le Monde*, 28 July.

Crumley, Bruce, 2012. The Fukushima effect: France starts to turn against its much vaunted nuclear industry. *Time*, 4 January.

Dehousse, Franklin, with Didier Verhoeven, 2014. The nuclear safety framework in the European Union after Fukushima. Egmont Paper 73. Brussels: Academia Press for Egmont, Royal Institute for International Relations.

Diesendorf, Mark, 2014. *Sustainable Energy Solutions for Climate Change*. Sydney: UNSW Press.

Economist, 2016. What's the (Hinkley) point? 25 February.

EDF (Électricité de France), 2014. Investisseurs et analystes. www.edf. fr/groupe-edf/espaces-dedies/finance/investisseurs-et-analystes/l-essentiel (accessed 1 February 2017).

EDF (Électricité de France), 2016. Net financial debt and cash flow: Change in net financial debt. 30 June. www.edf.fr/en/the-edf-group/dedicated-sections/finance/investors-analysts/credits/net-financial-debt-and-cash-flow (accessed 1 February 2017).

EDF (Électricité de France), 2017. La centrale nucléaire de Flamanville: Une production d'électricité au cœur de la Normandie. Dossier de Presse, January. www.edf.fr/sites/default/files/contrib/groupe-edf/producteur-industriel/carte-des-implantations/centrale-flamanville%203%20-%20epr/presentation/2017dpfla3.pdf (accessed 7 September 2017).

Energiewende Team, 2015. So far, so good? The French energy transition law in the starting blocks. Energy Transition, 29 July. energytransition. org/2015/07/french-energy-transition-law/ (accessed 1 February 2017).

Feiveson, Harold, Zia Mian, M. V. Ramana, and Frank von Hippel, eds, 2011. *Managing Spent Fuel from Nuclear Power Reactors: Experience and Lessons from Around the World*. Princeton, NJ: International Panel on Fissile Materials, Program on Science and Global Security, Princeton University.

Financial Times, 2016. UK under fire over fresh delay to Hinkley Point nuclear plant. 16 July.

Gerbault, Alain, 2011. Les atomes crochus des politiques avec le nucléaire. *Slate*, 14 March. www.slate.fr/story/35467/lobby-nucleaire (accessed 1 February 2017).

Goldberg, 2010. Combien coûtera le démantèlement du nucléaire? *L'expansion*, 7 January.

IAEA (International Atomic Energy Agency), 2014/2015. Energy, electricity and nuclear power estimates for the period up to 2050. Vienna: IAEA.

IAEA (International Atomic Energy Agency) and OECD/NEA (Nuclear Energy Agency), 2013. *INES: The International Nuclear and Radiological Event Scale User's Manual: 2008 Edition.* Vienna: IAEA. www.iaea.org/sites/default/files/ines.pdf (accessed 1 February 2017).

IEA (International Energy Agency), 2015. Projected costs of generating electricity 2015 edition: Executive summary. Paris: IEA/Nuclear Energy Agency.

INA (Institut national de l'audiovisual), 1975. Debat sur l'implantation des centrales nucléaires au conseil regional. 13 February.

INA (Institut national de l'audiovisual), 2011. La France et le choix du nucléaire. 16 March. fresques.ina.fr/jalons/fiche-media/InaEdu05226 /la-france-et-le-choix-du-nucleaire.html (accessed 1 February 2017).

International Panel on Fissile Materials, 2015. Global fissile material report 2015: Nuclear weapon and fissile material stockpiles and production. Princeton, NJ: International Panel on Fissile Materials, Program on Science and Global Security, Princeton University.

Ipsos, 2011. Global citizen reaction to the Fukushima nuclear plant disaster. June. www.ipsos.com/sites/default/files/migrations/en-uk/files/Assets/Docs/Polls/ipsos-global-advisor-nuclear-power-june-2011.pdf (accessed 7 September 2017).

Le Monde, 2015. Un campement contre la poubelle nucleaire. 4 August.

Le Monde, 2016. EDF refuse d'enclencher la fermeture de la centrale nucléaire de Fessenheim. 15 June.

Macron, Emmanuel, 2017. Programme: En Marche! en-marche.fr/ emmanuel-macron/le-programme (accessed 26 June 2017).

Mancini, Mauro, 2015. The divergence between actual and estimated costs in large industrial and infrastructure projects: Is nuclear special? In *Nuclear New Build: Insights into Financing and Project Management*, 177–88. Paris: Organisation for Economic Co-operation and Development and Nuclear Energy Agency.

Ministère du Developpement Durable, 2012. Résultat des stress tests concernant les installations nucléaires. Press release, 1 October.

Morice, Louis, 2011. Tchernobyl: Quand le nuage s'est (presque) arrete a la frontiere. Le Nouvel Observateur, 7 September. tempsreel. nouvelobs.com/planete/20110907.OBS9926/tchernobyl-quand-le-nuage-s-est-presque-arrete-a-la-frontiere.html (accessed 7 September 2017).

PricewaterhouseCoopers Advisory, 2011. Le poids socio-économique de l'électronucléaire en France. May.

Radio Free Asia, 2016. Safety fears cause concern amid delays to China's Taishan nuclear plant. 7 March. www.rfa.org/english/news/china/safety-03072016114147.html (accessed 1 February 2017).

Reuss, Paul, 2007. L'épopée de l'énergie nucléaire: une histoire scientifique et industrielle. Paris: EDP Sciences.

Roussely, François, 2010. Avenir de la filiere française du nucleaire civil: Synthèse du rapport. 16 June.

RTE (Réseau de transport d'électricité), 2016. Bilan électrique 2015. Paris: RTE.

Schlömer S., T. Bruckner, L. Fulton, E. Hertwich, A. McKinnon, D. Perczyk, J. Roy, R. Schaeffer, R. Sims, P. Smith, and R. Wiser, 2014. Annex III: Technology-specific cost and performance parameters. In *Climate Change 2014: Mitigation of Climate Change. Contribution of Working Group III to the Fifth Assessment Report of the Intergovernmental Panel on Climate Change*, edited by O. Edenhofer, R. Pichs-Madruga, Y. Sokona, E. Farahani, S. Kadner, K. Seyboth, A. Adler, I. Baum, S. Brunner, P. Eickemeier, B. Kriemann, J. Savolainen, S. Schlömer, C. von Stechow, T. Zwickel, and J. C. Minx, 1329–56. Cambridge: Cambridge University Press.

Schneider, Mycle, and Antony Froggatt, with Julie Hazemann, Todahiro Katsuta, M. V. Ramana, and Steve Thomas, 2015. *The World Nuclear Industry Status Report 2015*. Paris: Mycle Schneider Consulting Project.

Schneider, Mycle, and Antony Froggatt, with Julie Hazemann, Ian Fairlie, Tadahiro Katsuta, Fulcieri Maltini, and M. V. Ramana, 2016. *The World Nuclear Industry Status Report 2016*. Paris: Mycle Schneider Consulting Project.

Schneider, Mycle, and Yves Marignac, 2008. Spent fuel reprocesssing in France. Research Report No. 4. Princeton, NJ: International Panel on Fissile Materials, Program on Science and Global Security, Princeton University.

Soleymieux, Loïc, 2015. EPR de Flamanville: De plus en plus en retard, de plus en plus coûteux. *Le Monde*, 21 April.

Stothard, Michael, 2017. Green activist and minister Nicolas Hulot to test Macron over energy. *Financial Times*, 19 May.

Taylor, Simon, 2016. *The Fall and Rise of Nuclear Power in Britain: A History*. Cambridge: UIT Cambridge.

Topçu, Sezin, 2013. La France nucléaire: l'art de gouverner une technologie contestée. Paris: Ed. du Seuil. doi.org/10.14375/NP.9782021052701

Vie Publique, 2012. Débat télévisé entre MM. Nicolas Sarkozy, président la République et François Hololande, député PS, candidates à l'élection présidentielle 2012, le 2 mai 2012, sur les projets et propositions des deux candidats et sur le bilan du président sortant. discours.vie-publique.fr/notices/123000884.html (accessed 9 February 2017).

3

Energy subsidies: Global estimates, causes of variance, and gaps for the nuclear fuel cycle

Doug Koplow

Abstract

Subsidies to energy cost hundreds of billions of dollars per year, often skewing market decisions in ways detrimental to environmental quality and social welfare. Subsidy reform could provide large fiscal and environmental gains, although remains politically challenging to implement. Growing data collection by international agencies and others has expanded the fuels and countries captured in international subsidy figures. However, important gaps remain regarding credit support, liability subsidies, natural resource leasing, and state-owned enterprises (SOEs). In addition, data on subsidies to the nuclear fuel cycle remain particularly weak, with no global estimates at all and very few national ones. This is despite US$4.4 trillion in projected investments in new nuclear facilities by 2050, much of which will be state-guaranteed, and heavy government involvement with many parts of the nuclear fuel cycle. These hidden subsidies promote the expansion of nuclear facilities, infrastructure, and capabilities throughout the world even when other alternatives could provide the same energy services more quickly, at lower risk, and for less money.

Introduction

Hundreds of billions of dollars per year in energy subsidies slow the transition to cleaner forms of energy and divert limited public funds from health, education, and other social objectives. Reforming these subsidies, particularly in the fossil fuel sector, is now broadly accepted as an integral strategy to address climate change. As recognition of the fiscal and environmental benefits of subsidy reform has grown, data collection and compilation have also expanded. This is important progress, although subsidy elimination remains politically challenging and the pace of reform slower than needed.

Current beneficiaries continue to invest heavily to fight reform efforts, and remaining gaps in subsidy data impede effective opposition. Global subsidy estimates differ by more an order of magnitude, with valuation differences being a primary driver. Variance in geographic coverage, the types of subsidy programs evaluated, and the fuels assessed also contribute to the large range. Coverage is much worse for the nuclear fuel cycle than for renewables or fossil energy, making inter-fuel comparisons impossible. Industry estimates in all sectors tend to skew low, complicating policy debates.

Imperfect data will remain a challenge for years to come, yet does not justify slowing reform. There are broad areas of consensus regarding definitions and valuations. Further, even the lower bound estimates by international agencies indicate a problem massive in scale, and with significant negative side effects. Of the more than 35 countries for which the International Energy Agency (IEA) assessed fossil fuel subsidies in 2011 (IEA 2012), nearly half spent a larger share of their government budget on fossil fuel subsidies than on publicly funded health care (Koplow 2015a).

Although much attention on subsidies within developed countries has been focused on support to renewable energy, subsidies to conventional fuels such as oil, gas, coal, and nuclear are more entrenched and generally larger. The percentage depletion tax subsidy for oil and gas in the United States, for example, is nearly a century old; accident liability caps on nuclear reactors date to the mid-1950s. Subsidies extend the operating lives of inefficient and polluting coal plants and mask the real costs of nuclear power, helping to propel reactor sales to countries with no experience managing a nuclear sector. Subsidies to oil production tip many uneconomic sites into profitability, needlessly unlocking decades of new carbon (Erickson et al. 2017). Critical price signals that would encourage market participants to select lower-risk and cleaner energy paths are lost.

This chapter provides a brief introduction to the many forms of energy subsidies and how they are measured, the degree to which they crowd out other social spending, and their global scale. As a first step in addressing the dearth of data on nuclear subsidies, the chapter also introduces common approaches used to subsidise nuclear power worldwide.

What is an energy subsidy?

Subsidies are most commonly thought of as cash payments from the government to an individual or corporation. They are rarely that simple: a dizzying array of mechanisms are deployed to transfer value to, and risks from, particular forms of energy (see Table 3.1). These include tax breaks, subsidised credit or insurance, trade restrictions, price controls, and purchase mandates. While not every type of subsidy will be relevant to every situation, focusing only on cash grants will greatly understate the complexity and magnitude of the problem. Similarly, because some types of subsidies are much more important to one form of energy than to others, a full survey is needed to avoid skewing results.

Table 3.1 Governments transfer value to the energy sector in many different ways

Intervention category	Description	Captured in global estimates?	
		Inventory	Price gap
Direct transfer of funds			
Direct spending	Direct budgetary outlays for an energy-related purpose	Yes	Possibly[†]
Research and development	Partial or full government funding for energy-related research and development	Yes	Possibly[†]
Tax revenue foregone			
Tax*	Special tax levies or exemptions for energy-related activities, including production or consumption (includes acceleration of tax deductions relative to standard treatment)	As reported	Possibly[†]

Intervention category	Description	Captured in global estimates?	
		Inventory	Price gap
Other government revenue foregone			
Access*	Policies governing the terms of access to domestic onshore and offshore resources (e.g. leasing auctions, royalties, production sharing arrangements)	No	Possibly[†]
Information	Provision of market-related information that would otherwise have to be purchased by private market participants	Yes	No
Transfer of risk to government			
Lending and credit	Below-market provision of loans or loan guarantees for energy-related activities	No	No
Government ownership*	Government ownership of all or a significant part of an energy enterprise or a supporting service organisation; often includes high risk or expensive portions of fuel cycle (nuclear waste, oil security, or stockpiling)	No	Possibly[†]
Risk	Government-provided insurance or indemnification at below-market prices	No	No
Induced transfers			
Cross-subsidy*	Policies that reduce costs to particular types of customers or regions by increasing charges to other customers or regions	Partial	Possibly[†]
Import or export restrictions*	Restrictions on the free market flow of energy products and services between countries	Partial	Yes
Price controls*	Direct regulation of wholesale or retail energy prices	Some	Yes
Purchase requirements*	Required purchase of particular energy commodities, such as domestic coal, regardless of whether other choices are more economically attractive	No	Yes

Intervention category	Description	Captured in global estimates?	
		Inventory	Price gap
Regulation*	Government regulatory efforts that substantially alter the rights and responsibilities of various parties in energy markets or that exempt certain parties from those changes. Distortions can arise from weak regulations, weak enforcement of strong regulations, or over-regulation (i.e. the costs of compliance greatly exceed the social benefits)	No	No
Costs of externalities			
	Costs of negative externalities associated with energy production or consumption that are not accounted for in prices; examples include greenhouse gas emissions and pollutant and heat discharges to water systems	No	Generally not

* Can act either as a subsidy or as a tax depending on program specifics and one's position in the marketplace.

† Intervention may be partially captured in price gap calculations if it affects domestic prices to end-users or if (as with cross-subsidies) the transfers move across fuel types that are measured independently in the price gap analysis.

Sources: Koplow (1998); Kojima and Koplow (2015); main section headings from OECD (2011).

Many subsidies shift risks off energy producers

Markets are largely about how risks and rewards are partitioned amongst investors, producers, and consumers. Some subsidies directly increase the return to a specific party. Often, however, the subsidies work indirectly by changing the risk and reward profile of a particular activity or investment. Either approach boosts returns for some individuals, companies, or products while worsening the market position of competitors.

Investment, safety, price, geological, and regulatory risks vary by energy type, but are significant factors in energy markets overall. Because it is impossible to precisely predict future financial returns in an uncertain market environment, investors make guesses about the magnitudes and probabilities of the risks they will face to approximate *expected* returns. This is evaluated against their cost of capital, risk tolerance, and other business objectives in driving investment and corporate strategy. Interventions that shift core risks away from the private sector will alter these estimates, and can significantly distort energy market choices.

Government policies that limit or eliminate key down-side risks can be extremely valuable in turning unprofitable projects into profitable, investable ones. The down-side hedge is particularly material where there is a great deal of uncertainty over the ability to make a technology work, or on the timing and cost of delivery. Examples include very long-duration exposure to uncertain costs (as with nuclear waste disposal) and high capital costs that must be recovered through product sales over an extended period of time (as with coal plants deploying carbon capture and sequestration). The competitive environment can change dramatically during the development and construction period of high-risk projects, making them particularly uncomfortable for investors. Responses to this discomfort include a much higher cost of capital or withdrawal of funding entirely.

The challenge on the public policy side is twofold. First, risk-based subsidies do not actually *eliminate* risks; rather, they *transfer* them from the subsidy beneficiary to somebody else. Most often, the 'somebody else' is the government (and therefore taxpayers). However, other outcomes are also possible. Liability caps shift risks associated with uncompensated damages from accidents to plant neighbours or industries dependent on a common resource (as with contaminated fisheries after an oil spill). Purchase mandates shift price risk to customers. Second, comparable substitutes that can be developed and delivered without the same complications often exist, making public risk absorption hard to justify on policy grounds.

Subsidised insurance programs that socialise private risks are common: nuclear accidents, earthquakes, flooding, and dam failures are but a few examples. Not only do the subsidies harm competing products with lower economic or operational risks, but aggregate risks to society may actually rise. Critical decisions on where to locate a power plant or mine, or how heavily to fund operational safety and worker training, are made by the plant owners and shaped by a fear of financial losses. If government subsidies shift too much of this risk away from these private decision-makers, owners can make irreversible investment, siting, or operational decisions that increase the societal risks of future problems.

Grants are merely a starting point: An overview of subsidy types

Table 3.1 provides a comprehensive overview of the main types of transfer mechanisms, along with an estimate of how well current global estimates capture particular types of government interventions. Global subsidy estimates rely on two main techniques: measuring the degree to which domestic energy prices lag market prices for a fuel (the 'price gap' approach); and aggregating the subsidies associated with hundreds of individual government programs supporting energy (the 'inventory' approach).

The many categories within Table 3.1 underscore both the complexity of markets and the importance of tracking multiple venues of support in order to properly gauge their overall scale and impact. Although global estimates of subsidy magnitude run in the hundreds of billions of dollars annually, the real values are likely significantly larger. Current estimates have material gaps in terms of subsidy types, fuels, and geographic regions captured.

Another notable point in Table 3.1 is that quite a few of the transfer mechanisms can act as a tax or as a subsidy, depending on the program details and the associated market environment. If program rules or disbursements change over time, the direction of impact can shift as well. Fees levied on oil and gas, for example, are often earmarked to support industry-related site inspections and cleanup, or to fund infrastructure construction and maintenance. If the fees exceed these other costs, they may partially act as a tax; if they cover only part of the cost, a residual subsidy will remain. Subsidies to energy consumers can sometimes act as a tax on producers, and vice versa. Teasing out these interactions is a significant challenge of subsidy measurement.

Recipients frequently tap into as many subsidies as they can, capturing different types of support from multiple levels of government. This process is referred to as 'subsidy stacking', and applies to state-owned enterprises (SOEs) as well as to private firms or individuals. SOEs are common in many parts of the energy sector, including the nuclear fuel cycle.

Sometimes the support to SOEs is obvious, such as bailouts when a state firm runs out of cash. More often, however, the subsidies become evident only when compared to a free market baseline. SOEs borrow money and pay interest, for example, but not at a market rate. They earn break-even or even some net revenues on operations, but far less than needed to generate a reasonable rate of return on billions in invested taxpayer capital. Where competing resources are both less capital-intensive and privately owned, the lack of a required return on investment can be a large impediment to competitive energy markets. SOEs may pay no taxes, have inadequate insurance coverage relative to the riskiness of their operations, or receive below-market access to publicly owned minerals.

Energy subsidies as an 'investment': The opportunity costs of current patterns of support

Access to modern energy services has been clearly demonstrated to improve health, productivity, and welfare for recipient families (World Bank 2010: 19). Yet, billions of people lack access. As of 2013, 17 per cent of the global population had no electricity and a much higher proportion had limited or low-quality access. Nearly 40 per cent were without access to clean cooking fuels (IEA 2015: 101–6).

Intentionally or not, global energy subsidies represent a massive investment in the energy sector. It is reasonable to ask whether this spending helps to finance poverty reduction and the development of cleaner and better fuel sources for the world's poor. While there are clearly some benefits, studies suggest that the vast majority of support continues to be ineffective in reducing energy poverty.

Fossil fuel subsidies in the developing world have mostly been focused on keeping transport and cooking fuel prices below market levels. Although often justified as a poverty-reduction strategy, in practice much of this support 'leaks' to wealthier citizens. Higher income groups use much more energy per capita. They consume energy from power and gas networks that frequently do not even extend to the poorest areas due to the high installation costs and low purchasing power in those districts. Other subsidies rely on political connections to access, bypassing the poor. Low domestic prices also frequently trigger development of black markets that move subsidised supplies to other countries or illegal domestic sales at market prices or above.

Developing country surveys found that only 8 per cent of fossil fuel subsidies reached consumers in the poorest 20 per cent of the population (IEA 2011), and less than 25 per cent reached the poorest two quintiles (IEA, OPEC, OECD, and the World Bank 2010: 24).[1] Leakage rates for gasoline have been particularly high according to International Monetary Fund (IMF) analysis, with an estimated US$33 in subsidies to gasoline required for each US$1 that actually reached the poorest 20 per cent of society (Arze del Granado, Coady, and Gillingham 2010: 13).

The IEA estimated that US$13.1 billion in new capital was invested in 2013 to improve access to electricity and clean cooking fuels, an increase from the previous year (IEA 2015: 105). Still, the investment comprises a mere 2.5 per cent of the estimated half a trillion dollars in fossil fuel subsidies to consumers that year. Within the spending on energy poverty, the vast majority supported expansion of power generation or distribution networks. Less than 5 per cent focused on expanding options for clean cooking—despite significant negative health effects from the use of traditional cooking fuels.

Indeed, the huge fiscal burden of existing subsidies can absorb such a large portion of available government revenues that it crowds out spending in other welfare-enhancing areas. Table 3.2 compares country-level estimates of consumer subsidies to fossil fuels with gross domestic product (GDP), government revenues (as a proxy for the country's break-even public spending limit), and public spending on health care. Nearly one-sixth of the countries were spending more than 10 per cent of their GDP on fossil fuel subsidies, and more than half committed in excess of 10 per cent of government revenues. Most striking is the crowding out of other important social goals, with nearly half of the countries tracked by IEA spending more to subsidise fossil fuels than on health care.

Table 3.2 Subsidies to fossil fuel consumers crowd out other spending priorities

Country counts	Fossil fuel subsidy amount[1] as percentage of:		
	GDP[2]	Federal revenues[3]	Public spending on health care[4]
Total countries	37	38	37
Subsidies > 100% of metric	0	0	18

1   Countries surveyed were Angola, Bangladesh, China, India, Indonesia, Pakistan, the Philippines, South Africa, Sri Lanka, Thailand, and Vietnam.

Country counts	Fossil fuel subsidy amount[1] as percentage of:		
	GDP[2]	Federal revenues[3]	Public spending on health care[4]
Subsidies > 50% of metric	0	2	26
Subsidies > 25% of metric	0	5	32
Subsidies > 10% of metric	6	22	33

[1] Based on price gap subsidies to consumers in 2011 (IEA 2012).

[2] 2011 data from World Bank (2013a).

[3] 2012 estimates from CIA (2013).

[4] Based on World Health Organization data compiled by *Guardian* (2012). Population data used to scale per capita to national figures from World Bank (2013b).

Source: Extracted from Koplow (2015a); reprinted with permission from Oxford University Press.

The political economy of subsidies: Opacity helps the recipients

The political economy of subsidies helps explain the high leakage rates for many existing subsidy programs. Because subsidies entail the transfer of wealth from one group to another, tensions between these groups frequently arise. More transparency generally benefits those paying for the support either financially (primarily taxpayers) or competitively (makers of competing goods or services). In contrast, this visibility can hurt those receiving support (subsidy beneficiaries and the associated politicians). Risks of reputational damage (for example, this firm's products are not good enough to survive without subsidies) and economic loss (for example, once they see the subsidy, opponents will successfully mobilise to narrow or reverse the program) both increase for producers. For the political facilitators, attention from voters or competing industries creates similar reputational concerns, as well as risks to campaign contributions and re-election prospects.

Because benefits to subsidy recipients tend to be concentrated, while the groups paying for them are diffuse, recipients can more easily mobilise and fund efforts to create and protect subsidy programs. Complex subsidy mechanisms often extend that advantage, making it easier for recipients and associated politicians to avoid scrutiny. Data may be isolated in arcane reports outside of the main budget, or may not be reported at all. Even where reporting occurs, descriptions may be sufficiently vague to preclude linkage back to specific firms, industries, or elected officials. When risk-based subsidies turn bad (such as when credit recipients go bankrupt or natural disasters hit subsidised clients of government insurance programs), publicity is inevitable. Because these events often arise years after the subsidies were granted, however, political fallout is less likely; indeed, the sponsoring politician may no longer even be in elected office.

Limited reporting

Even discerning whether a subsidy was granted can sometimes be difficult, particularly in countries with strong central governments, state ownership of key industries, and a limited tradition of transparency. Levers potentially available to citizens to force political accountability may be greatly constrained. Free and fair elections, the right to sue firms or governments for redress, regulated securities markets with mandated disclosure, or statutorily protected rights to public information may exist in name only or not at all. In these types of settings, quantifying subsidies becomes extremely difficult. Absent an initiative from the central government itself, reform becomes nearly impossible.

Western democracies that pride themselves on open government are not immune from challenges either. Bias amongst regulators or legislators for a particular energy resource can influence policy in subtle ways, regardless of reporting. Further, although agency budgets are published regularly, publicly available, and usually independently audited, disclosures can be less complete and more fragile than often assumed. The existing rules often required a political battle to put in place, and remain subject to political pressure for bypass or reversal. The United Kingdom, for example, was recently forced to disclose that it had capped investor risk to nuclear waste management and reactor decommissioning costs at the planned new Hinkley Point C project. The subsidies only saw the light of day due to laws requiring disclosure of data under public records requests (Doward 2016).

Reporting in the United States provides another instructive example: disclosure of federal tax expenditures and credit subsidies occurs only due to statutory mandates. Even so, reporting is limited to a highly aggregated level. Attribution to specific recipients is impossible, and, depending on the subsidy, even allocation to particular industries can be difficult. Individual government insurance programs will often be evaluated, but risk subsidies across the government rarely get tallied or linked to specific subsidised beneficiary groups. Federal indemnification, where private liability is limited by statute rather than subsidised through a government-run insurance program, is rarely quantified at all.

Even the tax and credit subsidy data remain imperfect. Tax breaks are measured in terms of federal revenue foregone, but in some cases the breaks generate tax-exempt income to the recipient and are therefore more valuable than a similarly sized cash grant. Measurement of this second impact made comparisons of subsidy value across types of government support more accurate, but was discontinued in 2008. Metrics for US credit subsidies contain two important inaccuracies. First, the values exclude the cost of program administration. Second, and more importantly, interest rate subsidies are measured against the cost of funds to the US Treasury (the 'risk-free rate') rather than the estimated risk of the borrower, which would be much higher. Inadequate integration of risk in the pricing of sovereign credit and credit guarantees is an endemic problem around the world (Lucas 2013). Credit subsidy estimates for particularly high-risk energy enterprises (new energy technologies, nuclear power plants, carbon sequestration sites) will be disproportionately understated in official government data.

Reporting on subnational subsidies is generally weaker than that at the national level. State-level tracking even of tax subsidies in the US remained voluntary until a recent Governmental Accounting Standards Board ruling established standards and made reporting mandatory (Governmental Accounting Standards Board 2015). Subnational subsidies to credit, insurance, and project-related infrastructure are common, but rarely tracked. SOEs, particularly power plants, are also common at the state or municipal levels in many countries.

Challenging valuation

Valuation of many subsidy mechanisms requires 'counter-factual' baseline assumptions against which to compare current practice. What would a neutral tax, or an unsubsidised insurance or credit market, look like? The task is manageable, although challenging. Government-prepared assessments of tax breaks or credit subsidies may not all be using the same baseline assumptions, for example, making comparisons and aggregation more difficult. Similarly, when governments do not estimate subsidy values, developing them independently can be time consuming and complex, and may require related data that only the government has.

These impediments result in less frequent and less complete coverage for complex subsidy mechanisms than for grants. In most cases, even the Organisation for Economic Co-operation and Development's (OECD) inventory incorporates tax expenditure data only if the member state estimates it, and while plans are underway to incorporate credit subsidies, this important area has not yet been implemented.

A final complication is that industry can slow reform, or defuse political opposition to subsidy extension or expansion, by challenging the baseline assumptions adopted by governments. Indeed, beneficiaries routinely hire their own consultants to defend existing policies and to estimate subsidy costs using (more favourable) alternative assumptions. In comparison to analysis by government agencies or non-governmental organisations (NGOs), these assessments often report lower costs and bigger benefits of the status quo.

A notable parallel strategy to block reforms is to simply deny that key interventions are subsidies at all. Often, this involves claims that the subsidised treatment of one's own industry is part of the baseline tax system and not a diversion from it. The American Petroleum Institute, the largest oil and gas trade association in the United States, provides a textbook example of this approach. Its director of tax and accounting, Stephen Comstock, wrote: 'Contrary to what some in politics, the media and most recently, the president during the State of the Union, have said, the oil and natural gas industry currently receives not one taxpayer "subsidy," "loophole" or deduction' (Comstock 2014).

Objectively, this is a refutable statement, and his claims run counter to an impressive array of US federal agencies that have long viewed and valued these policies as clear subsidies to oil and gas.[2] While making the case that some government interventions constitute targeted subsidies to a particular industry is not always simple, Comstock contests even the most basic ones. His approach underscores the political challenges to reform: a sizeable portion of the electorate will accept the industry framing even if it is incorrect, and industry trade associations often have much larger budgets to promote their position than do NGOs or even the subsidy research staff at international organisations.

---

2   This includes the US Treasury, the Joint Committee on Taxation, the Congressional Budget Office, the Government Accountability Office, and the Congressional Research Service.

Global subsidies: Measurement strategies and magnitude

Assessing the implications of subsidies on key policy questions such as climate change or the cost structure of nuclear power requires aggregating data across subsidy types and from multiple levels of government into metrics of combined support. Subsidy measurement has focused on quantifying the value transferred to market participants from particular programs (program-specific or inventory approach) or on the variance between the observed and the 'free market' price for an energy commodity (price gap approach). The former captures the price effects, but does not identify the specific subsidies causing them. The latter tracks the individual subsidies, but does not delineate their pricing impacts (Koplow 2015b).

The price gap approach requires less data than the subsidy inventories, and is particularly useful in multi-country studies where government data are not readily available. However, this metric misses the many subsidies that boost industry profitability or allow marginal competitors to stay afloat, although without affecting equilibrium prices. Price gap estimates should therefore be viewed as lower bound (Koplow 2009).

The OECD's Total Support Estimate (TSE) metric captures both pricing distortions (net market transfers) and transfers that do not affect end-market prices (net budgetary transfers). The TSE tracks individual policies on producer and consumer sides of the market, allowing interactions to be evaluated. Government programs that support the general structure of a particular fuel market, although not a specific producer or consumer, are tracked separately. The OECD's approach is data intensive: its most recent review of government support to fossil fuels included more than 800 subsidies provided by a diverse array of government agencies—yet still contained important residual gaps in coverage (see OECD 2015).

A final differentiating factor across the commonly reported global estimates is the inclusion of 'normative' adjustments to subsidy value. The IMF valued fiscal subsidies (using IEA's price gap data plus producer subsidies as calculated by the OECD, less tax subsidies) into a 'pre-tax' measure of support. In addition, the IMF also developed a 'post-tax' estimate that was roughly 16 times its lower estimate, equal to nearly 7 per cent of global GDP (Coady et al. 2015). This higher value includes an imputed national consumption tax on fuel for countries that either had no existing tax of this type, or one that the IMF felt was inappropriately low; and negative externalities to both energy and to transport. The IMF's post-tax estimate remains contentious among subsidy practitioners because of these factors.

Table 3.3 provides a summary of the global estimates of subsidies to energy, highlighting a handful of key points. First, variation in coverage by policy type and country, in combination with the valuation approach used, results in very large differences in estimates across organisations. The factors sometimes work in opposite directions: the OECD captures a wider array of subsidy policies than does the IEA, but does not include countries such as Iran, Saudi Arabia, and Venezuela, which have very large subsidies to energy consumers. Second, data on renewables are lacking and that on nuclear subsidies non-existent. These are important holes to fill going forward. Third, even the lower-end estimates demonstrate that quantified subsidies are a material portion of global GDP. The drain on government resources from energy subsidies would be even higher were coverage of renewables, nuclear, and some of the missing subsidy types to improve.

Table 3.3 Global energy subsidy estimates: Massive scale, wide range

Fuel type	IEA	OECD	IMF (pre-tax)	IMF (post-tax)
	Measurement approach/billions of 2015 US$			
	Price gap	Total Support Estimate	IEA plus OECD, less tax breaks	IMF pre-tax plus tax breaks plus externalities
Fossil fuels	506	170	333	5,302
Nuclear electric	NE	NE	NE	NE
Renewables electric	112	NE	NE	NE
Biofuels, transport	23	NE	NE	NE
Total all fuels	641	170	333	5,302
% of world GDP	0.8	0.2	0.4	6.8

NE = not estimated

Data year: 2014 for IEA and OECD; mixed input years for IMF

Sources: Earth Track tabulation from IEA (2014, 2015); OECD (2015); Coady et al. (2015). GDP data from World Bank (2017).

Overview of global subsidies to the nuclear fuel cycle

Despite growing coverage of global subsidies to fossil fuels, there is no global estimate for support to nuclear power and very few national level estimates either. As there are far fewer nuclear facilities than oil and gas, estimating nuclear subsidies would seem a manageable task; a host of other attributes may help explain the lack of coverage. Worries about climate change, in combination with potential fiscal windfalls from subsidy reform, may have provided the needed focus and funding for fossil fuel subsidy research. The coverage by IEA of renewable subsidies is perhaps structural: purchase mandates (feed-in tariffs or renewable portfolio standards) form the bulk of support to renewables and show up well in price gap calculations.

In contrast, heavy government involvement in many aspects of the nuclear fuel cycle complicates subsidy tracking. Many reactors, nuclear construction firms, fuel cycle facilities, research programs, and (potential) waste repositories are partially or entirely government-owned. Subsidy evaluation for SOEs remains an area of particular weakness in current global data, and support to electrical power generation and transmission infrastructure has also been more challenging to estimate than subsidies to frequently traded fuels.

Further, governments may have little interest in compiling subsidy data on their nuclear sector. Key players in the new wave of reactor construction such as China and Russia view the industry as building strategic capabilities and international influence.[3] Profitability may also be a goal, but it is not the only, or perhaps even the main, one. Countries purchasing these reactors may have some mixed interests as well, such as the potential to gain skills and know-how that could be deployed in military initiatives. Indeed, in a recent review of the most likely pathways to a highly proliferated world, four of the six new post-2030 nuclear states are expected to use militarisation of their civilian programs to get there (Murdock et al. 2016: 16).

---

3   Thomas (2017) provides an overview of China's nuclear export plans, and some of the challenges the country is facing. Reuters (2016) notes the goal of the Russian government to use energy policy to achieve national political objectives.

Even current nuclear states see continued investments in the civilian sector as a necessary contributor to keeping their military capabilities active. A review of the United Kingdom's strong commitment to build a new reactor despite very large fiscal contributions from the state relative to the value of the power concluded that 'it is difficult fully to comprehend the persistent intensity of official UK attachments to nuclear power, without also considering the role of parallel commitments to maintaining national military nuclear submarine capabilities' (Cox, Johnstone, and Stirling 2016: 3). Continued investment in India's breeder reactor program is also focused on retaining a 'minimal credible deterrent' (Ramana 2016a). Making public the substantial subsidies to the civilian nuclear sector runs counter to these interests, while also increasing the political risks to sponsors if projects run into trouble.

Complexity is a final factor likely driving the lack of data. The most important subsidies to the nuclear fuel cycle involve complicated risk-shifting from industry onto taxpayers or plant neighbours: accident risks, cost overruns on decommissioning, nuclear waste, and plant financing. All of these areas entail a mixture of significant government involvement, long timescales, and a lack of systematic tabulation within countries. All also have enough uncertainty on potential outcomes to provide wide latitude for industry and nuclear boosters within governments to develop alternative narratives on the programs.

Regardless of the cause, the lack of global data on nuclear subsidies is both striking and likely to prove quite expensive to taxpayers and competing energy sources alike. The IEA and the Nuclear Energy Agency (NEA) recently called for US$4.4 *trillion* invested into new reactors by 2050 in order to stay within 2 degrees Celsius of warming (IEA/NEA 2015: 23). Much of this amount would involve direct or indirect (for example, via credit or power price guarantees) government support.

The industry has been pushing hard to position nuclear energy as a key strategy to address climate change as well. They have been supported by prominent scientists such as James Hansen, who wrote that 'nuclear power paves *the only* viable path forward on climate change' (Hansen et al. 2015, emphasis added). Yet, there are many ways to pull carbon out of our economy, both in the power sector and beyond. Nuclear industry proponents have been far more enthusiastic about proclaiming the centrality of nuclear in any climate change plan than they have been about exposing projects to real market tests. Large subsidies to trillions of dollars' worth of new, long-lived nuclear reactors will skew greenhouse gas reduction strategies away from quicker, cheaper, and lower-risk solutions.

Although a full review of nuclear subsidies around the world is not possible for this chapter, reviewing common patterns in government support to the nuclear fuel cycle around the world more generally is instructive (see Table 3.4). Additional discussion of risk-shifting and subsidisation in five key areas follows: financing new reactors or otherwise absorbing market price risk; socialisation of cost overruns and long-term management of high-level nuclear waste; shifting of financial risks associated with facility decommissioning; capping accident liability at levels well below likely damage; and enrichment.

Table 3.4 Common areas of subsidy to the nuclear fuel cycle

Intervention type	Description
Pre-production subsidies	
Government research and development	IEA member state spending on nuclear research and development (1978–2012) exceeded US$250 billion (IEA 2013). Nuclear research and development captured 51 per cent of total energy, nearly four times the next highest funded fuel cycle.
Funding, cost sharing on reactor design and licensing	No data on Russia and China, but likely mostly funded by the state. In the US, new reactor design supported by research and development; increasing cost-sharing on licensing.
Extraction subsidies	
	Some tax breaks and government support to uranium mining in the US; no readily available international data. Mining costs are a small portion of delivered cost of nuclear power; these interventions are relatively less important to economics of nuclear power.
Conversion subsidies	
Uranium milling and enrichment	Heavy government involvement in enrichment technology development, plant construction, and plant operation. Nearly 90 per cent of current enrichment capacity globally is state-owned. Historical remediation costs at US uranium milling sites exceeded the value of the ore mined (Koplow 2011: 61).
Power plant finance and construction	Capital-intensive plants, often plagued with construction delays and cost overruns, would have very high borrowing costs absent government subsidy. State involvement through loan guarantees, direct lending, tax-exempt bonds, and sometimes state ownership is common in most of the countries building new reactors today.

Intervention type	Description
Power plant operation	Direct subsidies being proposed for reactors in New York state to keep existing, already amortised plants from closing (Matyi 2016). Some production tax credits for new reactors in the US. Free or nearly free use of large quantities of cooling water by reactors common in the US; likely common in many other countries as well.
Accident risks	Liability caps well below probable damages from any major accident are common across most of the world. Incentive problems from inadequately low insurance cover sometimes compounded where state funds, rather than operator funds, are used to ensure compliance with particular levels of coverage. Use of retrospective premiums to cover the bulk mandated insurance payments (as in the US) subject to significant counter-party risks of non-payment.
Transportation and distribution subsidies	
	Transport of spent fuel and irradiated plant components during decommissioning is complex, although few of these trips have yet been completed. Nuclear transport is covered by liability caps in most countries. Significant share of electrical capacity in many countries; pro-rata beneficiary of any subsidies to power grid expansion or maintenance.
Consumption subsidies	
	Primarily in the form of purchase mandates, where customers must buy nuclear power at above-market rates. US: Construction work-in-progress (CWIP) rules allow significant cost overruns to be passed on to customers, often even if the reactor never successfully begins operation. Nuclear plant has historically been a high-cost supplier in regulated rate regions once capital costs are included. UK: Price guarantees through a 'contract-for-difference' scheme sets long-term price floors for new reactors well above current wholesale rates.
Post-production activity subsidies	
Post-production period for other fuels might last a few decades. For the nuclear fuel cycle, the metric could be centuries. Totally different technical and economic considerations; much larger probability of estimation errors.	
Decommissioning for reactors, fuel cycle facilities	Direct subsidies often provided through tax breaks on accrued funds and by state responsibility for costs or cost overruns. De facto subsidies arise where accrual for post-closure costs is not properly segregated from firm operations, and lost due to corporate restructuring, fraud, or bankruptcy.

Intervention type	Description
Nuclear waste management	Heavy state involvement with location, construction, and operation of nuclear waste repositories around the world, although no facilities currently operating.
	Long duration enterprise with substantial technological risks; where current funding exists, contributions are often capped at levels unlikely to cover real costs of the service provided.
	Socialised nuclear waste management likely to be priced at break-even (at best) with no return on invested taxpayer capital. Undermines alternative sources of energy.
Negative externalities	Close to zero carbon emissions, but negative externalities include radioactivity, accidents, and expansion of skills and facilities that can support or hide weapons proliferation activities.[1]

[1] A promising strategy to reduce proliferation externalities is to place uranium enrichment and reprocessing under international control. This would make illicit diversion much more difficult (Diesendorf 2014).

Source: Based on author's data.

Financing new reactors or absorbing market price risk on power

Nuclear power is a capital-intensive technology with long and often uncertain build times. These factors alone would contribute to a high-risk profile and elevated risk premiums from capital providers; a poor track record on cost and delivery times even on new projects compounds the issue. The fact that reactors produce a commodity product (electricity) in a segment undergoing rapid technical change and plausible disintermediation of centralised generation during the life span of the new capital further weakens the investment case for new construction.[4]

---

4   Nuclear energy has focused on differentiating itself in three main areas: low carbon, high-power density, and dispatchable energy. All three are important attributes. However, other sources of reducing carbon are far less expensive than fully costed nuclear. High-power density may not matter much given widespread grid connections throughout much of the developed world and benefits to mini-grids or distributed energy (both areas of weakness for nuclear) in the developing world where a combination of high costs and poor governance have limited grid expansion and often made centralised power unreliable. Dispatchable load to provide firm power capabilities is increasingly important with rising shares of intermittent renewables. However, in US markets that provide price premiums for firm capacity, even some existing reactors are uneconomic (*World Nuclear News* 2016b); the situation would be far worse for new reactors that must price to recover capital. Further, the timescale of the NEA's nuclear buildout, for example, is through 2050. Power storage markets are developing extremely fast, with battery research prodded by the massive market pressure of hundreds of millions of units per year from the portable devices and growing electric car segments. Reactor cost reductions have not materialised in the past, and are unlikely to outrace cost declines in power storage now.

Investors recognise that the market value of the power their new reactors produce could be much lower in five or 10 years when the reactor comes online than what they would have assumed, and that its value over the 50 or more years between the start of plant construction and when it closes is impossible to guess. These characteristics have led capital providers to largely abandon the private nuclear finance market.[5] Funding strategies have shifted largely to state-supported, non-competitive capital. Consider the IEA's and the NEA's take: 'Governments have a role to play in ensuring a stable, long-term investment framework that allows capital-intensive projects to be developed and provides adequate electricity prices over the long-term for all low-carbon technologies' (IEA/NEA 2015: 5). Approaches to provide this 'stable' framework all involve shifting core financial risks from the investor to the state. This includes favourable state-backed credit packages, long-term state-guaranteed price floors, and broadly worded allowances to shift advance financing and cost overrun risk to ratepayers. Direct state ownership of nuclear facilities is also common.

State credit packages

Of the 56 reactors under construction (according to IAEA PRIS 2016),[6] seven are in Russia and 20 in China. The Russian reactors are all state-financed and owned; Chinese state involvement in its domestic reactors is also pervasive. Rosatom, Russia's state-owned reactor developer, has bundled multibillion-dollar credit programs with proposed or pending reactor export deals as well, including in Bangladesh, US$12.7 billion (*World Nuclear News* 2016b); Egypt, US$25 billion (*Russia Times* 2015); Finland, €5 billion (Rosatom 2016); and Hungary, €10 billion (Rosatom 2016). Although many of these projects are unlikely to move to construction, the centrality of credit provision in deal structure is nonetheless instructive. Similarly, in China, '[t]he huge financial reserves of the Chinese government meant that its vendors would be able to provide finance as well as equipment, a big advantage in most potential markets'

---

5   Moody's Investors Service (2013: 20) noted that 'new construction of nuclear power plants is generally credit negative', that cost overruns on nuclear projects were more frequent than for other technologies, and that 'European power markets in their current state don't support new construction' of reactors.

6   The International Atomic Energy Agency (IAEA) Power Reactor Information System (PRIS) database listed 60 units under construction as of 1 November 2016. Two units have been under construction since 1999, and another two since the late 1980s; these have been excluded from the totals.

(Thomas 2017). This includes the UK's Hinkley Point C where Chinese state-owned entities are, along with France's Électricité de France (EDF), taking a one-third interest in the £18 billion project (Ruddick 2016).

Guaranteed price floors

To address the significant risk that electricity prices will be too low when a new reactor finally comes online, utilities often implement power purchase agreements for at least a sizeable portion of its capacity. The price and timing for nuclear kilowatt hours (kWh) is so uncertain that counter-parties on these transactions for new reactors are often the state. On the Hinkley Point C project in the UK, the government is guaranteeing project developers a minimum price three times the current wholesale price of power for 35 years. The cost to UK taxpayers of this price support has been estimated to be as much as £300 billion (*The Week UK* 2016; Ruddick 2016).

China set a wholesale floor price for power produced from all new nuclear projects in the country 'to promote the healthy development of nuclear power and guide investment into the sector' (WNA 2016a). Future upward adjustments may be needed, as some feel the current price floor is too low (WNA 2016a). A tender for a reactor project in the Czech Republic worth up to US$15 billion was scrapped in 2014 due to weakening prices for power and a refusal by the government to guarantee power prices (Lopatka 2014).

Ratepayer interest-free finance and exposure to cost overruns

The four new reactor projects in the United States (Summer 2 & 3 and Vogtle 3 & 4) are all in districts with favourable construction work-in-progress (CWIP) rules. CWIP allows reactor owners to increase rates prior to a new project entering production, effectively using ratepayers as a source of interest-free capital. In addition, CWIP shifts much of the risk of cost escalation from the investors and builders onto customers. While local rate boards can sometimes reject the pass-through of plant cost increases, in most cases customers end up paying.

Both attributes are enormously valuable to nuclear investors and plant owners. Vogtle 3 & 4 is nearly four years late and facing cost escalation of US$6 billion (43 per cent, to an expected total cost of more than US$20 billion). Still, all costs, including incremental financing costs on the delays, were deemed prudent and allowed to be billed through to ratepayers (SACE 2016). Interest-free pre-funding of the Vogtle reactors through October 2016 exceeded US$1.8 billion (SACE 2016). While not setting a price floor, plant investors have agreed to 'hell or high water' clauses in their wholesale power purchase agreements with Vogtle that do the same thing: they require their customers to buy power even if they do not need it in their districts, and at whatever price it is delivered even if that price is well above other supplier options (MEAG Power 2016: 10–12). The two Summer reactors are US$4 billion over initial cost estimates (40 per cent), and interest-free pre-funding of the project accounts for more than 16 per cent of customers' monthly bills (Wren 2016).

Socialisation of high-level nuclear waste management

High-level radioactive waste must be isolated and managed for thousands of years. Ongoing hazards include accidents, theft (if in accessible storage), and environmental contamination. Continued exposure to this unknown and long-duration risk is highly problematic for investors. Facilities capable of safely storing this waste are technically difficult to design, build, and operate. They are also politically unpopular. After 60 years of civilian nuclear power, there are no operating permanent repositories anywhere in the world (see Table 3.5). Delivery times continue to slip and the risks of massive cost escalation remain high. Andra, the French agency responsible for managing high-level nuclear wastes in that country, increased its cost estimate for the repository from at most €18 billion in 2005 to €35 billion in 2014 (*World Nuclear News* 2016a). In response to industry complaints, Andra responded that '[t]he evaluation of these costs is a particularly delicate exercise because it requires making assumptions on labour costs, taxation, materials or energy for more than 100 years' (*World Nuclear News* 2016a).

Table 3.5 Nuclear waste management largely unresolved, financial and technical risks borne by governments

Country	Percentage of global nuclear power capacity, 2015	Earliest high-level waste repository open date	Location chosen?	Accrued funding held outside firm	Ownership
United States	25.6	2048	Chosen site terminated	Yes	State
France	16.2	2025	Yes, although opposition remains	No	State

Country	Percentage of global nuclear power capacity, 2015	Earliest high-level waste repository open date	Location chosen?	Accrued funding held outside firm	Ownership
Japan	10.4	>2035	No	Yes	Utility
China[1]	7.8	>2050	'First priority' area selected	Yes[2]	State
Russia	7.0	No target date	No	Yes	State
South Korea	5.9	No target date	No	Yes	State
Canada	3.5	2035	No	Yes	State[3]
Germany	2.7	>2025	Opposition to target site	No	State
United Kingdom	2.3	No target date	No	Yes	State
Sweden	2.3	2028	Yes	Yes	Utility
Spain	1.8	No target date	No	Yes	State
India	1.6	No target date	No	Publicly funded[4]	State
Belgium	1.5	>2035	No	No	State
Switzerland	0.9	No target date	No	Yes	State
Finland	0.7	2023	Yes	Yes	Utility

[1] Published sources supplemented with additional information provided to the author by Hui Zhang, 1 December 2016.

[2] Funds held by government to finance spent fuel storage and reprocessing costs; financing of a permanent repository is not currently among the approved uses.

[3] Utility cooperative to handle nuclear wastes, but the utilities are state corporations.

[4] Most spent fuel considered a resource and transferred to a state-funded reprocessing infrastructure at no cost to the generator. Waste from light-water reactors appears to be an implicit liability of the state.

Sources: WNA (2016b; 2016d); NEA (2011a: 36, 37; 2011b: 4; 2011c; 2013a; 2013b: 5; 2014a: 1, 15; 2014b: 15–17; 2015: 13, 14; 2016a); US DOE (2013: 2); *Russia Times* (2016); Wang (2014); IAEA (2016: 16); Feiveson et al. (2011); Zhang and Bai (2015: 59); Zhou (2013); Ramana (2013, 2016b).

Lifecycle costs for the (now suspended) US nuclear waste repository were pegged at US$108 billion (in 2015 dollars) over the 100-year period evaluated (Cawley 2015). Estimation challenges caused by the extremely long time frame on top of project complexity are likely to be severe. Most countries have simply socialised the economic and technical risks of nuclear waste management. In return for a small (predictable) fee, responsibility for the long-term management of the nuclear waste—including developing workable technical solutions, and building and operating a repository—has been shifted to the state. In some cases, the state has taken on partial or total funding of these liabilities already. Even where rules require a polluter-pays system, shortfalls in financial accruals will be common. Some waste collections remain within firms, and therefore are at risk in a corporate action or bankruptcy. However, most countries now move accrued funds outside of corporate control.

Many of the waste-generating reactors will be shuttered long before repository funding shortfalls materialise. Post-closure fee adjustments will be impossible. Funding shortages are likely even in externally held waste funds. These financial problems will fall to taxpayers to deal with, and the problems are likely to be very large. Cost overruns in excess of 50 per cent on so-called 'megaprojects' are common (Flyvbjerg 2014: 9). Nuclear megaprojects had cost overruns nearly two-and-a-half times their initial budget, higher than all of the other sectors evaluated (Locatelli 2015: 11). Even assuming cost projections used to assess fees today are correct, the repositories are often assumed to operate at break-even, generating significant subsidies to nuclear. Simply applying the average return on investment for US utilities (a less complex, less risky investment than a nuclear waste repository) would have doubled the annual waste fee levied on US reactors (Koplow 2011: 97).

Subsidies to decommissioning costs

Closure costs are far more significant for nuclear than for other technologies. Decommissioning radioactive nuclear plants and fuel cycle facilities at the end of their operating lives is a major expense. The ability to delay decommissioning far into the future by leaving plants idle for many decades under some decommissioning strategies opens the door to a bit of financial engineering to make the problem seem to go away: large costs are pushed far into the future then discounted to the present using a variety of inflation and discount rate assumptions. The result is a manageable present value cost, which once adjusted to a cost per kWh generated can make decommissioning costs seem immaterial to plant economics.

Real-world outcomes are unlikely to be so predictable: the complexity, duration, and variability in the decommissioning process suggest that large costs to taxpayers are likely. The number of reactors decommissioned to date and thereby providing real cost experience is small. Since the regulatory and physical conditions that drive costs vary widely by installation, the available experience may not be indicative of future costs.

Some countries already subsidise accruals for decommissioning. Favourable tax treatment for decommissioning funds in the US, for example, was worth more than US$1 billion per year according to government estimates (Koplow 2011: 95). Further, although countries often try to ensure that operators accrue adequate funds for decommissioning over the facility's operating life, the practical challenges mirror those for nuclear waste disposal and management. Accrual rates may be too low to cover actual costs, shortfalls will not be known until the facility is already closed and unable to pay more, premature closure of plants due to market or other factors will trigger immediate decommissioning shortfalls, and assumptions on financial returns or period of fund growth may not match reality, with associated undershooting of funding targets.

In some cases, accrual rates may not be based on sound estimates of expected liabilities to begin with. For example, India's Comptroller and Auditor General noted that the decommissioning levy was last adjusted two decades earlier, and that the Atomic Energy Regulatory Board 'had not worked out the decommissioning cost formula in any of its documents' (Comptroller and Auditor General of India 2012: 66).

Collected funds of whatever scale can at least be protected from misuse or firm bankruptcy by mandating they be held in external trusts, and used only for the purposes for which they were set aside. Although this is required for nuclear power plants in the US, a review of practices within Europe found that only about half of the countries instituted this basic protection. Restricted external funds were even less common for other European fuel cycle facilities, such as uranium mining, milling and enrichment sites, and reprocessing plants (Wuppertal Institute for Climate, Environment and Energy and its Partners and Subcontractors 2007: 37).

These distinctions matter: £6 billion that was collected from ratepayers in the UK between 1990 and 1996 under the country's fossil fuel levy was presented to parliament as being for the purposes of decommissioning old and unsafe power stations. However, funds were not properly restricted and segregated, resulting in roughly half the collections being used to fund operations of Nuclear Electric. Most of the remainder was eventually appropriated by the UK Treasury following the nationalisation of nuclear decommissioning liabilities associated with British Nuclear Fuels Limited (Thomas 2007: 21–8). As of 2007, and despite 'more than 25 years of contributions, there [was] only about £800m in identified funds to pay for a liability of more than £75bn' (Thomas 2007: 1).

Capping accident liability at levels well below likely damage

Liability for off-site damage from nuclear reactors is set by domestic statute and supplemented by a series of international conventions that aim to standardise minimum liability levels. Under all of the existing frameworks, liability cover remains well below the likely damages from even a moderate-sized accident. Total coverage in the US (the largest accident pool in the world) is less than US$14 billion. In contrast, the cost of the Fukushima accident to Japanese taxpayers has already reached US$118 billion (Harding 2016), with significant costs left to be paid.

Rules on mandated coverage often contain different tiers of policies, with the first tier being most similar to conventional insurance where each reactor owner buys its own insurance policy to cover the initial portion of accident risk. Although titled 'operator liability' in summaries produced by the NEA (2016b), the first tier values sometimes mix in coverage from other reactors or backstopped by governments. The coverage paid for directly by the reactor owner provides the most direct price signals to operators on their risk levels; it would therefore be helpful to see cleaner data on that specific coverage. Directly funded policies can positively influence siting decisions as well as incentivise higher spending on facility and operator risk reduction.

A second tier of coverage often involves risk pooling amongst all reactors or additional coverage offered by the state. This approach dominates the insurance pool in the United States: retrospective premiums levied on all reactors should an accident exhaust the primary operator's coverage limit (US$450 million) comprise more than 95 per cent of available funding. The approach has some structural challenges, however. It is paid in annual installments over more than six years post-accident, bringing the present value of the coverage much lower than its face value. In addition, the pool declines as reactors close down, and payments are subject to counterparty risks should remaining reactors be in financial distress (fairly likely after a major accident). Concentrated reactor ownership worsens counterparty risks since multiple retrospective premium payments would be due at once. Exelon alone owns 23 reactors, or nearly one-quarter of the US nuclear fleet (Exelon Generation 2016).

Additional tiers of coverage are sometimes available from other countries under convention agreements, although the amounts are relatively small (less than US$100 million). Even tallying all reliable tiers of coverage together, the amount available can be surprisingly low. The NEA (2016b) indicates liability coverage in Russia of only US$5 million under formal conventions. Additional coverage may be available but, as with many other countries, would be at the discretion of the state or of legislative bodies. Even in a country with as much nuclear power as France, the NEA indicates available accident coverage of less than US$1 billion.

Uranium enrichment

Enrichment capability has long been a fraught area of the fuel cycle due to concerns that countries would use this capacity for both civilian and military purposes. Although much diminished today, historic involvement with civilian enrichment by the US government was both large and heavily subsidised (Koplow 1993). State ownership continues to dominate the sector today on a global basis, comprising nearly 90 per cent of the total enrichment capacity (see Table 3.6). This type of structure would normally be associated with over-investment in the sector (resulting in over-capacity and low prices to utilities), tax exemptions, and access to below-market rates on credit and insurance.

Table 3.6 Uranium enrichment is dominated by state actors globally

Company	2015 capacity	Percentage share of world capacity	Ownership	Percentage state-owned share of world capacity
	(thousands of separative work units)			
Areva	7,000	11.9	French state (85%); Kuwaiti state (5%); other (10%)	10.8
Urenco	19,100	32.6	⅓ UK government, ⅓ Dutch government, ⅓ German utilities	21.7
Japan Nuclear Fuel Limited	75	0.1	Mostly Japanese utilities	0.0

Company	2015 capacity	Percentage share of world capacity	Ownership	Percentage state-owned share of world capacity
	(thousands of separative work units)			
Tenex	26,578	45.4	State-owned (part of Rosatom)	45.4
China National Nuclear Corporation	5,760	9.8	State-owned	9.8
All others	87	0.1		0
Total	58,600	100¹		87.7

¹ Components may not sum to 100 per cent due to rounding.

Sources: Areva (2016); JNFL (2016); Tenex (2017); Urenco (2017); WNA (2016c); Zhang (2015).

Summary

Subsidies remain a large and pervasive element in energy markets worldwide. Despite increasing recognition of the political and economic benefits of reform, political impediments remain high and the pace of successful reform slow. Eliminating most subsidies to fossil fuels would properly align price signals with climate and environmental objectives, expediting the transition to cleaner fuels globally.

There are no global estimates of subsidies to nuclear energy. These widespread data gaps are increasingly important to close if investors, governments, and citizens are to properly weigh the low-carbon benefits of nuclear power against its real fiscal cost and fairly evaluate nuclear against other carbon reduction pathways. Key subsidies to the nuclear fuel cycle include low-cost financing, shifting market risks onto customers, and socialising complex portions of the fuel cycle including enrichment, decommissioning, accident liability, and nuclear waste management. These hidden subsidies promote the expansion of nuclear facilities, infrastructure, and capabilities throughout the world even when other alternatives could provide the same energy services more quickly, at lower risk, and for less money.

References

Areva, 2016. Shareholder structure December 31, 2015. www.areva. com/EN/finance-1166/shareholding-structure-of-the-world-leader-in-the-nuclear-industry-and-major-player-in-bioenergies.html? XTCR=1,58&XTMC=CATITAL%20STRUCTURE?XTMC=A LTERNA%20RADIOPROTECTION (accessed 30 January 2017).

Arze del Granado, Javier, David Coady, and Robert Gillingham, 2010. The unequal benefits of fuel subsidies: A review of evidence for developing countries. IMF WP/10/202. Washington, DC: International Monetary Fund.

Cawley, Kim, 2015. The federal government's responsibilities and liabilities under the Nuclear Waste Policy Act. Testimony before the Subcommittee on Environment and the Economy, Committee on Energy and Commerce, US House of Representatives, 3 December. Washington, DC: US Congressional Budget Office.

CIA (Central Intelligence Agency), 2013. Government revenues. *The World Factbook*. www.cia.gov/library/publications/the-world-factbook/fields/2056.html (accessed 6 March 2013).

Coady, David, Ian Parry, Louis Sears, and Baoping Shang, 2015. How large are global energy subsidies? WP/15/105. Washington, DC: International Monetary Fund.

Comptroller and Auditor General of India, 2012. Report of the Comptroller and Auditor General of India on activities of Atomic Energy Regulatory Board for the year ended March 2012. Report No. 9 of 2012–13, Performance Audit. New Delhi: Union Government, Department of Atomic Energy.

Comstock, Stephen, 2014. The truth on oil and natural gas 'subsidies'. *Energy Tomorrow*, 29 January.

Cox, Emily, Phil Johnstone, and Andy Stirling, 2016. Understanding the intensity of UK policy commitments to nuclear power. SWPS 2016–16. Brighton: Science Policy Research Unit, University of Sussex.

Diesendorf, Mark, 2014. *Sustainable Energy Solutions for Climate Change.* Sydney: UNSW Press.

Doward, Jamie, 2016. Secret government papers show taxpayers will pick up costs of Hinkley nuclear waste storage. *Guardian*, 30 October.

Erickson, Pete, Adrian Downs, Michael Lazarus, and Doug Koplow, 2017. Effect of government subsidies for upstream oil infrastructure on US oil production and global CO2 emissions. Working Paper 2017-02. Stockholm: Stockholm Environment Institute.

Exelon Generation, 2016. Exelon nuclear fact sheet.

Feiveson, Harold, Zia Mian, M. V. Ramana, and Frank von Hippel, eds, 2011. *Managing Spent Fuel from Nuclear Power Reactors: Experience and Lessons from Around the World*. Princeton, NJ: International Panel on Fissile Materials, Program on Science and Global Security, Princeton University.

Flyvbjerg, Bent, 2014. What you should know about megaprojects and why: An overview. *Project Management Journal* 45(2): 6–19. doi. org/10.1002/pmj.21409

Governmental Accounting Standards Board, 2015. Statement No. 77 of the Governmental Accounting Standards Board: Tax abatement disclosures. No. 353. Norwalk, CT: Governmental Accounting Standards Board.

Guardian, 2012. Datablog: Healthcare spending around the world, country by country, 30 June. Extract of health care spending data collected by the World Health Organization.

Hansen, James, Kerry Emanuel, Ken Caldeira, and Tom Wigley, 2015. Nuclear power paves the only viable path forward on climate change. *Guardian*, 4 December.

Harding, Robin, 2016. Japan taxpayers foot $100bn bill for Fukushima disaster: Costs shouldered by public despite government claims Tokyo Electric would pay. *Financial Times*, 6 March.

IAEA (International Atomic Energy Agency), 2016. Country profiles: India. Vienna: IAEA.

IAEA (International Atomic Energy Agency) PRIS (Power Reactor Information System), 2016. Country statistics: Number of power reactors by country and status. 2 November. www.iaea.org/PRIS/CountryStatistics/CountryStatisticsLandingPage.aspx (accessed 5 January 2017).

IEA (International Energy Agency), 2011. *World Energy Outlook 2011*. Paris: International Energy Agency.

IEA (International Energy Agency), 2012. *World Energy Outlook 2012*. Paris: International Energy Agency.

IEA (International Energy Agency), 2013. RD&D statistics database. www.iea.org/statistics/rddonlinedataservice/ (accessed 1 October 2013).

IEA (International Energy Agency), 2014. *World Energy Outlook 2014*. Paris: International Energy Agency.

IEA (International Energy Agency), 2015. *World Energy Outlook 2015*. Paris: International Energy Agency.

IEA (International Energy Agency)/NEA (Nuclear Energy Agency), 2015. Technology roadmap: Nuclear energy, 2015 edition. Paris: IEA/NEA.

IEA (International Energy Agency), OPEC (Organization of the Petroleum Exporting Countries), OECD (Organisation for Economic Co-operation and Development), and the World Bank, 2010. Analysis of the scope of energy subsidies and suggestions for the G-20 initiative: IEA, OPEC, OECD, World Bank joint report. Prepared for submission to the G-20 Summit Meeting (26–27 June), Toronto, 16 June.

JNFL (Japan Nuclear Fuel Limited), 2016. Corporate profile. April. www.jnfl.co.jp/en/about/company/ (accessed 30 January 2017).

Kojima, Masami, and Doug Koplow, 2015. Fossil fuel subsidies: Approaches and valuation. Policy Research Working Paper WPS7220. Washington, DC: World Bank Group.

Koplow, Doug, 1993. Federal energy subsidies: Energy, environmental and fiscal impacts—Report and appendices. Washington, DC: Alliance to Save Energy.

Koplow, Doug, 1998. Quantifying impediments to fossil fuel trade: An overview of major producing and consuming nations. Paper prepared for the OECD Trade Directorate.

Koplow, Doug, 2009. Measuring energy subsidies using the price-gap approach: What does it leave out? Geneva: International Institute for Sustainable Development.

Koplow, Doug, 2011. *Nuclear Power: Still Not Viable without Subsidies.* Cambridge, MA: Union of Concerned Scientists.

Koplow, Doug, 2015a. Global energy subsidies: Scale, opportunity costs, and barriers to reform. In *Energy Poverty: Global Challenges and Local Solutions*, edited by Antoine Halff, Benjamin K. Sovacool, and Jon Rozhon, 316–37. Oxford: Oxford University Press.

Koplow, Doug, 2015b. Subsidies to energy industries. In *Reference Module in Earth Systems and Environmental Sciences*, edited by Scott Elias, 1–16. Amsterdam: Elsevier. doi.org/10.1016/B978-0-12-409548-9.09269-1

Locatelli, Giorgio, 2015. Cost-time project performance in megaprojects in general and nuclear in particular. Presentation to the Technical Meeting on the Economic Analysis of HTGR and SMR, International Atomic Energy Agency, 25–28 August.

Lopatka, Jan, 2014. 2-CEZ scraps Temelin nuclear plant explosion, shares up. Reuters, 10 April.

Lucas, Deborah, 2013. Evaluating the cost of government credit support: The OECD context. Paper prepared for *Economic Policy*, Fifty-eighth Panel Meeting, Vilnius, 25–26 October.

Matyi, Bob, 2016. New York nuclear plant subsidies fuel controversy. *Platts*, 10 October.

MEAG Power, 2016. Annual Information Statement of Municipal Electric Authority of Georgia, MEAG Power, for the Fiscal Year Ended December 31, 2015. Atlanta, GA: Municipal Electric Authority of Georgia.

Moody's Investors Service, 2013. Nuclear generation's effect on credit quality: Moody's perspective on operating risks and new build. London: Moody's Investors Service.

Murdock, Clark, Thomas Karako, Ian Williams, and Michael Dyer, 2016. *Thinking about the Unthinkable in a Highly Proliferated World.* Lanham, MD: Rowman & Littlefield.

NEA (Nuclear Energy Agency), 2011a. Radioactive waste management programmes in OECD/NEA member countries: Switzerland. Profile. Paris: OECD.

NEA (Nuclear Energy Agency), 2011b. Radioactive waste management and decommissioning in the United Kingdom. Report. Paris: OECD.

NEA (Nuclear Energy Agency), 2011c. Radioactive waste management programmes in OECD/NEA member countries: United Kingdom. Profile. Paris: OECD.

NEA (Nuclear Energy Agecny), 2013a. Radioactive waste management programmes in OECD/NEA member countries: Spain. Profile. Paris: OECD.

NEA (Nuclear Energy Agency), 2013b. Radioactive waste management programmes in OECD/NEA member countries: Sweden. Profile. Paris: OECD.

NEA (Nuclear Energy Agency), 2014a. Radioactive waste management in Rep. of Korea. Report. Paris: OECD.

NEA (Nuclear Energy Agency), 2014b. Radioactive waste management programmes in OECD/NEA member countries: Russian Federation. Profile. Paris: OECD.

NEA (Nuclear Energy Agency), 2015. Radioactive waste management programmes in OECD/NEA member countries: Canada. Profile. Paris: OECD.

NEA (Nuclear Energy Agency), 2016a. Radioactive waste management programmes in OECD/NEA member countries: France. Profile. Paris: OECD.

NEA (Nuclear Energy Agency), 2016b. Nuclear operator liability amounts and financial security limits. Paris: OECD, November.

OECD (Organisation for Economic Co-operation and Development), 2011. *Inventory of Estimated Budgetary Support and Tax Expenditures for Fossil Fuels.* Paris: OECD.

OECD (Organisation for Economic Co-operation and Development), 2015. *OECD Companion to the Inventory of Support Measures for Fossil Fuels 2015*. Paris: OECD.

Ramana, M. V., 2013. *Power of Promise: Examining Nuclear Energy in India*. New Delhi: Penguin Global.

Ramana, M. V., 2016a. A fast reactor at any cost: The perverse pursuit of breeder reactors in India. *Bulletin of the Atomic Scientists*, 3 November.

Ramana, M. V., 2016b. Email communication with Doug Koplow, Earth Track, 30 November.

Reuters, 2016. Rosatom's global nuclear ambition cramped by Kremlin politics. 26 June.

Rosatom, 2016. Projects. www.rosatom.ru/en/investors/projects/ (accessed 1 November 2016).

Ruddick, Graham, 2016. China plans central role in UK nuclear industry after Hinkley Point approval. *Guardian*, 16 September.

Russia Times, 2015. Russia to loan Egypt $25bn for nuclear plant construction. 30 November.

Russia Times, 2016. 'Underground Chernobyl': French parliament OKs nuclear waste facility despite protests. 13 July.

SACE (Southern Alliance for Clean Energy), 2016. Proposed agreement would reward southern company for bungled, massively over budget and 45-month delayed Plant Vogtle reactors. 22 October.

Tenex (Techsnabexport), 2017. Company profile. www.tenex.ru/wps/wcm/connect/tenex/site.eng/company/ (accessed 30 January 2017).

The Week UK, 2016. Hinkley Point: Bold move or white elephant? 16 September.

Thomas, Steve, 2007. Final country report: United Kingdom. In *Comparison Among Different Decommissioning Funds Methodologies for Nuclear Installations*, edited by Wuppertal Institute for Climate, Environment and Energy and its Partners and Subcontractors. Final report on behalf of the European Commission Directorate-General Energy and Transport, H2, Service Contract TREN/05/NUCL/S07.55436.

Thomas, Steve, 2017. China's nuclear export drive: Trojan horse or Marshall Plan? *Energy Policy* 101, February: 683–91. doi. org/10.1016/j.enpol.2016.09.038

Urenco, 2017. Company structure, Urenco Ltd. www.urenco.com/about-us/company-structure/ (accessed 30 January 2017).

US DOE (Department of Energy), 2013. Strategy for the management and disposal of used nuclear fuel and high-level radioactive waste. January.

Wang, Ju, 2014. On area-specific underground research laboratory for geological disposal of high-level radioactive waste in China. *Journal of Rock Mechanics and Geotechnical Engineering* 6(2): 99–104. doi. org/10.1016/j.jrmge.2014.01.002

WNA (World Nuclear Association), 2016a. Nuclear power in China. 5 November.

WNA (World Nuclear Association), 2016b. Radioactive waste management: National policies and funding. September.

WNA (World Nuclear Association), 2016c. Uranium enrichment. September.

WNA (World Nuclear Association), 2016d. World nuclear power reactors & uranium requirements. 1 September.

World Bank, 2010. Addressing the electricity access gap. Background Paper. June.

World Bank, 2013a. Gross domestic product by country. Databank data series. databank.worldbank.org/data/reports.aspx?source=2&series= NY.GDP.MKTP.CD&country= (accessed 24 January 2017).

World Bank, 2013b. Total population, by country. Databank data series. databank.worldbank.org/data/reports.aspx?source=2&series=SP.POP. TOTL&country= (accessed 24 January 2017).

World Bank, 2017. Gross domestic product by country. Databank data series. databank.worldbank.org/data/reports.aspx?source=2&series= NY.GDP.MKTP.CD&country= (accessed 24 January 2017).

World Nuclear News, 2016a. French repository costs disputed. 12 January.

World Nuclear News, 2016b. Illinois rallies as nuclear plants fail in capacity auction. 25 May.

World Nuclear News, 2016c. Rosatom explains benefits of state backing to plant projects. 11 February.

Wren, David, 2016. Power surge: Cost overruns at nuclear plant a growing part of SCE&G customers' bills. *The Post and Courier*, 17 June.

Wuppertal Institute for Climate, Environment and Energy and its Partners and Subcontractors, eds, 2007. Comparison among different decommissioning funds methodologies for nuclear installations. Final report on behalf of the European Commission Directorate-General Energy and Transport, H2, Service Contract TREN/05/NUCL/S07.55436.

Zhang, Hui, 2015. China's uranium enrichment capacity: Rapid expansion to meet commercial needs. Cambridge, MA: Belfer Center for Science and International Affairs, Harvard University.

Zhang, Hui, and Yunsheng Bai, 2015. China's access to uranium resources. Cambridge, MA: Belfer Center for Science and International Affairs, Harvard University.

Zhou, Yen, 2013. China's nuclear waste: Management and disposal. Presentation, Managing the Atom Project, Belfer Center for Science and International Affairs, Harvard University, Cambridge, MA, 28 May.

Part II
Country studies

4

A new normal? The changing future of nuclear energy in China

M. V. Ramana and Amy King

Abstract

In recent years, China has reduced its goal for expanding nuclear power capacity, from a target of 70 gigawatts (GW) by 2020 issued in 2009 to just 58 GW by 2020 issued in 2016. This chapter argues that this decline in targets stems from three key factors. The first factor is China's transition to a relatively low-growth economy, which has led to correspondingly lower levels of growth in demand for energy and electricity. Given China's new low-growth economic environment, we argue that the need for rapid increases in nuclear power targets will likely become a thing of the past. The second factor is the set of policy changes adopted by the Chinese government following the March 2011 Fukushima Daiichi nuclear disaster in Japan. Since the Fukushima disaster, China's State Council has stopped plans for constructing inland nuclear reactors and restricted reactor construction to modern (third-generation) designs. The third factor is government responsiveness to public opposition to the siting of nuclear facilities near population centres. Collectively, these factors are likely to lead to a decline in the growth rate of nuclear power in China.

Introduction

In March 2016, China's National People's Congress endorsed its draft 13th Five Year Plan (2016–20), which set China the goal of developing 58 gigawatts (GW) of operating nuclear capacity by 2020, with another 30 GW to be under construction by then. At first glance, this goal appears ambitious, for it represents a doubling of China's current nuclear capacity of 29 GW (as of May 2016, according to the International Atomic Energy Agency's (IAEA) Power Reactor Information System (PRIS) database). Nevertheless, a closer look at this target, and the history behind it, tells a somewhat different story. Back in 2002, China's draft short- and medium-term plan for nuclear expansion called for China to build 40 GW of nuclear capacity by 2020. In 2009, the target figure was increased dramatically to 70 GW of nuclear capacity by 2020. Although large, there was an expectation that this 70 GW target would be met easily; for example, the director of science and technology at the China National Nuclear Corporation (CNNC)—one of the major state-owned enterprises (SOEs) involved in constructing and operating nuclear power plants—stated that 'reaching 70 GW before 2020 will not be a big problem' (Stanway 2009). So what happened between the announcement in 2009 of the 70 GW target and the announcement in 2016 of the lower 58 GW target? In this chapter, we argue that this decline in targets results in part from the policy changes and government responsiveness to public concerns following the March 2011 Fukushima Daiichi nuclear disaster, and in part from China's transition to a relatively low-growth economy with correspondingly lower levels of growth in demand for energy and electricity.

The first factor that affects nuclear power targets is the growth rate for electricity demand. Electricity demand has flattened in China as the country's economy has started to undergo structural changes, from being an economy primarily focused on export-led industrial production to one oriented towards the service sector and domestic consumption. This shift has led to what many have termed the 'new normal' (Green and Stern 2016; Hu 2015; Levi, Economy, and Rediker 2016), with a corresponding decrease in energy and electricity demand. The average amount of electricity produced by different kinds of power plants has also declined. Taken as a whole, power plants operated for 349 fewer hours in 2015 as compared to 2014; specifically, thermal plants operated 410 hours less than they did in 2014, whereas the number of hours that the average nuclear power plant fed electricity into the grid declined by

437 hours, the largest decline among all power plants (Wong 2016a). Some expect the decline to be a long-term trend (Ying 2016). Under these circumstances, it is likely that the need for rapid increases in nuclear power targets will become a thing of the past.

The second factor that has changed Chinese nuclear planning is the Fukushima disaster, which had an immediate dampening effect on the Chinese government's push to rapidly expand nuclear power. Five days after the accident at Fukushima started, the State Council, China's chief administrative authority, stated:

> We will temporarily suspend approval for nuclear power projects, including those that have already begun preliminary work, before nuclear safety regulations are approved … Safety is our top priority in developing nuclear power plants (Bristow 2011).

Since then, the Chinese government has verbally committed to ensuring nuclear safety, and has introduced a number of operational measures aimed at lowering the risk of accidents (as described in detail in King and Ramana 2015). Over and above these, the State Council made two important decisions: (1) to restrict construction of nuclear reactors in inland areas, and (2) to restrict the choice of reactor designs for construction to only the so-called 'third-generation' designs. At the time of writing, the State Council had also issued a third draft decision that may further restrict the expansion of nuclear power in China: namely, to require all nuclear developers to solicit local public opinion and undertake local 'social stability' impact reports before proceeding with new projects.

As we have argued elsewhere (King and Ramana 2015), these decisions, especially the one to stop construction in inland areas, are being shaped and contested by China's economic plans, national atmospheric pollution reduction plans, corporate economic interests, public opposition to nuclear power, and local government bureaucratic pressure. More generally, China's choices on nuclear power are shaped by China's system of governance, which affords a relatively high degree of political power to local authorities and SOEs, which dominate China's nuclear power sector (Xu 2008, 2014; Ramana and Saikawa 2011). Whether or not the decisions made by Chinese central authorities to enhance nuclear safety in the aftermath of Fukushima continue to be upheld in the future will be dependent on the interactions between, and priorities of, these multiple actors. Nevertheless, we argue, the two decisions made in 2011–12 have led to a lowering of the nuclear installation target for 2020. Had there not been a shift in the pattern of the economy, the post-Fukushima decisions might have only resulted in a temporary change in policy. But when combined with the shift in the nature of China's energy demand, and the government's growing concern about public opposition to nuclear energy, we anticipate that these changes in policy will shape deeply the future of Chinese nuclear capacity.

In the remainder of this chapter, we first elaborate on the decline in energy growth rates and their implications for nuclear power targets. Then we discuss, in turn, the ban on inland reactor construction, the implications of the State Council's restriction on the type of reactor designs that may be installed, and the implications of the government's responsiveness to negative public attitudes to nuclear power, such as its 2016 decision to cancel the building of a nuclear processing plant in Lianyungang, Jiangsu province, in the face of public protests. We conclude with the overall inferences of our arguments and some remarks on the political struggles that might shape the future of nuclear power in China.

The changing nature of energy demand in China

In the last couple of years, there have been significant shifts in China's economic growth rates and development strategy. After more than a decade of rapid growth in China's energy-intensive heavy industry sector between 2000 and 2013, China's economy has now begun transitioning towards less energy-intensive sectors such as services and advanced technology (Green and Stern 2016). This shift is intentional: the country's leadership has been trying deliberately to steer China away from an overwhelming focus on exports towards exploiting domestic demand for products, partly in the expectation that this shift will reduce the environmental impacts of large-scale industrial manufacture. Alongside this, there has been an emphasis on improving energy efficiency. For example, for the year 2016, the China National Energy Administration (CNEA) announced a target of reducing energy consumption per unit of gross domestic product (GDP) by at least 3.4 per cent (Xinhua 2016). All of these factors have contributed to a 'new normal', which has resulted in lower growth rates in energy consumption. Compared with energy consumption growth rates of around 8 per cent per year from 2000 to 2013, China's total energy consumption in 2014 grew by just

2.2 per cent, and by just 0.5 per cent in 2015 (Green and Stern 2016: 5; China Electricity Council 2016). Furthermore, as China's economy continues to transition away from heavy industry sectors such as steel and cement, it is estimated that energy consumption will grow by just 1.8 per cent per year out to 2025 (Green and Stern 2016: 10). Another estimate is from the oil and gas firm Exxon, which has forecast that China's annual energy demand will grow by just 2.2 per cent per year out to 2025, and predicts that 'the country's energy demand would plateau around 2030' (Groden 2016). To be sure, these estimates are for energy demand growth rather than electricity demand, which might be higher because of continued urbanisation and other trends. Nevertheless, if the energy demand growth rate has fallen by a factor of four, it is only to be expected that the electricity demand growth rate will decline as well, even if it may not be as precipitous.

There is a slow-growing realisation that these changes in the economy will impact electricity generation plans. In the words of Zhou Dadi, a senior research fellow with the Energy Research Institute of the National Development and Reform Commission (NDRC):

> A 'new normal' has been unfolding in China's power sector. It's marked by weakening demand and a contraction in output resulting from the industrial restructuring. We have been used to seeing annual electricity consumption increase by 8% or more but it's quite a different situation now (quoted in Ying 2016).

Analysts Jiang Lin, Gang He, and Alexandria Yuan (2016) used 20 years of provincial data on GDP and electricity consumption to deduce a plateauing effect of electricity consumption in the richest provinces, as the electricity demand saturates and the economy develops and moves to a more service-based economy. Their evidence suggests the emergence of a 'new normal relationship' for electricity use. Therefore, they warn:

> If the power system planning approach is not responsive to these emerging trends, there is a significant chance of overbuilding the power capacity in China, with hundreds of billions of dollars of investment potentially stuck as stranded assets (Lin, He, and Yuan 2016: 52).

Indeed, 'if all the coal power projects submitted for Environmental Impact Assessment (EIA) approval were put into operation in 2020' the available capacity could exceed the demand by 200 GW or more

(Yuan et al. 2016: 136). This should not be surprising since there are other areas where China seems to be building overcapacity; for example, in what have been termed 'ghost cities' (*Chinadialogue* 2015).

Excess capacity is not uniformly distributed, with some regions projected to have much greater mismatch than others (in part because of existing regional variations in power supply and demand). This has led to the imposition of cutbacks in electricity production from nuclear plants in specific provinces. Provinces that have seen nuclear operators being asked to cut back are Fujian, Hainan, and, most dramatically, Liaoning, 'where China General Nuclear's Hongyanhe nuclear plant has faced major curtailments' (Wong 2016b: 4). Looking ahead, if even some of the coal plants that are being planned are actually commissioned, then the situation would be further detrimental to the expansion of nuclear power since both coal and nuclear plants can be considered as competing for operation as a baseload supplier of electricity.

China's transitioning economy and the shift to lower rates of economic growth and energy consumption mean that the central government faces considerably less pressure to rapidly roll-out new nuclear power plants to meet the ambitious nuclear growth rates stipulated between 2002 and 2011. The impact of this lowered pressure is already apparent: provincial governments and those in the nuclear industry seem to be less persuasive when pushing to restart construction of inland nuclear power plants, as we shall see in the next section.

The inland construction ban and the consequent shortage of sites

One important constraint on nuclear expansion in China is the choice of sites. Prior to the Fukushima accident, China had plans for a vast expansion of nuclear power stations, not only at coastal sites where reactors had traditionally been sited, but also at new inland sites (Du 2010). Since Fukushima, however, the Chinese government has prohibited the construction of all inland reactors, with the State Council placing a ban on inland reactor construction for the duration of the 12th Five Year Plan (2011–15). The decision to ban inland reactors was made for safety reasons: all nuclear reactors need large quantities of water to cool their radioactive cores and lack of adequate cooling water could cause a severe accident. In the case of potential inland power plants, the only sources of water are rivers and lakes, both of which also serve many other needs, including water for irrigation and household consumption. Inland nuclear power plants therefore pose far higher risks to nearby water sources and people dependent on these resources than comparable coastal plants. As we describe below, ever since the decision was made in 2012, there has been much pressure from vested interests to lift this ban. But so far the central government has not succumbed to that pressure and the 13th Five Year Plan put off construction of inland reactors until at least 2020 (Yu 2016b).

These safety concerns have been echoed by Chinese researchers. One prominent critic of inland nuclear reactors, He Zuoxiu, is a leading theoretical physicist, who worked on China's first nuclear bomb. He has warned against building 'any nuclear power plant in the inland regions' due to his concerns about problems with water supply, and his calculation that a reactor accident in China by 2030 is 'highly probable' (He 2013; Xuyang 2012). Another key figure has been Wang Yinan, a policy researcher from a State Council subsidiary institute, who has questioned the construction of inland nuclear power plants on safety grounds (Yu 2016b). Other prominent critics include government officials from provinces that border potential inland nuclear projects. A good illustration of this is the Pengze nuclear power project in Jiangxi province, which was originally slated to host two AP1000 reactors (Wang 2009).

Pengze, along with Xianning in Hubei province and Taohuajiang in Hunan province, was among the first inland nuclear power projects proposed in the late 2000s. In 2010, Chinese nuclear officials expected that these sites would be 'ready for construction' by the end of the year (Zheng and Wu 2010). However, the Fukushima accident prompted strong opposition to the Pengze reactor, particularly in neighbouring Anhui province (Cui 2012). The Pengze plant is sited alongside the Yangtze River, which is a vital water supply for farmers in Anhui (Hook 2012). In July 2011, four retired government officials in Anhui province submitted a petition to local and central government agencies listing various problems with the clearance given to the project and calling for its halting. These concerns eventually reached the government of Wangjiang county in Anhui province, which is downstream from the proposed Pengze site. The Wangjiang government opposed the project and publicly accused the Pengze project of 'falsifying its EIA report', expressed concern about the high local population density and the risk of earthquakes, and objected to Jiangxi province's failure to consult its provincial neighbours before deciding where to site the plant (Wen 2014).

This individual case study points to the more general problem that conflicting interests can arise when making decisions about the siting of nuclear power plants (Aldrich 2008). The structure of benefits and risks is such that all of the economic gains from the nuclear power plant would accrue to the host province (Jiangxi), but accidents would also affect neighbouring provinces (such as Anhui) (Zhu 2014).

There are also very strong forces in China actively encouraging the resumption of inland construction, including local provincial and county-level governments who will host any future inland power plants, and the SOEs who will construct and manage the inland power plants. To understand these forces, consider the situation in Hunan, Hubei, and Jiangxi provinces where three inland nuclear power plants—Hunan's Taohuajiang power station, Hubei's Xianning Dafan power station, and Jiangxi's Pengze power station—were proposed prior to the Fukushima incident.

For the Hunan, Hubei, and Jiangxi provincial governments, the desire to restart planning for inland reactors and their eventual construction primarily stems from economic interests. Each of the provincial governments has a stake in the projects; for instance, the Hunan government holds a direct 5 per cent stake in the Taohuajiang project, while the Jiangxi government holds a more indirect stake of 40 per cent through its financial backing of a provincial energy firm, which has invested in the site (Yu 2015).

Provincial governments also stand to benefit economically once the reactors come into service, and local SOEs have taken great pains to remind provincial governments of these benefits. For instance, the general manager of the Hunan Taohuajiang Nuclear Power Company, Zheng Yanguo, told reporters in September 2014 that an investment of 70 billion yuan (US$11 billion) in the Taohuajiang nuclear power plant would return GDP growth of over 100 billion yuan to Hunan province, and generate annual tax income of around 15 billion yuan (*Zhongguo Jinggong Bao* 2014). These kinds of claims have led the Hunan Taohuajiang Nuclear Power Company, the Hunan provincial government, and the local Taojiang county government (where the Taohuajiang site is located) to exert joint pressure on the central government in Beijing to approve inland reactors, and to use the media and other public means of communication to raise expectations that inland reactors will soon be built (Xu 2014: 24; Securities Daily 2015). For instance, in September 2014, the Taohuajiang general manager informed reporters that 'preparatory work for the Taohuajiang nuclear power station will commence in 2016, in accordance with the schedule' (*Zhongguo Jinggong Bao* 2014). Likewise, at the March 2015 National Party Congress, delegations from Hunan and Hubei provinces called upon the central government to restart construction of inland power projects at the beginning of the 13th Five Year Plan, and reminded the central government that the Taohuajiang, Dafan, and Pengze projects were ready for construction (*Zhongguo Hedian Wang* 2015). Furthermore, there were also signs that new inland sites were being prepared in anticipation of a lifting of the ban. In July 2014, it was reported that the China General Nuclear Power Corporation (CGNPC) had 'agreed to invest 38 billion yuan ($6.1 billion) in two nuclear power plants in southwest China's Guizhou Province' (Xinhua 2014). Prior to the Fukushima nuclear disaster, there were no plans to construct a nuclear plant in that province (WNA 2010).

Provincial governments and those in the nuclear industry have also attempted to use the State Council's development targets as a way to pressure the central government to lift the ban on inland reactors. This is because there are few coastal sites available for new nuclear plants; raising nuclear capacity would thus require China to find new sites inland. According to a 2015 analysis in *Nuclear Intelligence Weekly*, a trade journal:

> The Mid-Long Term Nuclear Development Plan (2011–2020) ratified by the State Council in 2012 envisions 58,000 MW [megawatts] of operational nuclear capacity at the end of 2020, with 30,000 MW under construction for a total of 88,000 MW. This means that within the next six years, China would have to start six to seven 1,000 MW units on average each year, representing a combined capacity of roughly 40,000 MW. This can only be achieved by adding inland sites to the mix since available coastal sites are becoming fewer and fewer. Indeed, State Council data published back in 2007 listed more than 40 sites reserved in China for future nuclear power projects—and at least 31 of them are inland (Yu 2015).[1]

Subsequently, provincial governments and those in the nuclear sector have used these older targets for their lobbying purposes. For example, a feasibility study produced by the Hunan Taohuajiang Nuclear Power

---

1   Similar nuclear energy targets can also be found in the *Nengyuan hangye jiaqiang daqi wuran fangzhi gongzuo fang'an* [Energy industry work plan for strengthening the prevention and control of atmospheric pollution], which was jointly issued by the NDRC, Bureau of Energy, and Ministry of Environmental Protection in May 2014.

Company stated that due to the limitation in sites, there was a 'large gap' in China's ability to meet the central government's nuclear power targets with the power stations currently installed and under construction (*Zhongguo Jinggong Bao* 2014). Similarly, during the March 2016 Chinese People's Political Consultative Conference, the chairman of the CGNPC, He Yu, told reporters that the scale of China's current nuclear power is 'still too small' to achieve China's economic development goals (L. Wang 2016).

Despite this intense pressure, and the limited number of coastal sites still available to build new plants, the central government has so far not reversed its ban on inland nuclear construction. In December 2014, the State Council released its Energy Development Strategy Action Plan (2014–20) (hereafter the 'Action Plan'). A State Council circular discussing the Action Plan indicated that inland nuclear power still required further research and proof of safety (CNEA 2014b). Media reports around the time of the Action Plan's release also noted that 'there is still a lot of controversy around inland nuclear power in China', while officials from the CNEA told reporters that inland nuclear power 'must be proved and proved again' (CNEA 2014a).

Moreover, in March 2016, the NDRC deputy director and director of the CNEA, Nur Bekri, issued a clear retort to provincial governments and those in the nuclear industry who have used the central government's targets to lobby for the development of inland plants. Bekri stated that 'coastal nuclear power plants are sufficient to realize the nuclear power development targets [of 58 GW] contained in the 13th Five Year Plan' (Xie 2016). He went on to say that 'there is no timetable for restarting' inland nuclear power plants, and that the CNEA would recommend restarting inland projects only if safety could be 'absolutely guaranteed' (Xie 2016). This unusually strong language—strictly speaking, an absolute guarantee of safety is impossible—suggests two things: first, there are still significant political barriers to constructing nuclear plants away from the coast, and second, lower rates of energy demand make the NDRC and State Council better able to resist local government and industry pressure to restart inland construction. The second point is underscored by examining the regional distribution of the excess coal power capacity discussed earlier. If all the proposed coal plants are commissioned in the two inland regional electric grids of China—the Northwest and the Central China grids—then there would be 41,010 MW and 47,300 MW of excess capacity (Yuan et al. 2016: 142).

Problems with new reactor designs

The second constraint that results from the policy changes decreed by the State Council after its 2011–12 review relates to the requirement that '[n] ewly constructed power units must comply with third-generation nuclear power technology safety standards' (Wen 2012). The problem this poses to rapid nuclear expansion is that third-generation nuclear reactors, namely the latest designs, cost more and have taken longer to construct, both in China and elsewhere in the world (Schneider et al. 2015).

Prior to the Fukushima disaster, Chinese nuclear officials and policymakers had identified the development of Generation III reactor technology as a key goal for the Chinese nuclear sector. Among the priority areas for research and development identified by the 'National Medium and Long-Term Science and Technology Plan (2006–2020)', issued by the government in 2005, was the development of indigenous 'advanced large-scale pressurized water reactors' (Mu 2010: 380). Despite these goals, however, China had yet to develop indigenous Generation III reactor technology at the time of the Fukushima disaster. In 2011, the 27 reactors then under construction in China included the following reactor models: CNP-600, CPR-1000, AP1000, and the European pressurised reactor (EPR) (WNA 2011). Of these, the two Chinese-designed reactors—the CNP-600 and CPR-1000—were classified as Generation II, and their safety was considered less reliable than European or US models.[2] China's central government had mandated the setting up of a Chinese 'Nuclear Power Self-Reliance Program' to develop a domestic Generation III reactor using a modified Western design (in this case, the AP1000 reactor). In 2010, just one year before the Fukushima disaster, two academics from Tsinghua University observed that 'China has basically mastered the generation II of nuclear technology, but still lacks the *R&D required for generation III*' (Zhou and Zhang 2010: 4283, emphasis added). Therefore, it is clear that there was no indigenous capacity to come up with an independent Generation III design at the time of Fukushima.

---

2   The only reactors that were categorised as Generation III in 2010 were the AP1000, the advanced boiling water reactor (ABWR), the economic simplified boiling water reactor (ESBWR), the European pressurised reactor (EPR), and the water–water energetic reactor (VVER). Of these, the ESBWR had not received regulatory approval in 2010.

Nevertheless, in the wake of the Fukushima accident, the Chinese nuclear industry moved quickly to reassure the public that the continued expansion of nuclear power in China would be safe because it would be based on Generation III reactors. Lu Qizhou, general manager of the China Power Investment Corporation, for example, pointed out that the 'reactors in the Japanese nuclear power plants, which have been affected by the massive quake, are Generation II reactors and have to rely on back-up electricity to power their cooling system in times of emergency', whereas the 'AP1000 nuclear power reactors, currently under construction in China's coastal areas and set to be promoted in its vast hinterland, are Generation III reactors and have built in safety features to overcome such a problem' (Xinhua 2011).

Comments such as those by Lu Qizhou led to a flurry of competitive activity in the Chinese nuclear industry.[3] Because China did not yet have its own domestic Generation III capability, it needed to rely on imported Western models, such as the AP1000, if it wished to guarantee the use of Generation III technology. However, only one of the players in the Chinese nuclear power sector, the State Nuclear Power Technology Corporation (SNPTC), held government authority 'to sign contracts with foreign parties to receive … 3rd generation nuclear power technology' (SNPTC 2011). Given the domestic expectation after Fukushima that only Generation III reactors would be permitted in new power stations, by May 2011—just three months after Fukushima—the SNPTC had convinced officials at Westinghouse Electric Company that Westinghouse was going to dominate the Chinese reactor market (Li and Tranum 2011).

It did not take long for other players in the Chinese nuclear power sector to respond to this potential threat to their market share. Two of the other major players in the Chinese nuclear power sector, the CNNC and the CGNPC, responded by quickly producing their own reactor designs, which they described as being compliant with Generation III requirements.

In November 2011, the CGNPC announced that it had developed and held 'full intellectual property rights'—a key requirement for exports—over the newly designed ACPR1000, a reactor that it stated

---

3   News media also shared the impression that as 'China attaches more focus on the safety of nuclear technology, it is likely to adopt the third-generation AP 1000 technology developed by US-based Westinghouse Electric Co in its future plants' (Liu 2011).

had incorporated the lessons of Fukushima in 'meeting the standards of international third-generation nuclear power technology' (Pan 2011). A few months later, at the third Asia Nuclear Power Summit in January 2012, the CNNC unveiled its own ACP1000 reactor (Zhou 2012). Adding to this menagerie is the Hualong One, a Generation III design jointly developed by the CNNC and the CGNPC; in 2014, the Hualong One design was certified by the National Nuclear Saftey Administration (Hore-Lacy 2014). There are questions about whether the Hualong One is one design or if there are two separate designs developed by the CNNC and the CGNPC respectively, both of which are marketed under the same name (Yu 2016a; Thomas 2017).

These questions notwithstanding, the Hualong One design is now being promoted enthusiastically outside China as its most advanced reactor and, in February 2015, the CNNC signed a contract with Argentina for the export of this design (*World Nuclear News* 2015). In March 2016, the CGNPC and the CNNC set up a 50–50 joint venture to promote the Hualong design in overseas markets (*World Nuclear News* 2016). The Chinese nuclear industry has also begun to exploit Chinese President Xi Jinping's new 'Belt and Road Initiative' as a way to further expand the overseas roll-out of the Hualong One. In March 2016, CNNC chairman Sun Qin ambitiously claimed that the Hualong One was expected to obtain 20–30 per cent market share of the more than 40 countries within the 'Belt and Road' region that were seeking to develop nuclear power (CNNC 2016).

The speed with which all these new reactor designs were rolled out by the CNNC and the CGNPC raises serious questions about whether these reactors actually meet Generation III safety requirements. While the CNNC and the CGNPC were certainly talking about developing Generation III designs prior to Fukushima, in part to improve their chances of entering the reactor export market, the real momentum for the development of the ACPR1000, the ACP1000, and the Hualong One came only after the Fukushima disaster in 2011 (*World Nuclear News* 2010). For the sake of comparison, the Westinghouse AP1000 reactor, which was approved for construction in the United States in February 2012 (Hargreaves 2012), received approval only after 19 revisions to its reactor design were examined by the US Nuclear Regulatory Commission (NRC) (2011). Furthermore, the AP1000 was itself a modification of the AP600, which had been certified by the US NRC 13 years earlier in 1999 after a 'ten-year, multi-million dollar effort by the NRC staff, the US Department of Energy, the Electric Power Research Institute and supporting utilities, Westinghouse, its subcontractors and partners' (Westinghouse Electric Company 2000).

The relatively short period of time taken by Chinese corporations to develop their Generation III reactors suggests that these new Chinese models may be Generation III in name only. Indeed, analysts who have followed the development of these new reactors report that they are merely 'enhanced versions of the current CPR-1000' (Hinze and Zhou 2012), the Generation II reactor that was being constructed en masse in China prior to Fukushima. Even though the Hualong One has been certified, observers see the design as being at an 'early stage'.[4]

To the extent that there has been construction of more advanced reactor designs (imported from France and the United States) in China, these projects have been afflicted with significant cost overruns and delays. The EPR units 1 and 2 being built at Taishan were originally scheduled to 'be commissioned at the end of 2013 and in autumn 2014' respectively, and France's Areva had hoped 'to have started work on more reactors' by then (Thibault 2010). Neither of these expectations were met. In January 2016, Taishan-1 underwent its cold functional test (Taishan Nuclear Power Joint Venture Co. 2016), a pre-operational step that has to be completed before any fuel is loaded into the reactor. In March 2016, CGNPC officials projected that Taishan-1 would start up in 2017 (Chaffee 2016).

China's experience in building Generation III AP1000 reactors at the Sanmen and Haiyang sites has also been fairly troubled, with significant delays, cost escalations, and the identification of safety concerns (Stanway 2014; Yap and Spegele 2015; Lok-to 2016). With these reactors, the main source of problems, although not the only one, has been the reactor coolant pumps that were supplied by US manufacturer Curtiss-Wright Corporation. The reactor coolant pump forces water to circulate through the reactor and transfer the heat generated by the fission reactions in the reactor core. Problems with the reactor coolant pumps could have serious safety consequences and Chinese nuclear officials have expressed concern in the past about these problems. In 2013, for example, a former vice president of the CNNC complained: 'Our state leaders have put a high priority on [nuclear safety] but companies executing projects do not seem to have the same level of understanding' (Ng 2013). The result has

---

4   Personal communication, C. F. Yu, 9 March 2015.

been a very long series of delays. As of January 2017, the expectation was that the four AP1000s will go into operation before the end of 2017 (*World Nuclear News* 2017). If this were to happen as hoped for, electricity generation from the AP1000s would take place four years after schedule.

The slower pace of construction has not only resulted in targets for nuclear generation being set back but also in higher costs. Estimates by China's Nuclear Energy Agency suggest that the cost of constructing Generation III reactors is significantly higher (US$2,300 per kilowatt (kW) for the AP1000) than Generation II reactors (US$1,750 per kW for the CPR1000) (IEA/NEA 2010: 48). More recently, the Hualong One's deputy chief designer has estimated that the 'targeted construction cost of Hualong One … when production was scaled up' will actually be US$2,500 per kW. Adding the caveat 'when production was scaled up' means, by implication, that the cost of early units will be significantly higher and that US$2,500 per kW is only an aspirational goal for the future. And if China reproduces the pattern of cost increases and negative learning that has characterised nuclear plant construction in other countries (Boccard 2014; Grubler 2010; Koomey and Hultman 2007), the goal of US$2,500 per kW may never be reached. Offering further evidence for the expected higher costs of Generation III reactors is another newspaper article that has claimed implicitly that the cost of the Hualong One is US$3,000 per kW (Abe 2016).

The impact of this higher construction cost is that in the face of slower demand growth, it is possible that the Chinese government will choose to emphasise other, cheaper sources of energy over nuclear power. In particular, there is evidence of both rapidly increasing capacity of wind and solar energy, as well as declining costs of these sources. Thus, in comparison to the 2000–10 decade, one might expect that nuclear growth targets would be more modest.

Public opinion shapes nuclear policy

The third factor acting as a constraint on the roll-out of new nuclear power stations in China is the growing government responsiveness to public opposition to the siting of nuclear facilities. Since the Fukushima accident, there has been a significant increase in the Chinese public's perception of risk from nuclear facilities. Two surveys of residents living near the Tianwan nuclear power plant in Lianyungang, Jiangsu province, the closest nuclear plant to Fukushima, which were conducted in August 2008 and March–April 2011, found a dramatic decline in support for nuclear power (Huang et al. 2013). The percentage of respondents who agreed with the proposition 'Nuclear power should be used in our country' went down from 68 per cent to 32 per cent, and the fraction that agreed with the proposition 'We should quickly increase the number of nuclear power stations in China' declined from 40 per cent to 17 per cent. The percentage of supporters of building a nuclear power station in 'my city' declined from 23 per cent to 8 per cent, whereas those who were neutral came down from 64 per cent to 38 per cent. In contrast, the fraction of opponents increased from 13 per cent to 54 per cent. The surveys also found that perceived benefits of nuclear power and public trust in government had decreased significantly, whereas knowledge about nuclear power had increased significantly.

Similar studies in other locations also found significant levels of public concern about nuclear safety and reactor accidents (Sun, Zhu, and Meng 2016). China also had the fifth highest difference in the levels of acceptability of nuclear power before and after the Fukushima accident (Kim, Kim, and Kim 2013). More than half of all respondents in one poll felt that only 80 kilometres or more constitutes a safe distance between their homes and a nuclear reactor (He et al. 2014).

Since Fukushima, there has been increasingly prominent opposition to nuclear power plants in China (Buckley 2015; Lok-to 2016). In particular, August 2016 saw the outbreak of large-scale public protests in the city of Lianyungang. Lianyungang was being considered—along with five other sites—as the location for a 100 billion yuan (US$15 billion) nuclear reprocessing plant to be built by the CNNC using technology owned by the French company Areva (Green 2016). Thousands of people gathered on the weekend of 6–7 August 2016 to protest against the plant proposal, with protestors making extensive use of Chinese social media platforms such as WeChat to garner further public support (Liu 2016). In addition to their general opposition to nuclear power, protestors drew connections between the Fukushima accident and Lianyungang, arguing that storing radioactive material in a 'seismically active area' like Lianyungang was inappropriate and unsafe (Green 2016: 4). Protestors also cited their frustration with the lack of transparency surrounding the government's decision to site the reprocessing plant in their city. According to reports by the *South China Morning Post*, local residents only became aware of the plant's possible siting in their city following a press release by the Chinese

State Administration of Science, Technology and Industry for National Defence (Li 2016). As one resident stated, '[t]he government kept the project a secret. People only found out about it recently. That's why most people are worried' (Hornby and Lin 2016).

Despite attempts by the Lianyungang city government to reassure the public that no final decision on the plant's location had been made, protests continued on 8 and 9 August. Subsequently, on 10 August, the Lianyungang city government issued a dramatic turnaround in policy, announcing that it would 'suspend preliminary work on site selection for the nuclear recycling project' (Li 2016). At the same time, Lianyungang authorities also stepped up their efforts to halt the further spread of anti-nuclear protests. At least one individual was arrested for 'allegedly spreading rumours' about a forthcoming protest in Lianyungang, while workers at the Lianyungang Limited Harbor Holding Group—which runs the city's port—were forced to sign an agreement pledging not to 'believe rumours', 'spread rumours', or 'participate' in 'illegal assemblies' (Henochowicz 2016).

The central government also took a direct interest in the Lianyungang situation. On 11 August, central government authorities issued censorship instructions directing media organisations to delete and not republish an article by the Sohu media group entitled 'Cautiously welcoming the decision to suspend the Lianyungang nuclear waste project'. The Sohu article criticised the government's lack of transparency over the nuclear project, described the decision to suspend the project as the 'correct response', and warned the government not to 'underestimate the public's resolve in opposing nuclear waste' (Wade 2016).

The decision to stop the Lianyungang project is not the first cancellation of a proposed nuclear facility in response to public protests. Earlier cancellations include that of the Hongshiding nuclear power plant in Rushan in Shandong province, a nuclear fuel cycle plant in Jiangmen in Guangdong province, and the Hui'an nuclear plant project in Fujian province (Sheng 2014).

What might be more significant than these cancellations for the future of nuclear power in China is that, in September 2016, the Legislative Affairs Office of the State Council issued new draft 'Regulations on Nuclear Power Management' (*Hedian guanli tiaoli*), which for the first time required nuclear developers to consider public opinion in siting new nuclear projects. The announcement of these draft regulations followed shortly after the Lianyungang protests, with the Legislative Affairs Office stating that 'Japan's Fukushima accident once again created doubt about the safety of nuclear power among the public, and also caused feelings of fear and opposition to occur from time to time' (Stanway 2016). Developed by the NDRC and the CNEA, the draft regulations state that developers must work with provincial governments to undertake 'social stability' impact assessments on all new nuclear projects, and must actively seek out public opinion on new projects through public hearings. In addition, the draft regulations stipulate that citizens have the right to public disclosure of government information related to nuclear power (J. Wang 2016; Stanway 2016). Ultimately, the State Council argued, the draft regulations are designed to 'allow the public to participate more actively in the construction and supervision of nuclear projects' (Stanway 2016).

These draft regulations, coupled with the Lianyungang government's decision to cancel the proposed reprocessing plant, demonstrates that public opinion has become a third important constraint on the future development of nuclear power in China. Local and central government authorities in China are now increasingly concerned that growing public opposition to nuclear power could result in the eruption of large-scale protests across China. Local and central government authorities have responded to these concerns in part by using the traditional methods of censorship, arrest, and coercion of local labour. But they have also responded by heeding the public's concerns: by suspending the Lianyungang reprocessing plant and by proposing regulations that are designed to increase decision-making transparency and public participation in nuclear power decision-making processes. At the time of writing, it is still too soon to tell whether the Chinese public will actually be granted greater involvement in the decision-making process. However, the swift responses by the Lianyungang city government and the State Council serve as important examples of government sensitivity to public opinion. Given the growth in anti-nuclear sentiment in China since the Fukushima accident, we can expect that public opinion will continue to dampen the push for an expansion of nuclear power in China.

Conclusion

The future of nuclear power in China is not what it used to be. A decade ago, China had acquired the reputation of setting very ambitious targets for nuclear power in the country, meeting these targets, and then increasing the targets to even more ambitious heights. That is no longer the case. Because of the changing nature of the economy, the growth rate of energy demand has declined precipitously. The decline has been so sharp and swift that China today has a surplus in electrical generation capacity; consequently, in comparison with previous years, many power plants are being forced to run for fewer hours (as mentioned earlier). Further, due to the inertia in the system, many more power plants, including nuclear plants, are going to come online over the next several years. Thus, the mismatch between electricity demand and availability can be expected to become more severe in the coming years. As a result, a decline in the growth rate of nuclear power can also be expected.

These changes within China also play a part in the efforts by Chinese nuclear operators and reactor constructors to export reactors, with a specific focus on entering markets in Western Europe—for example, Hinkley Point C in the United Kingdom—by investing large amounts of money. At one level, this can be seen as the maturing of the Chinese nuclear industry and the development of its technical capacity. But, at a different level, this can also be seen as a response to the slowdown in growth in the domestic nuclear market; reactor exports, then, become a route for the continued expansion of reactor construction. However, entering the reactor export marketplace is not easy, and there are many questions about China's ability to supply reactors that perform adequately, especially with regard to safety (Thomas 2017). China also has little experience in executing nuclear projects anywhere except within its own borders. Chinese investment in the Hinkley Point C reactor is a way to address those concerns and this is being done, in part, by using China's financial clout. There is some expectation that, in exchange, the United Kingdom will allow for the construction of a Chinese reactor down the line at the Bradwell site.

One counter-argument might be that China is also in the process of reducing its reliance on fossil fuels, coal in particular and, as part of that process, it could shut down old coal plants, thus creating a demand for new power plants not based on fossil fuels, such as renewables or nuclear power. While this is certainly true, our arguments earlier suggest that the constraints on nuclear power plants—the lengthy time period for construction, the higher costs involved, the problems with imported nuclear reactor designs, and, last but not least, the limited number of coastal sites still available to build nuclear plants—make it more likely that the replacement for old fossil fuel plants will be renewables—wind and solar—rather than nuclear reactors.

A second counter-argument might be that nuclear energy is considered a baseload source of electricity, whereas solar and wind cannot perform in this fashion. While there are important differences between renewable electricity sources and nuclear power, this argument has less merit now because of the shift in the Chinese economy and the move away from high-energy consuming manufacturing industries—for example, steel or cement.[5] Industrial energy requirements dominate China's electricity demand pattern.

We do emphasise that our argument is suggestive not definitive. It is certainly possible that there may be a shift in the political balance of power between those who advocate inland reactor construction and those who resist such a push, thereby voiding the siting constraint for new nuclear plants. But the evidence so far makes that scenario seem unlikely. If, prior to the era of low-energy demand growth, advocates of siting reactors away from the coast could not overcome opposition, then it is less likely that they would be able to do so as energy demand growth slows down and as there is local opposition to the siting of nuclear power plants.

The import of these developments and shifts is not that China is moving away from nuclear energy, but neither is it likely to be the powerful engine for global nuclear expansion as had been assumed earlier.

---

5   The declining importance of baseload electricity generation is testified by many studies in various countries. In Great Britain, for example, one study found that 'with current patterns of electricity demand in GB [Great Britain], the need for baseload vanishes once the GB system secures an average of around 30% *of electricity generated* from wind, and 10% from PV [photovoltaics]' (Smith and Grubb 2016: 3, emphasis in original).

Acknowledgements

This chapter is based in part on King and Ramana (2015), but both the structure and the argument have been changed completely. The authors are grateful to Peter Van Ness for his invitation to the 'Nuclear Power in East Asia: The Costs and Benefits' workshop held at The Australian National University, Canberra, 12–14 August 2014, and for his valuable comments on a draft of this chapter. We would also like to thank Wei Peng, Bill Sweet, Stephen Thomas, C. F. Yu, and Derek Abbott for their valuable feedback.

References

Abe, Tetsuya, 2016. China nuclear industry: State-owned enterprises eye overseas power projects. *Nikkei Asian Review*, 5 January. asia.nikkei.com/Business/Deals/State-owned-enterprises-eye-overseas-power-projects (accessed 23 January 2017).

Aldrich, Daniel P., 2008. *Site Fights: Divisive Facilities and Civil Society in Japan and the West.* Ithaca, NY: Cornell University Press.

Boccard, Nicolas, 2014. The cost of nuclear electricity: France after Fukushima. *Energy Policy* 66(March): 450–61. doi.org/10.1016/j.enpol.2013.11.037

Bristow, Michael, 2011. China suspends nuclear building plans. BBC News, 17 March. www.bbc.co.uk/news/world-asia-pacific-12769392 (accessed 31 January 2017).

Buckley, Chris, 2015. China's nuclear vision collides with villagers' fears. *New York Times*, 21 November.

Chaffee, Phil, 2016. EDF faces British frustrations on Hinkley. *Nuclear Intelligence Weekly*, 24 March.

Chinadialogue, 2015. New 'ghost cities' typify out-of-control planning. 15 October. www.chinadialogue.net/article/show/single/en/8239-New-ghost-cities-typify-out-of-control-planning (accessed 23 January 2017).

China Electricity Council, 2016. Press release, 3 February. www.cec.org.cn/yaowenkuaidi/2016-02-03/148763.html (accessed 2 June 2016).

CNEA (China National Energy Administration), 2014a. Woguo Hedian Zhuangji Liang 2020nian Mubiao Bubian [No change in China's nuclear installation targets for 2020]. First published in Yicai. com, 20 November. www.china-nea.cn/html/2014-11/31338.html (accessed 23 January 2017).

CNEA (China National Energy Administration), 2014b. Guowuyuan Bangongting Guanyu Yinfa Nengyuan Fazhan Zhanlue Xingdong Jihua (2014nian-2020nian) de Tongzhi [State Council General Office circular concerning the publication of the Energy Development Strategy Action Plan (2014–2020)]. 3 December. www.nea.gov.cn/ 2014-12/03/c_133830458.htm (accessed 23 January 2017).

CNNC (China National Nuclear Corporation), 2016. Sun Qin Daibiao Tan: Zhan Zai Qianyan Lingyu Yinling He Gongye Fazhan [Talks with representative Sun Qin: At the forefront of the development of nuclear industry]. 1 March. www.cnnc.com.cn/publish/portal0/ tab664/info97004.htm (accessed 23 January 2017).

Cui, Zheng, 2012. Ex-officials battle plan to build nuclear plants. *Caixin Online*, 9 March. www.chinafile.com/reporting-opinion/ caixin-media/ex-officials-battle-plan-build-nuclear-plants (accessed 30 August 2017).

Du, Fenglei, 2010. Site selection for nuclear power plants in China. Presentation to Common Challenges on Site Selection for Nuclear Power Plants. Technical meeting. Vienna: International Atomic Energy Agency, 6–9 July.

Green, Fergus, and Nicholas Stern, 2016. China's changing economy: Implications for its carbon dioxide emissions. *Climate Policy* 16 March (online): 1–20.

Green, Jim, 2016. Protests against proposed reprocessing plant in China. *Nuclear Monitor* 829(August): 4–7.

Groden, Claire, 2016. Exxon cuts China energy demand growth forecast. *Fortune*, 26 January. fortune.com/2016/01/26/china-energy-demand/ (accessed 23 January 2017).

Grubler, Arnulf, 2010. The costs of the French nuclear scale-up: A case of negative learning by doing. *Energy Policy* 38(9): 5174–88. doi. org/10.1016/j.enpol.2010.05.003

Hargreaves, Steve, 2012. First new nuclear reactors ok'd in over 30 years. *CNNMoney*, 9 February. money.cnn.com/2012/02/09/news/economy/nuclear_reactors/index.htm (accessed 23 January 2017).

He, Guizhen, Arthur P. J. Mol, Lei Zhang, and Yonglong Lu, 2014. Nuclear power in China after Fukushima: Understanding public knowledge, attitudes, and trust. *Journal of Risk Research* 17(4): 435–51. doi.org/10.1080/13669877.2012.726251

He, Zuoxiu, 2013. Chinese nuclear disaster 'highly probable' by 2030. *Chinadialogue*, 19 March. www.chinadialogue.net/article/show/single/en/5808-Chinese-nuclear-disaster-highly-probable-by-2-3- (accessed 23 January 2017).

Henochowicz, Anne, 2016. Workers must pledge not to protest nuclear waste plant. *China Digital Times*, 16 August. chinadigitaltimes. net/2016/08/workers-pressed-pledge-nuclear-waste-plant-protest/ (accessed 23 January 2017).

Hinze, Jonathan, and Yun Zhou, 2012. China's commercial reactors. *Nuclear Engineering International*, February. belfercenter.ksg.harvard .edu/publication/21789/chinas_commercial_reactors.html?bread crumb=%2Fexperts%2F2342%2Fjonathan_hinze (accessed 23 January 2017).

Hook, Leslie, 2012. China nuclear protest builds steam. *Financial Times*, 28 February.

Hore-Lacy, Ian, 2014. China's new nuclear baby. *World Nuclear News*, 2 September. www.world-nuclear-news.org/E-Chinas-new-nuclear-baby-0209141.html (accessed 23 January 2017).

Hornby, Lucy, and Luna Lin, 2016. China protest against nuclear waste plant. *Financial Times*, 7 August.

Hu, Angang, 2015. Embracing China's 'new normal'. *Foreign Affairs*, May/June. www.foreignaffairs.com/articles/china/2015-04-20/embracing-chinas-new-normal (accessed 23 January 2017).

Huang, Lei, Ying Zhou, Yuting Han, James K. Hammitt, Jun Bi, and Yang Liu, 2013. Effect of the Fukushima nuclear accident on the risk perception of residents near a nuclear power plant in China. *Proceedings of the National Academy of Sciences* 110(49): 19742–7. doi. org/10.1073/pnas.1313825110

IEA (International Energy Agency)/NEA (Nuclear Energy Agency), 2010. *Projected Costs of Generating Electricity*. Paris: Nuclear Energy Agency, OECD.

Kim, Younghwan, Minki Kim, and Wonjoon Kim, 2013. Effect of the Fukushima nuclear disaster on global public acceptance of nuclear energy. *Energy Policy* 61: 822–8. doi.org/10.1016/j.enpol.2013.06.107

King, Amy, and M. V. Ramana, 2015. The China syndrome? Nuclear power growth and safety after Fukushima. *Asian Perspective* 39(4): 607–36.

Koomey, Jonathan, and Nathan E. Hultman, 2007. A reactor-level analysis of busbar costs for US nuclear plants, 1970–2005. *Energy Policy* 35: 5630–42. doi.org/10.1016/j.enpol.2007.06.005

Levi, Michael A., Elizabeth Economy, and Douglas Rediker, 2016. Can the world adjust to China's 'new normal'? *World Economic Forum*, 10 February. www.weforum.org/agenda/2016/02/can-the-world-adjust-to-china-s-new-normal/ (accessed 23 January 2017).

Li, Jing, 2016. Nuclear fuel plant on hold in eastern China after thousands protest. *South China Morning Post*, 10 August. www.scmp.com/print/news/china/policies-politics/article/2001726/nuclear-plant-scheme-halted-eastern-china-after (accessed 23 January 2017).

Li, Zhen, and Sam Tranum, 2011. Candris says Fukushima will help AP1000 in China. *Nuclear Intelligence Weekly*, 16 May.

Lin, Jiang, Gang He, and Alexandria Yuan, 2016. Economic rebalancing and electricity demand in China. *The Electricity Journal* 29(3): 48–54. doi.org/10.1016/j.tej.2016.03.010

Liu, Wen Xin (Cindy), 2016. City suspends nuclear project after thousands protest. *China Digital Times*, 11 August. chinadigitaltimes. net/2016/08/city-suspends-nuclear-project-thousands-protest/ (accessed 23 January 2017).

Liu, Yiyu, 2011. New nuclear plants may get green light soon. *China Daily*, 12 August. www.chinadaily.com.cn/cndy/2011-08/12/content_13097545.htm (accessed 23 January 2017).

Lok-to, Wong, 2016. Safety fears cause concern amid delays to China's Taishan nuclear plant. *Radio Free Asia*, 7 March. www.rfa.org/english/news/china/safety-03072016114147.html (accessed 23 January 2017).

Mu, Rongping, 2010. China. In *UNESCO Science Report 2010: The Current Status of Science Around the World*, 379–99. Paris: United Nations Educational, Scientific and Cultural Organization.

Ng, Eric, 2013. China nuclear plant delay raises safety concern. *South China Morning Post*, 7 October.

Pan, Wang, 2011. China rolls out new homegrown nuclear reactor. *People's Daily Online*, 18 November. en.people.cn/202936/7649438.html (accessed 23 January 2017).

Ramana, M. V., and Eri Saikawa, 2011. Choosing a standard reactor: International competition and domestic politics in Chinese nuclear policy. *Energy* 36(12): 6779–89. doi.org/10.1016/j.energy.2011.10.022

Schneider, Mycle, and Antony Froggatt, with Julie Hazemann, Tadahiro Katsuta, M. V. Ramana, and Steve Thomas, 2015. *The World Nuclear Industry Status Report 2015*. Paris: Mycle Schneider Consulting Project.

Securities Daily [Zhengquan Ribao], 2015. Shanghai Dianli Zhengshi Zhong Dian Tou Jituan Yu Guojia Hedian Jishu Gongsi Jiang Chongzu [Shanghai Electric Power confirms restructure of the CLP [China Power Investment] Group and the National State Nuclear Power Technology Corp]. East Money.com, 3 February. finance.eastmoney.com/news/1349,20150203474829119.html (accessed 23 January 2017).

Sheng, Chunhong, 2014. A look at anti-nuclear protests in China. *Nuclear Intelligence Weekly*, 11 April.

Smith, Andrew Z. P., and Michael Grubb, 2016. Hinkley Point C and other third-generation nuclear in the context of the UK's future energy system. CEE Briefing Note 20160915 AZPS1. London: RCUK Centre for Energy Epidemiology, University College London.

SNPTC (State Nuclear Power Technology Corporation), 2011. *Introduction of State Nuclear Power Technology Corporation.* Vol. 2011. 22 January. Beijing: State Nuclear Power Technology Corporation.

Stanway, David, 2009. China struggles to fuel its nuclear energy boom. Reuters, 10 December. www.reuters.com/article/2009/12/10/uranium-china-nuclear-idUSPEK20761020091210 (accessed 23 January 2017).

Stanway, David, 2014. China says first Westinghouse reactor delayed until at least end-2015. Reuters, 18 July. www.reuters.com/article/2014/07/18/china-nuclear-ap-idUSL4N0PT0T820140718 (accessed 23 January 2017).

Stanway, David, 2016. China nuclear developers must seek public consent: Draft rules. Reuters, 20 September. www.reuters.com/article/us-china-nuclear-safety-idUSKCN11Q18K (accessed 23 January 2017).

Sun, Chuanwang, Xiting Zhu, and Xiaochun Meng, 2016. Post-Fukushima public acceptance on resuming the nuclear power program in China. *Renewable and Sustainable Energy Reviews* 62: 685–94. doi.org/10.1016/j.rser.2016.05.041

Taishan Nuclear Power Joint Venture Co., 2016. Taishan Unit 1 CFT completed successfully. 1 February. en.tnpjvc.com.cn/n1623/n1624/c1235803/content.html (accessed 23 January 2017).

Thibault, Harold, 2010. Construction schedule on Chinese third-generation nuclear plants races ahead of European models. *Guardian*, 28 December.

Thomas, Steve, 2017. China's nuclear export drive: Trojan Horse or Marshall Plan? *Energy Policy* 101: 683–91. doi.org/10.1016/j.enpol.2016.09.038

US NRC (Nuclear Regulatory Commission), 2011. AP1000 Design Certification Amendment. *Federal Register*, 30 December. www.federalregister.gov/articles/2011/12/30/2011-33266/ap1000-design-certification-amendment#h-13 (accessed 23 January 2017).

Wade, Samuel, 2016. Minitrue: Delete article on nuclear project suspension. *China Digital Times*, 11 August. chinadigitaltimes.net/2016/08/minitrue-delete-article-lianyungang-nuclear-suspension/ (accessed 23 January 2017).

Wang, Jiayuan, 2016. Liang Bumen Ni Guiding: Hedianchang Xuanzhi Deng Shixiang Ying Zhengqiu Gongzhong Yijiang [Two departments draft rules: Must seek public opinion on nuclear power plant site selection]. *Sina*, 19 September. finance.sina.com.cn/china/2016-09-19/doc-ifxvyqwa3505314.shtml (accessed 23 January 2017).

Wang, Lu, 2016. Neilu Fazhan Hedian Yuqi Shengwen Nengyuan Ju Huiying: Chongqi Wu Shijian Biao [Inland nuclear power development expected to heat up energy agency response: No restart schedule]. *Jingji Cankao Bao* [Economic Information Daily], 7 March. news.xinhuanet.com/fortune/2016-03/07/c_128778398.htm (accessed 23 January 2017).

Wang, Qiang, 2009. China needing a cautious approach to nuclear power strategy. *Energy Policy* 37(7): 2487–91. doi.org/10.1016/j.enpol.2009.03.033

Wen, Bo, 2014. Inland provinces: Nuclear at crossroads. *China Water Risk*, 13 August. chinawaterrisk.org/opinions/inland-provinces-nuclear-power-at-crossroads/ (accessed 23 January 2017).

Wen, Jiabao, 2012. Wen Jiabao Zhuchi Zhaokai Guowuyuan Changwu Huiyi [Wen Jiabao chairs executive meeting of the State Council]. Central People's Government of the People's Republic of China website, 24 October. www.gov.cn/ldhd/2012-10/24/content_2250357.htm (accessed 23 January 2017).

Westinghouse Electric Company, 2000. Westinghouse AP600 receives design certification from US NRC; Company to aggressively market advanced/passive reactor throughout the world; Technology offers improved safety features. health.phys.iit.edu/extended_archive/0001/msg00154.html (accessed 4 September 2017).

WNA (World Nuclear Association), 2010. Nuclear power in China. London: World Nuclear Association.

WNA (World Nuclear Association), 2011. Nuclear power in China. London: World Nuclear Association.

Wong, Kimfeng, 2016a. Coal loses more market share to nuclear, renewables. *Nuclear Intelligence Weekly*, 19 February.

Wong, Kimfeng, 2016b. A radical solution to loosen coal's grip. *Nuclear Intelligence Weekly*, 5 August.

World Nuclear News, 2010. China prepares to export reactors. 25 November.

World Nuclear News, 2015. Hualong One selected for Argentina. 5 February.

World Nuclear News, 2016. Hualong One joint venture officially launched. 17 March.

World Nuclear News, 2017. Construction milestones at new Chinese units. 5 January. www.world-nuclear-news.org/NN-Construction-milestones-at-new-Chinese-units-0501175.html (accessed 6 February 2017).

Xie, Wei, 2016. Guojia Dian Tou Dongshi Zhang Wang Binghua: Neilu Hedian Yao Jian, Dan Yao Bawo Shiji [State Power Investment Chairman Wang Binghua: We must build inland nuclear power, but we must grasp the opportune moment]. *Zhongguo Jingji Zhoukan* [China Economic Weekly], 14 March. www.ceweekly.cn/2016/0314/144250.shtml (accessed 23 January 2017).

Xinhua, 2011. China not to change plan for nuclear power projects: Government. 12 March. news.xinhuanet.com/english2010/china/2011-03/12/c_13774519.htm (accessed 31 January 2017).

Xinhua, 2014. CGN invests $6b on nuclear power in Guizhou. *China Daily*, 14 July. www.chinadaily.com.cn/china/2014-07/11/content_17736464.htm (accessed 23 January 2017).

Xinhua, 2016. China sets energy use target for 2016. 1 April. news.xinhuanet.com/english/2016-04/01/c_135244392.htm (accessed 23 January 2017).

Xu, Yi-Chong, 2008. Nuclear energy in China: Contested regimes. *Energy* 33(8): 1197–205. doi.org/10.1016/j.energy.2008.03.006

Xu, Yi-Chong, 2014. The struggle for safe nuclear expansion in China. *Energy Policy* 73: 21–9. doi.org/10.1016/j.enpol.2014.05.045

Xuyang, Jingjing, 2012. Not in my backyard. *Global Times*, 17 February. en.people.cn/90882/7731890.html (accessed 23 January 2017).

Yap, Chuin-Wei, and Brian Spegele, 2015. China's first advanced nuclear reactor faces more delays. *Wall Street Journal*, 15 January.

Ying, Li, 2016. China's power sector and the economic 'new normal'. *Chinadialogue*, 25 January. www.chinadialogue.net/article/show/single/en/8558-China-s-power-sector-and-the-economic-new-normal- (accessed 23 January 2017).

Yu, C. F., 2015. Inland nuclear developers await policy change. *Nuclear Intelligence Weekly*, 2 January.

Yu, C. F., 2016a. CNNC and CGN launch Hualong JV. *Nuclear Intelligence Weekly*, 8 January.

Yu, C. F., 2016b. Construction on inland plants unlikely before 2020. *Nuclear Intelligence Weekly*, 1 April.

Yuan, Jiahai, Peng Li, Yang Wang, Qian Liu, Xinyi Shen, Kai Zhang, and Liansai Dong, 2016. Coal power overcapacity and investment bubble in China during 2015–2020. *Energy Policy* 97: 136–44. doi. org/10.1016/j.enpol.2016.07.009

Zheng, Xiaoyi, and Qi Wu, 2010. China advances in independently tapping nuclear power. Xinhua, 18 February. old.csr-china.net/en/second.aspx?nodeid=ddd0b45c-b7c4-4947-b2e3-e20374708733&page=contentpage&contentid=23c48153-ebb7-4af5-855e-fff6e8ec4b67 (accessed 11 September 2017).

Zhongguo Hedian Wang, 2015. Hunan, Hubei Huyu Chongqi Neilu Hedian Xiangmu Anquan Xing Rengyou Zhengyi [Hunan, Hubei call for the restarting of inland nuclear power projects – Safety still controversial]. ChinaPower.com, 16 March. np.chinapower.com.cn/201503/16/0044739.html (accessed 23 January 2017).

Zhongguo Jinggong Bao, 2014. Xiang E Gan Li Tui Hedian Neilu 'shou He' 2016 Nian Huo Kaizha' [Hunan Hubei Jiangxi push to open the 'first nuclear' inland power station, probably in 2016]. 9 September. www.heneng.net.cn/index.php?mod=news&action=show&article_id=31982&category_id=9 (accessed 23 January 2017).

Zhou, Sheng, and Xiliang Zhang, 2010. Nuclear energy development in China: A study of opportunities and challenges. *Energy* 35(11): 4282–8. doi.org/10.1016/j.energy.2009.04.020

Zhou, Yun, 2012. China's nuclear energy industry, one year after Fukushima. Policy Brief, Belfer Center's Technology and Policy Blog, 5 March.

Zhu, Yue, 2014. China's nuclear expansion threatened by public unease. *Chinadialogue*, 23 September. www.chinadialogue.net/article/show/single/en/7336-Chinese-protesters-threaten-nuclear-expansion (accessed 23 January 2017).

5

Protesting policy and practice in South Korea's nuclear energy industry

Lauren Richardson

Abstract

Japan's March 2011 (3/11) crisis spurred a revival in anti-nuclear activism around the globe. This was certainly the case in South Korea, Japan's nearest neighbour, which was subject to some of the nuclear fallout from Fukushima. This chapter examines the puzzle of why the South Korean anti-nuclear movement was apparently powerless in the face of its government's decision to ratchet up nuclear energy production post-3/11. It argues that its limitations stem from the highly insulated nature of energy policymaking in South Korea; the enmeshing of nuclear power in the government's 'Green Growth Strategy'; and certain tactical insufficiencies within the movement itself. Notwithstanding these limitations, the movement has successfully capitalised upon more recent domestic shocks to the nuclear power industry, resulting in a slight, yet significant, curtailing of the South Korean government's nuclear energy capacity targets.

Introduction

The March 2011 (3/11) earthquake in northeastern Japan and ensuing nuclear meltdown at the Fukushima Daiichi plant had profound reverberations for the global nuclear industry. In the wake of the disaster, countries as far-reaching as Germany and Switzerland brought their

133

nuclear energy programs to a complete halt. Closer to the source of the calamity, the Taipei government initiated a gradual phase-out of its nuclear reactors and suspended plans for the construction of a fourth nuclear plant. These policy shifts were precipitated by nationwide anti-nuclear demonstrations, which erupted in response to the Fukushima crisis. Somewhat surprising, however, was that Japan's nearest neighbour, South Korea, reacted to the complete contrary. Despite the fact that Korean territory was subject to some of the nuclear fallout from Fukushima (see Hong et al. 2012), the South Korean government proceeded to ratchet up its nuclear energy program post-3/11 and pushed ahead with plans to become a major exporter of nuclear technology. Indeed, within only months of Japan's disaster, South Korean President Lee Myung-bak reiterated his administration's goal of doubling the number of domestic reactors, and reaffirmed nuclear technology as a primary export focus.

This response was puzzling for a number of reasons. First, similarly to the cases of Germany, Switzerland, and Taiwan, the South Korean anti-nuclear movement expanded to unprecedented proportions in the aftermath of Fukushima, yet ostensibly to no avail. This expansion was driven by a marked decline in public trust in the safety of nuclear reactors, and witnessed activists mounting a formidable challenge to nuclear energy policy. Moreover, since overthrowing the nation's long-standing authoritarian regime in the late 1980s, South Korean civil society has evolved to wield powerful influence across a variety of policy domains; activists, though, were apparently powerless in the face of their government's decision to increase nuclear-generating capacity. This is somewhat perplexing given that, in the very same year of the Fukushima calamity, South Korean civic groups contributed to undercutting a proposed security accord between Seoul and Tokyo, and 'comfort women' victims compelled their foreign ministry to pursue compensation from Japan more vigorously on their behalf—to name but two realms of policy influence.

Why then was South Korea's anti-nuclear movement unable to subvert the South Korean government's nuclear energy policy? Does the movement's lack of evident success suggest that it exerted no tangible influence on nuclear energy development in South Korea? What factors have served to impede its effectiveness? This chapter addresses these questions through an analysis of the movement's campaign to alter policy and practice in the South Korean nuclear energy industry from the late 1980s to 2016. As the challenges encountered by the movement stem in part from the structural development of nuclear energy in South Korea, the chapter begins by outlining the evolution of this process. It proceeds to assess the efficacy of the anti-nuclear movement in pre- and post-Fukushima contexts, with reference to its aims and pressure tactics. It then assesses the reasons behind the government's lack of responsiveness to the movement, before finally examining two emergent encumbrances to nuclear energy policy.

The chapter advances three broad arguments. First, the anti-nuclear movement has had considerable success in preventing the construction of nuclear waste disposal sites; this endeavour has been more fruitful than strategies that sought to undermine the establishment of new nuclear power plants. Second, the movement's inability to abort nuclear energy production stems from the highly insulated nature of energy policymaking in South Korea, the enmeshing of nuclear power in the government's 'Green Growth Strategy', and certain tactical insufficiencies in the anti-nuclear movement. Third, notwithstanding these limitations, the movement has capitalised upon recent domestic shocks to the nuclear power industry, resulting in a curtailing of the government's nuclear energy capacity targets.

The evolution of South Korea's nuclear energy policy

Since its post-Korean War (1950–53) inception, energy policy in South Korea has been driven by the need to spur economic growth, minimise dependence on imports, and ensure long-term energy security. In the late 1950s, the South Korean government opted to develop a nuclear power program as a means to fuel the restoration of its war-shattered economy. Officials presumed that nuclear reactors would provide a stable source of energy, facilitate export-oriented growth, and reduce the nation's reliance on costly oil, coal, and gas imports. Toward this end, Seoul joined the International Atomic Energy Agency (IAEA) in 1957, and thereafter enacted *Framework Act No. 483 on Atomic Energy* (1958) and established an Office of Atomic Energy (1959).

Under the iron grip of a succession of authoritarian leaders from the 1960s to the late 1980s, nuclear energy legislation proceeded mostly unhindered by public resistance. Indeed, the Park Chung-hee dictatorship (1961–79) was quick to charge would-be demonstrators with violating anti-communism and national security laws, and resorted to barrages of tear gas and martial law to restrain them. It was against this backdrop that the nation's first reactor, a small research unit, was brought to criticality in 1962. Some 10 years later, the Park government commissioned the construction of the Kori nuclear power plant in the port city of Busan, and this began generating in 1978 (Hwang and Kim 2013: 196).

In addition to the authoritarian milieu, South Korea's alliance with the United States constituted a further driving force in its development of nuclear energy. Once Seoul embarked on its nuclear power program, a confluence of interests emerged between the American nuclear industry, business conglomerates (*chaebol*), and officials in South Korea. Nuclear power companies in the US had a specific agenda to promote the advancement of nuclear technology in non-communist countries, and thus viewed South Korea as an attractive business prospect. In fact, the American firm Combustion Engineering (later incorporated into Westinghouse Electric) supplied South Korea with its first nuclear reactor in 1978—the Kori-1 unit—and thereupon imparted technological know-how to the fledgling industry.

The US government, meanwhile, sought a degree of control over its ally's nuclear energy policy; this was predicated on dissuading South Korea from developing an indigenous nuclear weapons capability. Prompted by mounting military pressure from Pyongyang and the withdrawal of thousands of US troops from South Korea in 1971, Park started harbouring aspirations of nuclear weapons development and proliferation (Hayes and Moon 2011). Through the enactment of the Agreement for Cooperation between the Government of the United States of America and the Government of the Republic of Korea Concerning Civil Uses of Atomic Energy in 1972, Washington attempted to curb these ambitions by pledging to provide nuclear materials and technology to Seoul on the condition that they be used exclusively for energy production purposes. The terms of the agreement further undermined Seoul's nuclear weapons potential by prohibiting uranium enrichment and limiting its fuel cycle options and raw material supply. When the Korea Atomic Energy Research Institute attempted to circumvent these terms by purchasing reprocessing plants from Belgium in the mid-1970s, the US and Canadian governments thwarted the deal by exerting financial leverage vis-à-vis Seoul, and Washington further threatened to cut off support for its ally's nuclear power program (Hayes and Moon 2011: 51–3). Under the weight of this pressure, Park eventually abandoned his weapons development and proliferation plans at the end of the decade.

Throughout the early to mid-1980s, the expansion of South Korea's nuclear energy capacity proceeded mostly unencumbered by civic dissent. This was largely owing to the preoccupation of the populace with achieving democratisation (Leem 2006). In this context, the state-owned Korea Electric Power Company (KEPCO) oversaw the construction of an additional eight reactors, through the assistance of American nuclear firms. By the end of the decade, South Korea's nuclear energy industry had evolved to supply 45 per cent of the nation's energy needs and had virtually attained technical self-reliance. Nuclear power thus became closely correlated with South Korea's rapid industrialisation and economic rise.

The bottom-up movement against nuclear energy

As the transition to democracy began in the late 1980s, however, the nuclear energy industry began to encounter significant social resistance. After a decade of sustained civil uprisings against the authoritarian leadership, South Korean citizens started to question Park's development model, in particular its driving force of nuclear energy. This questioning, which was fueled by increasing political liberalisation, gradually gave rise to a nascent anti-nuclear movement. In its early stages, this movement remained fairly localised around nuclear reactor sites. Yet the Fukushima crisis served to galvanise and encourage its transnational expansion. Although the movement's overarching objective of achieving a nuclear-free South Korea ultimately proved abortive, it did succeed in stymieing the construction of a number of nuclear waste disposal sites. This section examines the movement's opposition tactics before and after 3/11.

Phase 1: Pre-Fukushima

The South Korean anti-nuclear movement emerged as an amalgamation of various environmental and other civic-minded groups. Spurred in part by the numerous nuclear power plant–related accidents that had occurred by the end of the 1980s, including the Chernobyl disaster, citizens joined forces to prevent further environmental damage and curb the nation's steadily increasing pollution. As a first step, they jointly established the National Headquarters for Nuclear Power Eradication, and thereupon launched a bottom-up campaign against nuclear energy.

One of the first major rallying points of the movement was the matter of radioactive waste disposal. Given that close to 50 per cent of the nation's electricity was being derived from nuclear power by the 1980s, spent fuel repositories were reaching capacity and the storage of radioactive waste had begun to pose a formidable challenge. Activists perceived this state of affairs as a potential environmental disaster. When the government first announced its candidate sites for nuclear waste disposal in 1986—and every instance thereafter—impassioned civic resistance thus followed. Brandishing messages about the dangers of nuclear materials, citizens staged large-scale protests at government complexes and proposed waste sites. These early grassroots efforts met with overwhelming success: over a period of eight years, the anti-nuclear movement thwarted the construction of 12 nuclear waste disposal sites (Sayvetz 2012).

In an attempt to circumvent further public obstruction, the South Korean government began targeting remote locales to play host to waste depositories. In the mid-1990s, officials designated Gulup Island, a small landmass off South Korea's western coast, as a potential site. This plan was instigated without public consultation and when news of it was leaked to the public, anti-nuclear activists rallied in anger. The Korean Federation for Environmental Movements (KFEM) elected to head a campaign to prevent the site's construction. Boasting a membership of more than 13,000, the KFEM worked in tandem with various civic groups to advocate for the Gulup Island residents, who were strongly averse to the prospect of a nuclear waste dump in their residential vicinity (Sayvetz 2012). In a show of broad-based consensus against the proposed site, the KFEM convened mass rallies and filed an oppositional petition that attracted thousands of signatures.

When the government belatedly agreed to convene a public hearing regarding the site, representatives from a number of civic groups voiced their concerns about the presence of a geological fault on the island. Their apprehensions, however, ostensibly fell on deaf ears. Public pressure thus continued to mount and, in the spring of 1995, over 300 residents in the nearby Deokjeok Island—who were also fearful of the site's potential consequences—staged a protest in front of the Ministry of Science and Technology in Seoul. Faced with this unrelenting opposition, government officials were impelled to solicit experts from the IAEA to conduct a survey on the proposed site. Their findings revealed the presence of a fault, confirming residents' suspicions that the site was particularly perilous for the storage of nuclear waste. In light of this development, the central government decided to abort the Gulup Island plan in November 1995.

The movement continued to challenge the construction of radioactive waste sites throughout the 1990s and into the early twenty-first century. These attempts tended to remain localised in nature and dissipated once a proposal was successfully undermined.

Phase 2: Post-Fukushima

Following the meltdown of the three reactors in Fukushima, South Korea's anti-nuclear movement underwent somewhat of a resurgence. This was characterised by the mobilisation of a broader spectrum of activists and an increase in the breadth of the movement's anti-nuclear activities. As images of the triple meltdown at the Fukushima Daiichi plant filtered through South Korean media outlets, various religious groups, unions, co-ops, professional associations, non-governmental organisations, academics, and parents groups joined the appeal for a nuclear-free future. Moreover, the 3/11 crisis spurred the South Korean movement to transnationalise its anti-nuclear efforts through joining forces with like-minded activists in the region. This was instigated by a group of Catholic South Korean dioceses who pledged to form an East Asian civil society network with anti-nuclear activists in Japan and China; their objective was to present a united front of opposition to the nuclear power industry regardless of the tensions between their respective countries. As described in their initial prospectus, 'the more we share information on the dangers on nuclear power and spread technology and wisdom regarding natural energy, the more East Asia will become the center of peace, not conflict; of life, not destruction' (East Coast Solidarity for Anti-Nuke Group 2012). Under the nomenclature of the East Coast Solidarity for Anti-Nuke Group, the group debuted on the first anniversary of the Fukushima disaster with a declared membership of 311 citizens, signifying that the South Korean movement was no longer a domestic phenomenon localised around nuclear waste sites.

In accordance with the expansion of its constituents, the movement increased the scope of its anti-nuclear efforts in the aftermath of Fukushima. Moving beyond the initial focus of countering the construction of new waste storage sites and plants, activists began to advocate more broadly for the cessation of nuclear energy production; accordingly, they targeted existing plants. The logic driving the movement's post-Fukushima campaign was essentially fourfold: (1) uranium sources will eventually be exhausted, and therefore nuclear energy is not a viable permanent energy source; (2) most of the developed countries around the world are no longer constructing new nuclear reactors and, since Fukushima, are seriously rethinking their nuclear energy policies; (3) when factoring in the social costs, nuclear energy cannot be considered cost-effective; and (4) as the mining and processing of uranium produces carbon dioxide (CO_2) emissions, nuclear power cannot be conceived of as an environmentally friendly source. Meanwhile, the overarching logic informing the movement was that Japan's 'March 11 disaster has proven that nuclear power plants are not safe' (Nagata 2012).

First among the anti-nuclear movement's post-3/11 objectives was to nullify the lifespan extensions of the nation's two oldest nuclear reactors—Kori-1 and Wolsong-1. The former unit, which was already running beyond its technological lifespan, had experienced a number of technical problems in the spring of 2011, and was consequently temporarily shut down. Yet shortly thereafter, nuclear officials declared it suitable for operation and allowed it to resume power generation. Likewise, the latter unit, which began operating in 1983 at a plant in North Gyeongsang province, was taken offline for extended maintenance in June 2009. As its operating license was due to expire in 2012, Korea Hydro & Nuclear Power (KHNP) spent ₩560 billion (US$509 million) on refitting the unit with the hope of prolonging its lifespan. Ultimately, the reactor was cleared for restart in June 2011.

These decisions by nuclear energy officials were made in close succession to the Fukushima disaster, and thus aroused fears among local residents of a similar catastrophe occurring in their own vicinity. Under the banner of a group called Collective Action for a Nuclear Free Society, residents demanded that the life extensions of the reactors be nullified. Toward this end, they staged protests in front of the Nuclear Safety and Security Commission (NSSC) in Seoul, where officials deliberated the fate of the reactors, and chanted anti-nuclear slogans. In spite of these objections, however, nuclear officials permitted Kori-1's continued operation. And although they agreed to shut down Wolsong-1 at the conclusion of its lifespan in November 2012, they later backtracked, granting permission for it to restart in February 2015 and operate for a further 10 years. These two decisions constituted a major setback for the movement.

In addition to focusing on aged reactors, the anti-nuclear movement continued on its mission to abort the construction of new nuclear power plants. Activists concentrated on the candidate sites of Samcheok and Yeongdeok, two cities on the east coast of South Korea in which the government proposed to build eight new reactors (four at each site). The local government of Samcheok had originally agreed to host a nuclear power plant in 2010. Yet following the Fukushima disaster, anti-nuclear sentiment swept throughout the city, culminating in the formation of the Pan-Citizen Alliance for Cancelling the Samcheok Nuclear Power Plant. To signal their changed stance on nuclear power to the central government, the city residents elected a new mayor, Kim Yang-ho, who had campaigned on an anti-nuclear platform. In order to elicit a collective anti-nuclear expression, Kim held a referendum in October 2014. As he anticipated, the majority of citizens indicated their opposition to the plant's construction: among the 69.8 per cent of the voting population who participated in the referendum, 85 per cent voted against the proposed site. Due to the fact that the referendum was not legally sanctioned, however, the national government declared it non-binding and thus ignored the result.

In the second candidate city of Yeongdeok, a similar outcome transpired. Being a rural and coastal county with a dwindling population and struggling economy, Yeongdeok's residents had initially been enthused about the prospect of economic revitalisation that a nuclear power plant would offer. Not only would it bring much-needed employment opportunities, but the South Korean government had pledged to provide ₩1.5 trillion (US$1.35 billion) over a 60-year period, to compensate for any potential associated dangers. Having lost their earlier (2005) bid to host a storage site for low-level radioactive waste, the citizens of Yeongdeok were particularly keen to secure the nuclear power plant venture. Their enthusiasm quickly dissipated, however, in the face of Japan's 3/11 disaster. Indeed, residents had not foreseen the possibility of tsunami damage to the plant when originally submitting their host bid. In the aftermath of Fukushima, local citizens thus called for a county referendum to overturn the plan. In this instance, the mayor was unwilling to support the initiative and therefore residents organised it on their own accord. Perhaps owing to this lack of official backing, the referendum failed to attract the requisite one-third of voters for it to hold legal sway (Kim 2015). In any case, national officials dismissed both the Samcheok and Yeongdeok voter outcomes on the grounds that central government projects are not subject to local referenda results.

Evidently, the pressure tactics of the South Korean anti-nuclear movement have produced mixed results. Early protests were successful in undermining nuclear waste site proposals and plans for the construction of a small number of nuclear power plants. Yet in the post-Fukushima period, the movement largely failed in its aims to abrogate the lifespan extensions of aged reactors and reverse site selection decisions for new nuclear power plants.

Explaining the limited policy change

Despite the magnitude of the Fukushima crisis and ensuing tide of pressure from the anti-nuclear movement, Seoul's nuclear power policy showed no immediate signs of deceleration—at least on the surface. The disaster only prompted limited government measures aimed at counteracting potential contamination from Japan's meltdown, and enhancing the safety of domestic nuclear installations. In the two months following 3/11, all 30,000 passengers that entered South Korea from Japan (by ship or aircraft) were screened for radioactivity; only two people, however, required decontamination (Korean Government 2011). Over the same two months, the central government ordered nuclear officials to carry out a special safety inspection of all nuclear power plants throughout the country, yet, ultimately, no abnormalities were detected. Finally, in June 2011, the South Korean National Assembly passed a bill to establish the NSSC, a regulatory body tasked with protecting public health and safety.

Together these measures constituted the extent of the South Korean government's responsiveness to 3/11 and the subsequent pressure from the anti-nuclear movement. South Korea continues to stand as the sixth largest consumer of nuclear energy in the world, second in Asia only to Japan. There remain 24 nuclear reactors operating nationwide, with another five under construction. Government officials continue to emphasise the safety and low-cost efficiency of nuclear power, while largely eschewing the development of renewable energy sources. Expanding the nuclear energy industry is still a national strategic priority, as exemplified in the Ministry of Science and Technology's (2006) Third Comprehensive Plan for Nuclear Energy Development (2007–11). The government predicted in this report that the nation would derive 59 per cent of its electricity from nuclear power sources by 2030.

In addition to these domestic ambitions, nuclear energy technology has evolved to become a major export industry for South Korea. The Ministry of Knowledge Economy intends to export another 80 reactors, worth a total of US$400 billion, by 2030. The nation secured its first major international contract in 2009, when KEPCO signed a US$40 billion deal to construct four nuclear reactors for the United Arab Emirates (UAE). Undeterred by the Fukushima meltdown, President Lee embarked on an official visit to the UAE on 13 March 2011—a mere two days after Japan's crisis began to unfold—to reaffirm his plans for future energy cooperation. Besides the UAE deal, Seoul has secured a US$173 million contract to build a nuclear research reactor in Jordan, and to construct several reactors in Saudi Arabia worth a total of US$2 billion. Other target export countries for South Korea's nuclear industry include China, Finland, Hungary, Indonesia, Malaysia, Turkey, and Vietnam.

What explains the failure of the anti-nuclear movement to subvert the development of nuclear energy in South Korea? Pressure tactics cannot singularly account for the limited policy change. Rather, a combination of three factors have served to militate against substantial nuclear power reform: (1) the highly insulated and top-down nature of nuclear energy policymaking in South Korea—this has restricted the number of legislative handles around which activists can mobilise to influence policy decisions; (2) the centrality of nuclear energy to the South Korean government's Green Growth Strategy, a factor that has legitimated its continued expansion; and (3) shortcomings in the anti-nuclear movement's pressure strategy, specifically, its laxness in articulating a feasible alternative energy strategy to nuclear power.

The insularity of nuclear power policymaking

The primary hurdle faced by the movement has been the elite-driven nature of policymaking on nuclear energy. In contrast to the many other policy domains in South Korea that allow for substantial input from citizens, decisions on nuclear energy continue to be formulated exclusively by government officials and technocrats, in a highly insulated environment. The key actors engaged in this process include the Ministry of Commerce, Industry and Energy; the Ministry of Trade, Industry and Energy; the Ministry of Science, ICT and Future Planning; the NSSC; and various *chaebol* and bureaucratic authorities. Each of these institutions is in turn informed by pro-nuclear politicians and technocrats, producing an iron triangle of decision-making that excludes civil society. This triangular structure was particularly reinforced with the installation of Lee—a former *chaebol* leader (Hyundai executive)—as South Korean president in 2008.

As a corollary of this elite-driven process, nuclear energy policy is implemented through a top-down dynamic. This has been characterised by a 'decide-announce-defend' sequence (Norman and Nagtzaam 2016: 250), whereby the central government enacts a policy, proceeds to impose it on local government and citizens, and then seeks to placate any objections by offering financial rewards and other incentives. This sequence was vividly evinced in the Gulup Island fiasco. However, as this strategy has proved abortive on a number of occasions, the government has attempted since 2004 to move toward a slightly more consultative mechanism that incorporates citizens' preferences. Activists continue, though, to face significant barriers in shaping the nuclear energy agenda. The elite-driven and top-down dynamic of the policy process has in fact steered their pressure tactics away from government lobbying, toward the more viable strategy of obstructing policy implementation.

Nuclear power as 'green' energy

A further inhibiting factor for the movement has been the enmeshing of nuclear power in the South Korean government's Green Growth Strategy. Essentially, this has added another layer of insularity to nuclear energy policy in South Korea.

As a consequence of South Korea's rapid industrialisation over the last few decades, its greenhouse gas emissions virtually doubled between 1990 and 2005—an increment exceeding most of the Organisation for Economic Co-operation and Development (OECD) countries. At the same time, Seoul's annual mean temperature increased by 1.5 degrees Celsius, surpassing the global average of 0.7 degrees Celsius (von Hippel, Yun, and Cho 2011). These developments, coupled with an emergent international consensus on the need to address climate change, forced the South Korean government to consider ways to curtail its CO_2 emissions. Being at once low-carbon and cost-effective, nuclear energy was seized upon by South Korean officials as a convenient solution to the nation's environmental and climate woes, and also as a means to deal with rising energy demands. In 2009, the Lee administration announced a national Green Growth Strategy premised on three major objectives: reducing fossil fuel use, tracking greenhouse gas emissions, and establishing several new nuclear power plants. Renewable energy was relegated only a marginal status under the plan.

This linking of nuclear power to the national environmental and climate strategy was institutionalised through the government's Five Year Plan for Green Growth (2009–13), and the *Framework Act on Low Carbon, Green Growth* (2010). As a result of this process, the political opportunity structure surrounding nuclear energy became less favourable to activists. The discursive framing of nuclear power, as both a means to reduce carbon emissions and promote energy independence, enabled the South Korean government to legitimise its plans to expand nuclear power domestically and export nuclear technology abroad. Indeed, Lee boasted to his constituencies that the planned export of four reactors to the UAE would equate to '40 million tons of carbon mitigation' (Lee 2010: 11–12).

To challenge this stance of the government, the anti-nuclear movement has attempted to counter-frame nuclear power as an environmentally unfriendly energy source. As previously mentioned, activists have argued that the mining and processing of uranium produces CO_2 emissions. The movement has furthermore underscored the clause of the South Korea–US atomic energy agreement, which prohibits the reprocessing of spent fuel, and thus renders the necessity of environmentally hazardous radioactive waste sites. As many of South Korea's nuclear power plants are located in coastal areas that are subject to occasional earthquakes, activists have also raised the possibility of the occurrence of a Fukushima-style disaster. This counter-frame, however, has yet to tip the cost–benefit analysis of nuclear energy by the wider populace. Indeed, there remains an overriding belief within South Korean society that nuclear power holds the key to combating climate change, as argued by the government.

Tactical insufficiencies in the anti-nuclear movement

The limited policy change in nuclear energy development can further be attributed to insufficiencies in the tactics of the anti-nuclear movement. Throughout their campaign against nuclear power, activists have neglected to formulate a feasible alternative energy source. Instead of demanding new policies (Hermanns 2015: 276), they have tended towards the reactionary tactics of undercutting policy implementation and emphasising the hazards inherent in nuclear energy. Given that South Korea is lacking in natural resources and its economy is structured around manufacturing, this approach of the movement has been problematic for the offsetting of nuclear power. In the absence of a strategy delineating how the nation's energy needs might otherwise be met—accounting both for energy security issues and projected industrialisation—it is improbable that the South Korean government would eschew nuclear power as a major energy source. Formulating such a strategy is all the more necessary in light of the nation's dense population, relatively small landmass, and mountainous terrain, all of which render certain forms of renewable energy—such as wind farms—less conceivable than in other countries.

And while the anti-nuclear movement has significantly increased in scope since Fukushima, its pressure tactics have not resulted in a marked change in public opinion vis-à-vis nuclear power. According to annual polls conducted by the Korea Nuclear Energy Agency, South Korean citizens have upheld consistent views about the importance of nuclear-generated energy throughout recent years, with national support for nuclear power plants hovering between 80 per cent and 90 per cent—even after Fukushima. This has served to further bolster the government's mandate to expand its nuclear energy program. The 3/11 disaster did, however, result in lowered perceptions regarding the safety of nuclear reactors and radioactive waste management in South Korea, with 39 per cent and 24 per cent of survey respondents expressing their confidence in these respective realms. Additionally, polls conducted one year prior to and one year after Fukushima indicated a decline of 8 per cent (from 28 per cent to 20 per cent) in local acceptance of nuclear power (Dalton and Cha 2016). These statistics reflect the fact that opposition to nuclear power is highly localised to rural areas—where nuclear power plants and waste sites are concentrated—while support for nuclear power rests with the larger cities, such as Seoul, where the power-brokers reside and nuclear power plants are a rare sight.

In effect, the downturn in local approval of and confidence in the safety of nuclear reactors has complicated the policy implementation process in South Korea. At the same time, though, the sustained broad-based support for nuclear power generation has functioned to attenuate the pressure tactics of the anti-nuclear movement.

New challenges to South Korea's nuclear energy industry

Notwithstanding the limitations of the anti-nuclear movement in shaping energy policy in South Korea, recent years have seen the emergence of two new challenges to the government's nuclear power strategy. Manifesting both endogenously and exogenously, effectively these have sent shockwaves throughout the industry, forcing Seoul to curb its generating capacity ambitions. For its part, the anti-nuclear movement has seized upon these shocks as opportunities to whip up further opposition to nuclear energy among South Korea's populace.

Corruption scandals

The first of these challenges manifested as a series of corruption scandals implicating nuclear officials, and a consequent erosion of public trust in nuclear energy regulation. As part of Seoul's bid to expand its nuclear-generating capacity, 11 new reactors had been planned for construction in the period 2012–21. This proposal was derailed, however, when it was found—during a routine inspection—that the plant manager had covered-up a reactor power failure (KHNP 2012). When the reactor in question had lost power, the emergency diesel generator failed to start, signalling a host of potential dangers. The plant manager refrained from reporting the mishap due to a fear of inciting a public backlash and 'worsening the plant's credibility' (IAEA–NSNI 2012: 3).

Given Kori's location in South Korea's second-most populous city of Busan, this act of cover-up provided ample opportunity for the anti-nuclear movement to stoke public concerns about regulatory practices. Thus, amidst the controversy, the KFEM and the No Nukes Busan Citizen Countermeasure Commission simulated a radioactive leak (on the scale of the Chernobyl disaster) at the plant, to determine the probable effects. The results were published in a report, and predicted that such an accident would produce roughly 900,000 casualties in Busan, and ₩628 trillion (US\$533 billion) worth of property damage (Yi 2012). This scenario, which was reminiscent of the safety regulatory failure at the Fukushima Daiichi plant, struck widespread fear in the minds of residents. While a panel of experts from the IAEA proceeded to declare the two reactors as safe, their assurances failed to allay the concerns of local citizens who were quickly losing trust in nuclear officials (IAEA 2012).

On the heels of this incident a second corruption scandal occurred, further highlighting the lack of transparency in the regulation of nuclear power plants in South Korea. This unravelled in November 2012, when regulators discovered that at least 5,000 small reactor components at the Yeonggwang nuclear power plant lacked proper certificaiton, and that at least 60 of the quality assurance certificates for these components were fake. After launching an official investigation, the KHNP announced that between 2003 and 2012, the plant had been supplied with a total of 7,682 items with forged quality certificates (LaForge 2013–14). In light of these revelations, the KHNP was compelled to shut down two of the plant's six reactors until the dubious reactor components were replaced. As citizen protests erupted over the controversy, nuclear authorities were prompted to inspect the components of all 23 reactors nationwide. This led to the discovery of copious forged safety certificates for reactor parts at the Kori and Wolseong plants. Consequently, the Kori-2 and Shin Wolseong-1 units were shut down in June 2013, and Kori-1 and Shin Wolseong-2 were ordered to remain offline while the unauthorised parts were refitted. In the ascription of culpability for these scandals, 100 people were indicted on bribery charges, including a former chief executive of the KHNP and a vice president of KEPCO (LaForge 2013–14).

Once again, these events triggered an upsurge in anti-nuclear ferment in South Korea. Citizens attributed the corrupt practices in safety certification to the culture of secrecy shrouding the nuclear energy industry. These sentiments were evinced in protests that erupted in response to the shut down of the Yeonggwang reactors, which attracted as many as 2,500 citizens. Calling for an overall safety review of South Korea's nuclear power plants, participants burned effigies of the KHNP and brandished placards claiming, 'We feel uneasy!' To placate the public outcry, Cho Seok, the chief executive officer of the KHNP, issued a public apology in September 2013, conceding that the corruption scandals constituted the 'utmost crisis' ever faced by the nuclear sector, and vowed to reform South Korea's corporate culture.

Together these controversies engendered a loss of overall public trust in the government's capacity to regulate nuclear energy production. This outcome was inevitably reinforced by the parallels that citizens drew between the regulatory shortcomings at Fukushima Daiichi and that of their national nuclear power plants.

Cyber-attacks on nuclear power plants

The second formidable challenge to South Korea's nuclear energy program emerged in the form of a cyber-attack. This occurred in December 2014, when a hacker leaked the partial blueprints and operating manuals for three domestic nuclear reactors, in addition to the personal data on 10,000 KHNP employees (Baylon, Livingstone, and Brunt 2015). The material was first published online via a blog, and then on a Twitter account under the profile 'president of the anti-nuclear reactor group'. The hacker, whose identity was unknown (the South Korean government suspected Pyongyang), issued a threat to the effect that unless three specific reactor units—Kori-1, Kori-3, and Wolseong-2—were shut down by Christmas, they would systematically be destroyed and further data would be published online. 'Will you take responsibility when these blueprints, installation diagrams and programs are released to the countries that want them?' the hacker threatened in Korean. The three nuclear reactors at the centre of the controversy had long been targeted by the anti-nuclear movement, given their close proximity to populous areas.

Despite having accessed the reactors' blueprints and manuals, however, the hacker was unable to obtain critical technical data pertaining to the nuclear facilities; indeed, this information is stored securely within the KHNP's control monitoring system, which is separate from its internal network. The attacks nevertheless prompted the government to raise its cyber-crisis alert level to 'attention'—the second on a five-step scale—and to run a series of cyber-warfare drills on its various nuclear power plants. More worrisome for government and nuclear officials was that the cyber-attack and its attendant threats provided further fuel for the anti-nuclear movement and stirred greater social unrest among residents in the Kori and Wolseong plant vicinities. In the eyes of local citizens, the susceptibility of the KHNP's internal server to cyber-attacks constituted yet another danger associated with nuclear energy production. These apprehensions were buttressed by the hacker's pronouncement that anyone living in proximity to the plants should vacate their homes immediately (McCurry 2014).

What was the combined impact of these challenges on South Korea's nuclear energy program? In short, the rise in anti-nuclear sentiment in relation to the scandals essentially reined in the government's nuclear power aspirations. Faced with unprecedented criticism over the safety standards and regulatory practices at domestic nuclear power plants,

South Korea's Ministry of Trade, Industry, and Energy was compelled to drastically lower the national nuclear energy capacity target. Whereas the initial goal was to attain 59 per cent capacity by 2030, in the aftermath of the scandals, this was reduced to a more modest 22–29 per cent (by 2035) (Ministry of Trade, Industry, and Energy 2014: 40). The justification provided for this revision was the need to avoid 'excessive expansion' of nuclear energy and, doubtlessly, was premised on the increasing concerns of citizens. As a further ramification, the KHNP agreed to permanently shut down the Kori-1 reactor in June 2017 on the advice of the central government, rendering it the first of South Korea's nuclear power units to enter the decommissioning phase. The controversies moreover necessarily imposed a significant financial burden on the KHNP: a congressional hearing in October 2013 estimated this cost to be as high as US$2.8 billion (Cho 2013).

The overarching effect of the scandals is that South Korea's nuclear energy industry has been rendered more accountable to the public. This status quo is being reinforced by the recent corruption charges levelled against the Park Geun-hye administration, and the consequent presidential impeachment proceedings. As allegations emerged that President Park—daughter of Park Chung-hee—had colluded with a confidante in the embezzlement of large sums of public funds, over a million South Korean citizens took to the streets in protest. Their refusal to accept their president's apology and to continue to call for her resignation is stark evidence of society's diminished tolerance for government malfeasance.

Conclusion: The post-Fukushima legacy of the South Korean anti-nuclear movement

The Fukushima disaster of March 2011 was a vivid reminder for the world that nuclear power plants can cause catastrophic damage. A number of governments accordingly aborted or considerably slowed the pace of their nuclear energy programs, taking heed of rising concerns about the safety of nuclear reactors among their populaces. Yet, as we have seen, South Korea conversely pushed ahead with its ambition to become a foremost nuclear powerhouse after 3/11. This was in spite of the anti-nuclear movement gaining significant traction and mounting a concerted effort to alter policy and practice in the industry. The aim of this chapter has been to explain the limited effect of the movement through an examination of its anti-nuclear campaigns in pre- and post-Fukushima contexts.

It found that, owing to the fact that nuclear power became firmly ensconced in Seoul's energy policy long before the advent of anti-nuclear activism, the movement faced formidable structural obstacles from its incipient stages. This entrenchment of nuclear energy occurred as a consequence of decades-long dictatorial rule, the US–South Korea alliance, and the export-oriented development model installed by former President Park Chung-hee. Early collaboration among activists on opposing nuclear energy was hampered primarily by two factors: the dictates of authoritarian leadership and the preoccupation of the South Korean citizenry with achieving democratisation.

Once the anti-nuclear movement eventually materialised in the late 1980s, it proceeded to challenge various facets of nuclear energy policy with mixed results. In the earlier stages of its campaign, activists attained a degree of success in thwarting the construction of new nuclear power plants and radioactive waste disposals. They largely failed, though, in their post-Fukushima objectives of countering the lifespan extension of reactors due for decommissioning, and overturning county-level agreements (enacted pre-3/11) to host new nuclear power plants.

This limited policy change, it was argued, cannot solely be understood in terms of deficiencies within the movement. Rather, a combination of factors have served to constrain the opportunity structure for activists, including the insulated and top-down nature of nuclear energy policymaking in South Korea, and the integrality of nuclear power to the government's Green Growth Strategy. For its part, the movement has neglected to formulate a viable alternative to nuclear energy, which has long constituted a driving force of economic growth for the nation.

While the anti-nuclear movement failed to achieve a phase-out of nuclear power in South Korea, it would be imprecise to conclude that its efforts have been ineffectual. In fact, activists have succeeded in politicising nuclear energy and weakening its public support base. This process was facilitated by the recent revelations of endemic corruption within the industry (and government writ large), as well as the cyber-attacks targeting the more notorious nuclear reactors in the country. The movement capitalised upon these scandals to mobilise further anti-nuclear sentiment, and to fuel public mistrust in the regulation of nuclear energy. As a result, the South Korean government's policy of expanding nuclear energy is now subject to an increasingly hostile domestic atmosphere, which stands in sharp contrast to the earlier authoritarian era. Furthermore, the movement partially eroded the government's monopoly over nuclear energy, by compelling the industry to enhance its transparency, improve the safety of existing reactors, and to conform to greater public scrutiny. But perhaps the most significant legacy of the movement thus far is that it helped to persuade the government to scale back its target for nuclear power generation by as much as 30 per cent.

Nevertheless, South Korea remains on track to cement its status as a nuclear power stronghold. In order to change this status quo, the anti-nuclear movement will need to exert constant pressure, citing the lessons of Fukushima, and to formulate a feasible alternative to nuclear energy. This, in turn, will help the South Korean government to resolve its dilemma of being reliant on nuclear reactors to sustain economic growth and reduce CO_2 emissions, on the one hand, and subject to rising anti-nuclear views from its electorate, on the other.

If Seoul continues to pursue the further development of nuclear power without establishing a consultative mechanism that adequately incorporates the views of South Korean citizens, effectively it will only add greater fuel to the anti-nuclear movement. As surmised by Yeon Hyeong-cheol of the KFEM:

> Nuclear power plants are directly connected to the lives of the residents, yet the government has ignored citizens' opinions and insisted on a policy in favour of expanding nuclear power plants. Now that we have confirmed the [anti-nuclear] thoughts of the citizens, we will actively engage in movements to close down old nuclear power plants and to oppose the construction of new nuclear power plants nationwide (Choi 2014).

References

Baylon, Caroline, David Livingstone, and Roger Brunt, 2015. Cyber security at civil nuclear facilities: Understanding the risks. Chatham House Report. London: Royal Institute of International Affairs.

Cho, Mee-young, 2013. Stung by scandal, South Korea weighs up cost of nuclear energy. Reuters, 28 October.

Choi, Seung-hyeon, 2014. Referendum on Samcheok nuclear power plant ends in overwhelming opposition, a true victory for citizen autonomy: Expected to accelerate anti-nuclear movements in other regions. *Kyunghyang Shinmun*, 10 October.

Dalton, Toby, and Minkyeong Cha, 2016. South Korea's nuclear energy future. *The Diplomat*, 23 February.

East Coast Solidarity for Anti-Nuke Group, 2012. Pamphlet. Seoul.

Hayes, Peter, and Chung-in Moon, 2011. Park Chung Hee, the CIA, and the bomb. *Global Asia* 6(3): 46–58.

Hermanns, Heike, 2015. South Korean nuclear energy policies and the public agenda in the 21st century. *Asian Politics & Policy* 7(2): 265–82. doi.org/10.1111/aspp.12179

Hong, G. H., M. A. Hernández-Ceballos, R. L. Lozano, Y. I. Kim, H. M. Lee, S. H. Kim, S.-W. Yeh, J. P. Bolivar, and M. Baskaran, 2012. Radioactive impact in South Korea from the damaged nuclear reactors in Fukushima: Evidence of long and short range transport. *Journal of Radiological Protection* 32(4): 397–411. doi.org/10.1088/0952-4746/32/4/397

Hwang, Hae Ryong, and Shin Whan Kim, 2013. Korean nuclear power technology. In *Asia's Energy Trends and Developments*, vol. 2, edited by Mark Hong and Amy Lugg, 193–204. Singapore: World Scientific. doi.org/10.1142/9789814425582_0010

IAEA (International Atomic Energy Agency), 2012. IAEA completes expert mission to Kori 1 nuclear power plant in Republic of Korea. Press release, 11 June.

IAEA (International Atomic Energy Agency)–NSNI (Division of Nuclear Installation Safety), 2012. Report of the expert mission to review the station blackout event that happened at Kori 1 NPP on 9 February 2012, Republic of Korea. 4–11 June. kfem.or.kr/wp-content/uploads/2012/07/1419321177_zkGfGz.pdf (accessed 6 February 2017).

KHNP (Korea Hydro & Nuclear Power), 2012. IAEA completes expert mission to Kori 1 nuclear power plant in Republic of Korea. Press release, 13 June. www.khnp.co.kr/eng/board/BRD_000201/boardView.do?pageIndex=6&boardSeq=1604&mnCd=EN0501&schPageUnit=10&searchCondition=0&searchKeyword= (accessed 6 February 2017).

Kim, Se-jeong, 2015. Referendum on nuke plant turns invalid. *Korea Times*, 13 November.

Korean Government, 2011. Policy Issue 0: Report of the Korean government response to the Fukushima Daiichi nuclear accident. www.oecd-nea.org/nsd/fukushima/documents/Korea_2011_08Policy00GovernmentResponsetoFukushimaAccident.pdf (accessed 25 January 2017).

LaForge, John, 2013–14. Defective reactor parts scandal in South Korea sees 100 indicted. *Nukewatch Quarterly* Winter: 7.

Lee, Myung-bak, 2010. Shifting paradigms: The road to global green growth. *Global Asia* 4(4): 8–12.

Leem, Sung-Jin, 2006. Unchanging vision of nuclear energy: Nuclear power policy of the South Korean government and citizens' challenge. *Energy & Environment* 17(3): 439–56. doi.org/10.1260/0958305506778119425

McCurry, Justin, 2014. South Korean nuclear operator hacked amid cyber-attack fears. *Guardian*, 23 December.

Ministry of Science and Technology, 2006. Je 3-cha wonjaryeok jinheung jonghap gyehoek [Third comprehensive plan for nuclear energy development]. Seoul: Ministry of Science and Technology.

Ministry of Trade, Industry, and Energy, 2014. Je 2-cha eneoji gibbon gyehoek [The second national energy plan]. Seoul: Ministry of Trade, Industry, and Energy.

Nagata, Kazuaki, 2012. Fukushima puts East Asia nuclear policies on notice. *Japan Times*, 1 February.

Norman, Andrew, and Gerry Nagtzaam, 2016. *Decision-Making and Radioactive Waste Disposal*. Abingdon: Routledge.

Sayvetz, Leah Grady, 2012. South Koreans stop plan for nuclear waste dump on Gulup Island, 1994–95. Global Nonviolent Action Database. nvdatabase.swarthmore.edu/content/south-koreans-stop-plan-nuclear-waste-dump-gulup-island-1994-95 (accessed 25 January 2017).

von Hippel, David, Sun-Jin Yun, and Myung-Rae Cho, 2011. The current status of green growth in Korea: Energy and urban security. *Asia-Pacific Journal* 9(44)4: 1–15.

Yi, Whan-woo, 2012. Potential nuclear risk. *Korea Times*, 21 May.

6

Control or manipulation?
Nuclear power in Taiwan

Gloria Kuang-Jung Hsu

Abstract

Over the last three decades, the development of nuclear energy in Taiwan has shifted from a secret weapons program to civilian applications, from an expansion of nuclear power towards a nuclear-free future. But some countries may still wish to gain weapon capabilities through civil programs, as Taiwan did many years ago. Even though the days of military involvement have long gone, the past still casts a long shadow over those in the field of nuclear energy. An old nuclear culture persists, and advocating nuclear power is still more important than safety regulation. Unless nuclear safety can be restored to its rightful position, continuing current practices are likely to threaten operational safety and risks, placing waste management in disarray. Over the years, a number of nuclear-related incidents have occurred in Taiwan, demonstrating the importance of having a system of checks and balances, strictly enforced through domestic and international transparency.

Introduction

Taiwan's Chin San nuclear power plant's Unit-1 has been idle since December 2014, pending legislative hearings on the broken handle of a fuel assembly. Of six operating nuclear reactors in Taiwan, four have full spent fuel pools, with no space for a full core removal in an emergency.

The reactors are nonetheless allowed to continue to operate. Interim spent fuel dry storage programs, which have been criticised for their lack of basic safety features, have been put on hold by the local government.

On 20 May 2016, Ms Tsai Ing-Wen of the Democratic Progressive Party (DPP) was inaugurated as the president of Taiwan. Before the presidential election, Tsai promised that Taiwan would become nuclear free by 2025. Two days after Tsai won the election in January 2016, the Taiwan Power Company (Taipower) swiftly updated its projection on future power shortages from high risk to little risk if Taiwan becomes nuclear free (Lin 2016a). Two months later, Taipower chair Huang Chong-Chiou denied that Taipower had ever made any such U-turn in its electricity projection, and said he could not guarantee adequate electricity supply without nuclear power (Huang 2016). Taipower's apparently contradictory statements were criticised at the time as a political reflex, devoid of professional judgement.

As if trying to prove Huang's point, in May 2016, Taipower began issuing warnings of a possible power shortage (Lin 2016b). Seemingly manipulated by Taipower, Premier Lin Chuan announced his intention to restart troublesome Unit-1 of the Chin San nuclear power plant to fill the electricity gap, only two weeks after Tsai's inauguration. Lin's remarks immediately provoked vehement criticism from civil society and from DPP legislators who were outraged by this betrayal of President Tsai's nuclear-free promise. Premier Lin retracted his words the next day.

This incident exposes the administration's limited understanding of the intricate relationship between Taipower and the Atomic Energy Council (AEC), and their role in Taiwan's energy policy over the last four decades. Unless the government is willing to seek advice from outside the establishment, many risky and urgent nuclear-related problems will remain unresolved. 'Nuclear free' will remain only a slogan.

In this chapter, I describe the early secret nuclear weapons program in order to help understand the power distribution and the psychology behind the scenes. Next, I describe the devolution to peaceful applications, the emerging significance of Taipower, the fourth (Lungmen) nuclear power plant controversy and its current status, and future challenges. I then outline nuclear waste problems, the discovery of numerous radioactive buildings, the low-level waste storage facilities on Orchid Island, and recent spent fuel reprocessing issues. In the final section, I discuss reasons for a series of unfortunate incidents, and why new Taiwanese policymakers should try hard to prevent history from repeating itself.

Taiwan's early nuclear weapons program

The Israeli connection and US opposition

The story begins with an arms race between the ruling Kuomintang (KMT, Nationalist Party) and the People's Republic of China. In March 1962, President Chiang Kai-Shek learned from US intelligence that China was developing a nuclear weapons program in northwest China. Not wishing to fall behind the Communists, Chiang decided to pursue a nuclear weapons program. Dr Ernst Bergmann, Chair of the Israeli Atomic Energy Commission, in response to an invitation by General Tang Chun-Po, paid a secret visit to Taiwan in 1963 and spent three days at Sun Moon Lake Resort with President Chiang and General Tang (Wang 2010).

In spring 1964, the establishment of the military-controlled Chung Shan Science Institute (CSSI) for nuclear energy, rocket propulsion, and electronics was formally announced. General Tang served as chair of its preparatory bureau and Dr Bergmann as its foreign advisor. To facilitate the nuclear weapons program, the Defense Ministry immediately started sending talented military personnel overseas for advanced science and technology degrees.

In October 1964, China successfully detonated its first atomic bomb. Taiwan's nuclear weapons program, the Hsin Chu project, was initiated with Bergmann's assistance, and included a heavy-water reactor, a heavy-water plant, and a reprocessing plant. The CSSI began its work in July 1969; it had three research departments, including the Institute of Nuclear Energy Research (INER). President Chiang nominated the renowned physicist Wu Da-You as chair of the National Science Council in 1967, hoping for his assistance on the nuclear weapons program. Instead, Dr Wu submitted a 10,000-word written statement forcefully repudiating the idea. In the meantime, Western intelligence started to suspect that something underhand was afoot. The Israeli newspaper *Haaretz* first reported visits of atomic scientists from Taiwan in December 1965.[1] It was learned later that Dr Bergmann was their contact person.

---

1   US Embassy Tel Aviv, Nationalist Chinese atomic experts visit Israel. Airgram 793, 19 March 1966, in Burr (1999).

In early 1966, the Taiwanese government approached West Germany for a 50 megawatt (MW) heavy-water reactor from Siemens. As this was the first major German nuclear equipment export, the German government favoured the deal, on the condition that sensitive parts would be secured under International Atomic Energy Agency (IAEA) safeguards. Taiwanese officials stated that the reactor would aid research for an economic feasibility study and would be operated by the Union Industrial Research Institute of the Ministry of Economic Affairs, instead of the state-owned Taipower.[2] Taiwan's representatives repeatedly claimed that there was no relationship between the reactor purchase and nuclear weapons research. The United States remained unconvinced and strongly opposed the sale of the German reactor to Taiwan.

Also in 1966, four experts from the IAEA travelled to Taiwan to help with site selection for two 450 MW nuclear power plants, one in the north and one in the south of the island. During the site selection process, a Taipower representative requested an additional site for a 200 MW reactor, as a pilot plant, in either Hsin Chu or near Shimen Dam. The Taipower representative said that this 200 MW reactor would be sponsored by a 'consortium' of universities and other government institutes. The US Embassy immediately suspected military involvement.[3] Archival evidence from electronic briefings to the US Embassy in Taipei confirms that Taiwan intended to proceed with nuclear weapons development.[4]

The Hsin Chu program was aborted in 1969, probably due to a combination of domestic and international pressure (Albright and Gay 1998). Dr Wu strongly opposed the weapons program for being too costly and too close to the population centre. The Taiwanese authorities also worried that the international community could deny Taiwan access to all nuclear resources and that a direct confrontation with the United States was possible.

---

2   US Embassy Taipei, GRC plans for purchase of 50 megawatt (MW) heavy-water nuclear power plant. Airgram 566, 30 April 1966, in Burr (1999).
3   US Embassy Taipei, GRC request to IAEA team for advice on location of reactor for possible use by military research institute. Airgram 813, 8 April 1966, in Burr (1999).
4   US Embassy Taipei, Indications GRC continues to pursue atomic weaponry. Airgram 1037, 20 June 1966, in Burr (1999).

A deceptive shift of focus

After terminating the Hsin Chu program, Bergmann persuaded President Chiang to modify Taiwan's nuclear strategy by considering civilian applications of nuclear power. To dilute its ability to function militarily, the INER was reassigned, from the military-controlled CSSI to a new position affiliated with the AEC, as proposed by Dr Wu. The president of National Taiwan University was named as head of the CSSI. Soon after, genuine civilian programs were initiated. The Executive Yuan approved the first nuclear power plant project in August 1969. Two General Electric (GE) light-water reactors (LWRs) were acquired in 1970. Two more boiling water reactors were considered for purchase in 1974. These civilian nuclear activities persuaded some United States intelligence officers that Taiwan had shifted its focus: 'This type of reactor is not by any means an optimum choice with regard to producing plutonium for weapons use', a US Embassy official noted.[5]

Meanwhile, Taiwan launched another secret nuclear program codenamed 'Tao Yuan'. But Taiwan's attempt to purchase a reprocessing facility from the United States was vetoed by the Richard Nixon administration in 1969. Instead, the INER acquired a 40 MW heavy-water reactor from Canada, which became critical in April 1973. Combining equipment acquired from the United States, France, Germany, and other countries, the Taiwanese developed a small reprocessing facility, a plutonium chemistry laboratory, and a plant to produce natural uranium fuel (Albright and Gay 1998). The plan to purchase a reprocessing plant from France failed due to an exorbitantly high price and/or pressure from Beijing.[6] Washington learned that the Taiwanese government had turned to a West German firm for parts for a reprocessing plant in 1972. But it being a time when the Sino-US relationship was normalising, and not wanting to agitate either Taiwan or China, the US did nothing.

Washington, nevertheless, steadily increased pressure on Taiwan to forego its nuclear military program. The United States offered to support Taiwan's reprocessing of spent fuel in the US or other countries so that the

5　US Embassy Taipei, ROC nuclear intentions. Cable to State Department No. 2354, 20 April 1973, in Burr (1999).
6　US State Department, German inquiry regarding safeguards on export of parts to ROC reprocessing plant. Memorandum of Conversation, 22 November 1972, in Burr (1999).

Taiwanese could save resources.[7] In January 1973, despite US opposition, Taiwan signed a deal with the West German firm UHDE for a spent fuel reprocessing facility. In heated exchanges with the US Ambassador to Germany, Martin J. Hillenbrand, Taiwan's Foreign Minister Shen Chang-Huan maintained that Taiwan had not made a decision about the reprocessing issue and denied the existence of a nuclear weapons program. Under pressure from the West German Foreign Office and the United States, UHDE backed out of the deal in February. The next day, the Taiwanese foreign minister informed the US ambassador that Taiwan had decided not to purchase the reprocessing plant.

In March 1973, the AEC Secretariat's Victor Cheng told the US Deputy Assistant Secretary of State for Far Eastern Affairs, Richard Sneider, that Taipei did not keep 'any nuclear secrets from its friend'. Cheng presented a progress report on building a laboratory scale reprocessing facility at the INER, with a potential capacity to produce approximately 300 grams of separated plutonium per year.[8] The US estimates came out differently, showing that the Canadian research reactor could generate enough plutonium in one year for a nuclear weapons test if the reactor was running at optimal capacity. A visit by IAEA inspectors in 1976 led to suspicion that the INER may have been secretly diverting spent fuel for reprocessing. In September 1976, the United States made a formal diplomatic request through Ambassador Leonard S. Unger that Taiwan should renounce the development of nuclear weapons. On 17 September 1976, Premier Chiang Ching-Kuo and his cabinet issued a public statement solemnly declaring that Taiwan had no 'intention to use its human and natural resources for the development of nuclear weapons' or to obtain technology to reprocess spent fuel (*United Daily News* 1976).

In April 1977, the United States learned that the INER had been in touch with a Dutch firm regarding reprocessing technology. The chief US concerns were about heavy-water production and the 'hot laboratory' at the INER. An IAEA inspector discovered an unsafeguarded exit port at the fuel pond in March 1977, but no spent fuel diversion was found. The United States demanded that Taiwan terminate all fuel cycle activities, reorient facilities to peaceful applications, and transfer all plutonium to

---

7   US State Department, Proposed reprocessing plant for Republic of China. Cable 2051 to Embassies in Bonn, Brussels, and Taipei, 4 January 1973, in Burr (1999).
8   US State Department, ROC nuclear research. Cable 51747 to Embassies in Taipei and Tokyo, 21 March 1973, in Burr (1999).

the United States. A team of US experts arrived, tore down all of the suspected facilities, and destroyed the Tao Yuan program. At the same time, AEC Secretariat Cheng visited Washington to discuss the licensing of the first nuclear power reactor. Despite repeated intervention from the United States, suspicion continued. In September 1978, US Secretary of State Cyrus Vance sent a letter to President Chiang concerning suspected activities in the CSSI. President Chiang was obviously annoyed that 'Taiwan's vulnerability and its unique relationship with the US [should] allow the latter to treat [the] ROC [Republic of China] in a fashion which few other countries would tolerate' (*United Daily News* 1976).

On 12 January 1988, the deputy director of the INER, Chang Hsien-Yi, defected to the United States carrying a large amount of sensitive material. Chang appeared at a closed-door hearing in Washington, arranged by the Central Intelligence Agency (CIA), a few days later. The Taiwanese authorities only learned of Chang's disappearance after this. President Chiang died on 13 January 1988. Two days later, an expert team from the United States visited the INER. They extracted all of the heavy water and thoroughly demolished all nuclear weapon–related facilities. Chang may have been recruited by the CIA and worked in secret for the United States for 20 years before the incident. His defection led to the closure of Taiwan's nuclear weapons program and was highly praised as one of the few covert operations in which the CIA was successful (Weiner 2007). But many Taiwanese view Chang as a traitor.

The civilian applications

New nuclear power plants

The organisational framework of nuclear development was established in 1955. Human resources were cultivated at the Institute of Nuclear Science of the National Tsing Hua University. The Institute of Nuclear Science was the only institute in the university, newly relocated from the Chinese mainland. Taiwan signed the Agreement for Cooperation between the Government of the United States of America and the Government of the Republic of China Concerning Civil Uses of Atomic Energy on 18 July 1955, to ensure the transfer of nuclear technologies and materials. The national regulator, the AEC, was established in the same year, on a provisional organisation status, with the majority of its personnel seconded from other ministries. The Department of Nuclear Engineering at the National Tsing Hua University was established in 1964, and began recruiting undergraduates. The timing coincided with the start of the Hsin Chu project and the establishment of the CSSI.

Site selection for nuclear power plants, mentioned earlier, began in 1964. In 1969, with help from IAEA experts, Chin San and Yen Liao were found to be suitable and the former was preferred (CEPD 1979). Bechtel Corporation was contracted to provide support for site selection, machinery, and instrument preparation, in addition to technical and economic feasibility studies. Two GE LWRs, 636 MW each, were recommended for the Chin Shan nuclear power plant project, and received formal approval from the Executive Yuan in August 1969. It was one of Taiwan's 'ten major infrastructure projects', an achievement of Premier Chiang Ching-Kuo, who later succeeded his father as president (Small and Medium Enterprise Administration, Ministry of Economic Affairs n.d.).

The second (Kuosheng) and third (Maanshan) nuclear power plants went forward without dispute. Total installed nuclear capacity reached 5,144 MW when the Maanshan plant was in full operation in 1985. The fourth nuclear power plant project, Lungmen, was approved in September 1980, with the intention of connecting to the grid by 1994.

Since nuclear power plants were meant to be a distraction from weapons development, the Taiwanese authorities paid little attention to the actual trend of electricity demand. The reserve margin for electricity—the percentage of installed capacity that is not needed even during periods of annual peak demand—suddenly jumped to 55 per cent in 1985 when the Maanshan plant began feeding electricity to the grid (Control Yuan 2012). It was estimated that over 70 per cent of electricity-generating capacity was left idle for most of the year. The addition of two more huge nuclear reactors was completely unnecessary. Fifty-five KMT legislators appealed for an emergency motion to halt the Lungmen program (Legislative Yuan, Atomic Energy Council 1985). Their concerns, including operational risk, nuclear waste management, and energy security, were presciently similar to current anti-nuclear rhetoric. The premier dutifully complied and indicated that there was 'no need to start until all suspicions are cleared', so the Lungmen program was retracted (Legislative Yuan, Atomic Energy Council 1985).

A shifting line of command and a partisan political issue

The defection of the deputy director of the INER, the death of President Chiang Ching-Kuo, and the total destruction of Taiwan's nuclear weapons program all happened within one week in early 1988. Not only were these inconceivable shocks for Taiwan's nuclear weapons proponents, but huge political changes were also expected. Perhaps too embarrassed, or too busy with power struggles to care about the future of nuclear energy, military personnel have cut their involvement with nuclear power since then. Without any challenge from other agencies, Taipower with its hefty cash coffer conveniently took over the decision-making role.

In the following years, installed electricity capacities mysteriously shrank; as much as 3.8 gigawatts (GW) in estimated capacity was taken offline between 1986 and 1991 (calculations based on open data provided by Taipower n.d.a). As a result, the reserve margin fell from 55 per cent to 4.8 per cent. Unexpected power outages and rolling blackouts became routine and occurred with increasing frequency (Central News Agency 1991a, 1991b, 1992a, 1992b). The Lungmen nuclear power plant program was revived as the most viable option to rescue Taiwan from power shortages. Analyses by the Chung-Hua Institution for Economic Research (Wang 1991) showed the error in this thinking—they found that the majority of blackouts were caused by malfunctions of nuclear plants. Taipower was criticised for deliberately creating power shortages in order to gain support for nuclear power expansion (Central News Agency 1994, 1995). Despite this, the authority planned to install eight more nuclear reactors by 2020 (Gi 1998), including the two in Lungmen.

After the Chernobyl nuclear disaster in 1986, people began to realise the risks associated with nuclear power. When the Lungmen nuclear power plant program was revived in 1990, it met with much public suspicion, as more people spoke out about nuclear safety issues. The DPP, Taiwan's first real opposition party, was formed in September 1986, and had anti-nuclear party guidelines to 'oppose any program to install a new nuclear reactor; encourage the development of alternative energy resources; and set [up] a timetable to close all nuclear power plants' (Democratic Progressive Party Principles and Guidelines n.d.: 13). When, in June 1992, the KMT majority in the Legislative Yuan hammered through an eight-year budget for the nuclear project, the Lungmen nuclear power plant project became a partisan issue, with the DPP strongly opposed. DPP member Chen

Shui-Bian won the presidential election in May 2000 and, keeping his campaign promise to residents near the planned Lungmen site, terminated the program on 27 October 2000.

The termination announcement was made after Chen met with Lien Chan, chair of the KMT opposition. Lien was not informed of Chen's decision beforehand and felt that he had been publicly humiliated. The KMT not only launched an all-out campaign against Chen's decision, it also initiated an impeachment process in the KMT-dominated Legislative Yuan. President Chen bowed to sustained pressure, and the Lungmen program resumed in February 2001.[9] The failed attempt to stop Lungmen was seen as a major setback for the ruling DPP. Subsequently, key politicians have been reluctant to get involved in issues related to the Lungmen program.

The problematic history of the Lungmen nuclear power plant

Construction flaws

Construction of the Lungmen nuclear power plant revealed the strained relationship between the regulator, the AEC, and the operator, Taipower. The three existing nuclear power plants were completed under the supervision of US consulting firms Ebasco and Bechtel. With very little experience, Taipower undertook the construction of the Lungmen nuclear power plant using GE blueprints. The equally inexperienced AEC set up a regulatory committee in January 1997 to monitor Lungmen's quality and progress. The AEC began publishing short monthly monitoring reports in 2002, when the real work started. Many of the flaws identified during the early stages of the construction were quickly corrected. The first major discovery was triggered by anonymous tips, pointing to the lower-than-required-strength welding applied to the reactor base frame (AEC 2002). Follow-up by the AEC in April 2002 confirmed the problem, so the base frame was rebuilt.

---

9   Premier Chang Chun-hsiung and Legislative Yuan Speaker Wang Jin-pyng signed an agreement to resume construction, and the decision was announced the day after.

An increasing number of flaws were identified as construction progressed. Major problems listed in the AEC's reports included reinforced tendons for a containment anchor being accidentally cut (AEC 2007), and careless contractors repeatedly setting working platforms directly on top of installed pipes and tubing, causing rust, obvious dents, and even punctures (AEC 2008a, 2008b, 2010). Workers' logs were completed with clearly too much work that could be finished in a single day (AEC 2009). Moreover, many joints inside the Lungmen reactor building were inadequately sealed with Teflon tape (AEC 2011).

However, more serious allegations raised by an insider were categorised by the AEC as 'not safety related' (AEC 2008c). These included headline-grabbing design alterations and the systematic cutting of corners on materials (Wang and Wei 2008). It was found that Taipower had made 395 alterations to the Lungmen nuclear power plant design, including support for an emergency cooling system, without consulting the AEC or GE. In addition, Taipower knowingly used neoprene gaskets instead of carbon fiber gaskets in pull box and conduit fittings, despite the fact that the specification clearly precludes using such gaskets. The former can easily be ignited with a cigarette lighter, whereas the latter can endure temperatures of up to 1,000 degrees Celsius. It was also found that hot dip–galvanised zinc steel was replaced by zinc-electroplated steel: the coating on the former is 25 times thicker than that on the latter, and can last more than 50 years in coastal areas. In reply to questions from journalists, the Taipower Lungmen site manager said that a nuclear power plant is not a humid environment, zinc-electroplated steel is adequate, and neoprene releases toxic fumes when it catches fire. Since no one could survive such high temperatures, this concern seemed spurious (Wang and Wei 2008).

The AEC imposed a fine of NT$500,000 (about US$16,700) on Taipower in April 2008, and insisted that Taipower re-evaluate the safety of altered items and make no more alterations without the AEC's consent.[10] A couple of months later, the AEC discovered that Taipower had made about 700 additional alterations without the AEC's knowledge. A total fine of NT$3.5 million (about US$117,000) was imposed.[11] Yet again, more alterations without authorisation were discovered in mid-2011

---

10   The AEC website lists the penalties and fines of confirmed violations and irregularities related to nuclear plant operation and construction.
11   Violation nos 0970020065 and 0970020065, issued 19 November 2008.

(Lee 2011). This time, the AEC not only imposed a higher fine of NT$15 million (US$500,000), but also announced that it would take culpable Taipower executives to criminal court.[12] Apparently, Taipower holds little respect for the AEC.

Shared irresponsibility

Although Taipower is a state-owned utility monopoly, few government administrations have had firm control of its management. Magazine interviews with several Taipower executives in June 2008 revealed the rationale behind the alterations at Lungmen (Lee 2008). They blamed 'GE's over-conservative design' for all the problems.[13] The excessive GE design, the executives said, required 'tens to thousands of times more [materials] than the project really needed', making 'construction difficult' and 'inflating the costs'.[14] Also, GE designs apparently could not be trusted since the United States had not constructed a new nuclear power plant in 30 years, during which 'GE lost a major part of its nuclear capability'. Taipower executives claimed that they had found 'numerous contradictions' during construction, and therefore 'had no choice but to make improvised changes in order not to delay the whole project' (Lee 2008).

The AEC had overlooked some important issues. In the short inspection reports in May and August 2007, the AEC mentioned only in passing the poor cement jobs in both reactor containments. Reports described threaded steel, cigarette butts, and plastic bottles found in the wall of the reinforced concrete containment vessel, with no photos attached. Some places had steel bars partially exposed. Also found in Unit-1 were workers chipping away at the newly built containment, with over 40 tendon steel bars cut, to make room for the spent fuel pool. It was not until a photograph that showed a plastic bottle in the containment wall was leaked to the press in April 2013, that people began to realise how potentially catastrophic and dire the situation was.

---

12 Violation no. 1010001075, issued 16 January 2012.

13 Interview with Shih Hung-gee, who served as general manager of the Maanshan nuclear power plant, chief engineer of the Lungmen nuclear power plant, and deputy chair of Taipower. After retiring in 2009, Shih stayed on as an advisor for Lungmen (Lee 2008).

14 Explained by Lin Jun-long, head of the Lungmen nuclear power plant Progress Oversight Unit at Taipower (Lee 2008).

According to the AEC, a fine of NT$400,000 (about US$13,000) was imposed, the plastic bottle was removed, and the holes were filled with equal-strength concrete. The AEC assured the public that the strength of both containments was not compromised even after modifications (AEC 2014). Less than two weeks later, however, reports were published of a failed integrated leak-rate test and structure integral test for Unit-1 between 26 February and 5 March 2014 (Tang and Chien 2014). Leaks were substantial, but difficult to locate. Suspected causes ranged from more unseen plastic bottles in containments, second-hand valves, and construction short-cuts on the penetration seal within the nuclear island. In addition, records showed that as many as 197 items had been moved from Unit-2 to Unit-1 to replace broken parts, probably as a result of inadequate handling.

As construction at Lungmen began, scandals came to light from time to time, but public reaction was rather mild. In 2008, Ma Ying-Jeou of the pro-nuclear KMT won the presidential election, and began earnestly to pursue the task of making the Lungmen nuclear power plant operational. Nuclear energy was framed as an indispensable part of the climate mitigation program, and more nuclear power was suggested (Ho 2008). Although grid connection time was postponed repeatedly (Central News Agency 2014), work nevertheless continued.

Lungmen nuclear power plant's demise

The Fukushima disaster of March 2011 completely changed the situation. People suddenly realised how much Taiwan and Japan had in common, especially regarding seismic vulnerability. Many were bewildered as to how a prudent society with such advanced technology could become so helpless, and what would become of Taiwan in a similar situation. Immediate responses from the AEC were anything but reassuring. Without any evaluation, and just two days after the Fukushima disaster, the AEC deputy chair boasted that 'all nuclear power plants in Taiwan are as sturdy as Buddha sitting on his platform' (Now News 2011). Neighbouring countries, such as the Philippines, Vietnam, and China, all had detected radioactive materials from Fukushima, but the AEC insisted that nothing was detected until 31 March 2011.[15] The sensitivity of the AEC's instruments was questioned by the public (Yen et al. 2011).

---

15   In response to a legislator's question about whether radioactive material would come to Taiwan, deputy chair Huang Ching-tong of the AEC said, 'No radioactive substances will reach Taiwan, since the Japanese have not had enough for themselves' (Tung Sen News 2011).

In February 2013, the KMT's premier proposed holding a referendum to settle the future of Lungmen nuclear power plant. The *Referendum Act* of Taiwan requires more than 50 per cent voter turnout, plus an absolute favourable majority vote, in order for the referendum to be legally binding. Since the law passed in 2006, six national referenda have been held and all were rejected because turnouts were between 26 and 45 per cent. Under the current law, framing of the referendum question determines the outcome. The KMT's proposal was as follows: 'Do you agree that the construction of the Fourth [Lungmen] Nuclear Power Plant should be halted and that it not become operational?' Having the intended ballot date set at the end of 2013, the administration calculated that few would come to vote solely for the referendum, thus legitimising the project.

Meanwhile, the AEC requested that the European Union (EU) perform a Taiwan 'stress test', to be completed one month before the planned voting date. A well-received international assessment certainly would win more public support. Some concluded that the Taiwan stress test was a propaganda exercise and not for nuclear safety. Non-governmental representatives discovered that geological information in the AEC-prepared Taiwan stress test national report was much shorter than what was already known. Distances between nearby faults and nuclear reactors were either overlooked or completely absent in the report (AEC 2013; Tsai 2013; Hsu 2013). In the end, the AEC received a polite and lukewarm assessment report (ENSREG 2013). But waves of demonstrations occurred nationwide, including an anti-nuclear protest on 9 March 2013, which drew a crowd of more than 200,000 people.

Pressure from the electorate forced KMT legislators to withdraw the referendum proposal (Shih 2013). But a controversial service trade agreement with China that KMT legislators passed very quickly renewed widespread demonstrations in March 2014. On 22 April 2014, Lin Yi-Hsiung, former DPP chair and long-time anti-nuclear activist, went on a hunger strike, calling for the termination of the Lungmen nuclear power plant. Under all this pressure, President Ma Ying-Jeou reluctantly made concessions on the project, including ceasing construction and sealing of Unit-1 pending a later decision, and completely stopping construction of Unit-2 (M.-S. Huang 2014). The latter decision was probably made because the administration was clearly aware that the possibility of Unit-2 becoming operational was very slim. Lin ended his hunger strike on 30 April 2014.

Radioactive-contaminated buildings

Checking whether property for sale is radioactively contaminated may not be common elsewhere, but in Taiwan it is a regular service provided by real estate agents. Since 1992, more than 300 buildings in Taiwan, including 1,600 apartments, have been identified as being contaminated with elevated levels of radiation. The widespread radioactivity in housing resulted from repeated negligence and a cover-up by AEC personnel for over a decade. The AEC first learned about the existence of radioactive steel in January 1983. Steel bars bought by Chin San nuclear power plant were found to be highly radioactive, with radiation levels of 70 microSievert per hour (μSv/h, approximately 700 times background levels of around 0.1 μSv/h).[16] The supplier, Chin San Steel, bought steel ingots from a steel factory in Taoyuan. The AEC learned that Chin San Steel had sold two tonnes of steel bars to Chin Shan nuclear power plant and 29.9 tonnes to Jien-Kang Construction Company for the Tien Mo dormitory of the International Commercial Bank of China. Readings on unused steel bars on the construction site were about 50 μSv/h, and those of the half-built dormitory ranged from 1 to 5 μSv/h in March 1983.

The AEC swiftly demanded that the construction company remove all installed radioactive steel bars. On 26 March 1983, the AEC made Jien-Kang Construction and Chin San Steel promise to safeguard both 17.2 tonnes used and 12.7 tonnes of unused radioactive steel bars in the Chin San Steel warehouse, with no transfer to be made without the AEC's permission. However, an AEC inspector visited the Chin San Steel warehouse on 24 May 1984, and found that all of the unused radioactive steel bars were gone, allegedly sold without prior consent from the AEC. In his reply to the AEC's inquiry, the owner of Chin San Steel claimed that the entire supply of steel was far too rusty to be useful and therefore it was buried on-site. The AEC seemed to be persuaded, and it did not follow-up on this issue.

On 15 August 1992, the public first learned about a 'radioactive' villa in the *Liberty Times* (Chang, Chan, and Wang 1997). A tip-off from a disgruntled AEC employee led a reporter to a building called the Ming Shan Villa. Radiation levels as high as 600 μSv/h were found, emanating from the building frame, about 300 times higher than the limit for

---

16 The permissible limit in Taiwan for workers in nuclear power plants is 100 milliSieverts (mSv) in five years. Six hundred μSv/h is equivalent to 5.26 sieverts (Sv) per year (5260 mSv/year).

workers in nuclear facilities. It was soon discovered that the AEC had known about the incident since March 1985 when an AEC contractor was sent to evaluate the X-ray machine installation in a new dental clinic (Wang 1996). With the X-ray machine switched off, the radiation level reading was 280 µSv/h. The inspector quickly realised the readings came from the building beams. Saying nothing to the dentist, he reported back to the AEC with his sketch of the radiation distribution. Fearing that the radioactive steel beams were those that had disappeared in the commercial bank case one year earlier, the AEC chose to cover it up.[17] It issued the dentist a regular licence and decided not to carry out any future sampling in that building. The AEC also extracted promises from everyone, including the contractor and the X-ray machine dealer, not to reveal those measurements to the public. The dental clinic received AEC licence renewals several times subsequently.

After seeing the *Liberty Times* report, the AEC tried to downplay its seriousness by referring to the radiation level found in Ming Shan Villa as 'slightly above background level' (Central News Agency 1992c). Only after more details were leaked to major media outlets did the AEC reluctantly admit that the radiation level was 1,000 times higher than background from radioactive cobalt-60 with a half-life of 5.2 years (Central News Agency 1992d). A frantic nationwide radiation monitoring effort began. Identified radioactive-contaminated buildings included business offices, kindergartens, schools, and residential homes. Around 13,300 residents have been exposed to radiation. Meagre support and multiple criteria set by the AEC for rebuilding[18] have resulted to date in less than 7 per cent of buildings being rebuilt, and 15 per cent adding lead shields or having radioactive steel bars replaced. About 80 per cent of the contaminated buildings have been left intact, including Ming Shan Villa. Twenty years after the news emerged, 3,600 dwellers still reside in radioactive-contaminated apartments.[19] It seems that the AEC wishes to let nature solve this messy travesty.

17   Many dispute this because the reading in Ming Shan was much higher than that in the International Commercial Bank of China case.
18   Apartments with an annual dosage above 25 mSv are acquired by the government; those living where the annual dosage is 5–25 mSv receive compensation of NT$200,000 (about US$6,667). No compensation is provided where the annual dosage is below 5 mSv. Only buildings with more than 20 per cent of apartments above 5 mSv per year are eligible for better floor–area ratio if rebuilt.
19   With a half-life of 5.2 years, radiation levels reduced to 6.7 per cent of the original level 20 years ago. Therefore, the original radiation level is 2.54 times higher than seven years ago, when the AEC first learned about the problem.

Sources of radioactive materials

Where did these radioactive materials come from? Most radioactive steel bars were produced between 1982 and 1983 by Hsin Jung Steel Company, in Taoyuan county, and derived from scrap metal. Hsin Jung Steel claimed that no imported scrap metal was used during that period. An AEC investigatory report pointed a finger at the Army Chemical Infantry School, not far from Hsin Jung Steel (Lee 1984). The school reported that one of the cobalt-60 sources of 23.8 Curie was missing in September 1982, and the radioactive steel bars were first found on the market one month later. The AEC thus conveniently suspected that the school might have lost more than one of the cobalt-60 sources of 23.8 Curie, and specifically ruled out the possibility that radioactive materials might have come from its Institute of Nuclear Energy Sciences, which is closer to Hsin Jung Steel. An Army spokesman denied the accusation (Lin 1984).

Radiation Safety Improvement Organisation, a non-governmental organisation, challenged the official report.[20] According to its calculation, a couple of radiation sources can generate no more than a few hundred tonnes of steel, far less than the 7,000 tonnes that the AEC had already identified. Among leaked AEC confidential files were the inventory books of Hsin Jung Steel dating from 1982 to 1983, which the AEC seized in its 1985 internal investigation of the Ming Shan Villa incident. Records showed that Hsin Jung Steel had purchased 604 tonnes of scrap metal from Taipower on 29 October 1982. In November 1982, Hsin Jung Steel sold the steel to various construction companies; all were later identified as radioactively contaminated. Approximately 6,000 tonnes of scrap steel were collected after annual maintenance at the three operating nuclear power plants. Taipower denied that the scrap steel sold was radioactive plumbing, but offered no explanation as to where the scrap steel went.

Residents of Ming Shan Villa filed a petition with the government's Control Yuan against the AEC. In June 1994, the Control Yuan passed a resolution condemning the AEC bureaucrats (Control Yuan 1994). The accused were sent to the Public Functionary Disciplinary Sanction Committee, and the Supreme Prosecutor's Office was asked to investigate possible administrative and/or criminal responsibilities. However, judges

---

20 Wang Yu-lin is chair of the Radiation Safety Improvement Organisation.

found the AEC bureaucrats innocent on the grounds that 'steel bars are not under AEC jurisdiction, nor is the radioactive contaminated steel' (Control Yuan 1994).

Low-level nuclear waste on Orchid Island

Orchid Island (Lan Yu), where the aboriginal Tao tribe resides, is situated in southeastern Taiwan. A low-level nuclear waste[21] temporary storage facility was constructed there in 1978, disguised as a fish cannery. It began operating in May 1982. The initial plan was to dump the waste into deep ocean trenches adjacent to Orchid Island. But the 1993 Amendment to the Convention on the Prevention of Marine Pollution by Dumping of Wastes and Other Matter added low-level nuclear waste to the black list, and the dumping plan was abandoned (*LanYu BiWeekly* 1996).[22]

However, low-level nuclear waste kept being sent to the island. It was alleged that workers were permitted to release liquid radioactive waste into the surrounding environment. Since the original plan was to dump the nuclear waste into the ocean, drums were made of ordinary steel. In this hot, humid, and salty environment, about one-third of the waste barrels showed clear signs of rusting by early 1995. Locals complained about increasing numbers of cancer-related deaths and cases of children with learning disabilities. Taiwan's National Health Statistics indicated that Orchid Island has the highest cancer death rate in Taiwan (Chiu, Wang, and Liu 2013). Feeling deceived and abandoned by the government, the Taos began protests in 1988, which soon gathered momentum. On 27 April 1996, the Taos successfully prevented Taipower's low-level nuclear waste shipments from docking. No more waste shipments to Orchid Island have occurred since then. By that time the total number of nuclear waste barrels stored was 97,672.

Taipower first promised that a permanent low-level nuclear waste disposal site would be identified in 1996, and that all nuclear waste would be removed from Orchid Island by 2002. This target was also one of the

---

21 According to the AEC's definition, all radioactive materials except spent fuel are categorised as low-level waste, including equipment and materials that have had direct contact with fuel.

22 This international convention was established in 1972 and entered into force on 30 August 1975. The United States and Japan agreed to ban marine dumping of low-level nuclear waste after learning that the Soviet Union dumped about 900 tonnes of low-level nuclear waste into the Sea of Japan prior to 1993. A global ban on all dumping of radioactive waste at sea came into force in 1994.

conditions attached to the approval of Lungmen nuclear power plant's Environmental Impact Assessment (EIA). However, in an EIA revision submitted in July 2001, Taipower argued that permanent low-level waste storage was relevant neither to the operations of Lungmen nuclear power plant, nor to its environmental impact. Taipower successfully had the condition removed. The revision took place immediately after the KMT lost the presidential election, after five decades in power. In the ensuing political chaos, Taipower's revision received little attention.

Taiwanese authorities made a number of attempts to locate a permanent low-level nuclear waste site. A search panel for Taipower first identified Hsiao Chiu in Kingmen county as the most suitable site in 1998. After the DPP won the general election in 2000, the new AEC chair objected to the location as too close to China, potentially causing unwanted tensions (Liu 2002). Several legislators even suggested paying the aboriginals to leave Orchid Island, and to use it as a permanent nuclear waste site. Officials quickly denied having such a plan (Lo 2002). Exporting the waste was also seriously considered. In 1997, Taipower signed an agreement with North Korea for the latter to store 60,000 barrels of low-level nuclear waste from Taiwan. Nothing came of this proposal, since the North Korean facility was not ready in time, and South Korea issued strong protests.

There was speculation that China offered help on low-level nuclear waste storage, and that the waste might be sent to Solomon Islands, or the Marshall Islands. Nevertheless, no substantive solution materialised. Commencing in 2000, Taipower began paying NT$220 million (US$7.3 million) every three years to rent the waste storage site (Taipower n.d.b). An Orchid Island Storage Facility Relocation Committee was established under the Executive Yuan in May 2002. Meanwhile, the condition of the drums on Orchid Island continued to deteriorate.

A repackaging program was begun in 2008. Most canisters were found to be rusted, broken, and some even shattered. Although all the waste had been on Orchid Island for more than 10 years, the readings on sampled barrels were 2–4 milliSievert (mSv) per hour. A couple of hours of exposure would exceed the maximum annual dosage allowed for nuclear workers. However, the majority of the repackaging work was done by unskilled temporary labourers who were only provided with a simple dust-free suit, not radiation-protective gear. Instead of using a negative pressure chamber, everything was carried out in the open air. Workers wore regular clothing when applying new paint to the repacked canisters. Taipower's reason was that there was no 'dust' in the vicinity. The existence of the poor working conditions were made public during a Legislative Yuan hearing. The AEC chair first mocked legislators for using outdated information (CTI 2012). After enough evidence was presented, the AEC chair admitted that he was wrong (AEC 2012).

It is not clear whether the AEC was just too lax and incompetent to be aware of the poor working conditions, or whether it knowingly tried to cover-up this issue. Either way, serious negligence occurred at Orchid Island. There are 100,277 barrels after repackaging. Thirteen years have passed and promised completion dates have consistently failed to be met. The Taos are stuck with the low-level waste. Their experience provides a vivid example of problems that can be encountered when allowing nuclear waste to be stored in communities.

High-level nuclear waste: Conflict of interest

Over the past three decades, all spent fuel from nuclear power plants has been stored in the spent fuel pools next to the reactors. To date, the four reactor spent fuel pools have only enough room for an additional 9, 7, 34, and 146 fuel assemblies (AEC 2016). It is thus impossible for those reactors to have a whole core removal in case of an emergency. After the Fukushima disaster, both the AEC and Taipower understand very well the dire situation of having four fully loaded spent fuel pools. Somehow, they have let those reactors continue operating.

Taipower anticipated the lack of storage space for the spent fuel two decades ago. Four types of dry casks for spent fuel interim storage were listed in its EIA reports for Interim Dry Storage of Spent Fuel from Chin Shan and Kuosheng nuclear power plant, submitted in March 1995, and permission was received in June 1995. According to Taipower, the dry cask contract has been open for international tenders four times since 1995. It failed to find qualified tenders at a 'reasonable' price. Therefore, the contract was given to the INER, a deal which the AEC argued complied with the law. The INER is Taiwan's only nuclear energy research institute and does not work on nuclear regulation. All of the INER staff involved in the dry cask project are excluded from making an assessment, leading the AEC to argue that no conflict of interest exists (INER 2015).

In 2005, Taipower submitted its environmental change and countermeasures report, a revision of an earlier EIA,[23] and cement casks were chosen for both Chin Shan and Kuosheng nuclear power plant (Taipower 2005). Permission was granted in September 2008. Instead of having enough dry casks to store all possible spent nuclear fuel generated, the revised plans cover less than one-fifth of the overall spent nuclear fuel.[24] The change strengthened the suspicion that the interim dry cask program was intended only to prolong reactor lifetime (Wei 2005), rather than for reactor decommission. An application for lifetime extension of the Chin Shan nuclear power plant was submitted in 2009 (Chung 2009).

The cement cask that Taipower chose consists of one stainless steel container, 1.59 cm thick 304L steel, plus two outer layers of reinforced concrete. According to Taipower, the dry casks will stay intact for 40 years in open air, next to the ocean. Taipower is so confident that it only plans to install temperature sensors on each cask and one radiation detector for the whole storage area of 30 units. All processes were evaluated using computer simulations. Requests for a few experimental validations were rejected as unnecessary. Also deemed unnecessary are back-up plans and related facilities. Local residents worry that these interim storage sites may become permanent high-level waste dump sites. In response to concerns of its constituency, the New Taipei City government has held the required Land and Water Conservation License for the dry cask storage since 2013 (Lai 2013).[25] Meanwhile, the spent fuel pools in Chin Shan and Kuosheng nuclear power plants keep piling up.

Unexpectedly, the United States recently offered Taipower another option, making use of the 1955 nuclear cooperation agreement. That agreement, based on Section 123 of the US Atomic Energy Act, included nine non-proliferation criteria, prohibiting Taiwan from operating sensitive nuclear

---

23 The law requires that projects that pass EIA but do not start within three years must send an environmental change and countermeasures report for reassessment.

24 Environmental Protection Administration, Taiwan (2008) reduced the number of spent fuel bundles to be stored from 8,448 to 1,680; Environmental Protection Administration, Taiwan (2009) reduced the number of spent fuel bundles to be stored from 13,840 to 2,400 bundles.

25 Required by law, the prospective dry cask storage site has to complete all land and water conservation requirements set by local government. Taipower submitted a licence request to the New Taipei City after finishing ground preparations for dry cask in Chin San nuclear power plant in July 2013 (Lai 2013).

facilities and from any activities involving sensitive nuclear technologies.[26] The agreement was renewed on 6 January 2014 and entered into force on 22 June 2014.[27] For the first time, the agreement allows irradiated source material or special fissionable material to be transferred from Taiwan to France, or other countries or destinations for storage and reprocessing. A similar provision was recently included in Section 123, and in an agreement between the United States and the United Arab Emirates.

With the backing of the new US–Taiwan 123 Agreement, as it is called, Taipower then announced a tender invitation for reprocessing on the last working day before the lunar New Year holiday (17 February 2015), fully aware that no budget was allocated for this project. Taipower's behaviour met objections from both sides of the Legislative Yuan. A legislative resolution forbids tender soliciting unless a reprocessing budget passes the review. However, a tender request for reprocessing project bids remained on the public procurement site with a due date of 9 April 2015. Only after mounting public uproar did Taipower and the Ministry of Economic Affairs retract the announcement on 1 April 2015.

Generally speaking, a legally binding gold standard in the nuclear cooperation agreement will help establish a global precedent that enrichment and reprocessing are not necessary for a civil nuclear program. However, a substantial amount of high-level waste will remain even after reprocessing, so it must be dealt with carefully. Reprocessing does not solve the current high-level waste dilemma; in a global perspective, reprocessing only adds to the existing stockpile of fissile materials, increasing the risk of their becoming prey to terrorists and intensifying global vulnerability. The only advantage of reprocessing is to postpone high-level waste problems by two to three decades; the odious problems are then offloaded to future generations.

---

26 Under the 1955 agreement, sensitive nuclear technology is defined as any information or facility 'designed or used primarily for uranium enrichment, reprocessing of nuclear fuel, heavy water production, or fabrication of nuclear fuel containing plutonium'. These processes have the potential to be used not only for civil power generation but also for building a nuclear weapon.

27 The agreement is between the American Institute in Taiwan, which represents US interests in the absence of an embassy, and the Taipei Economic and Cultural Representative Office (TECRO) in the United States. Under the *Taiwan Relations Act 1979*, the United States concludes executive agreements such as Section 123 with TECRO.

Conclusion

The aforementioned absurdities and travesties of decision-making—the apparent abandonment of assigned responsibility by the regulator and the over-zealous pursuit of nuclear energy of the operator—are reminiscent of failed weapons programs. In the early 1960s, nuclear engineering was a glorified and prestigious discipline, in which only the brightest students had the chance to participate. There were ample job opportunities waiting for them in research institutes, government, and utility companies. The government's move away from military applications cut job opportunities. Nuclear accidents in Chernobyl and Three Mile Island turned many prospective students away. To boost its undergraduate recruits, the Department of Nuclear Science at the National Tsing Hua University had its name changed twice, in 1995 and 1997, finally becoming the Department of Engineering and System Sciences.

Most of the senior staff in the AEC and Taipower were the top students, who at the start of their careers were full of enthusiasm for nuclear energy. Their career advancement was unfortunately stopped by domestic and international circumstances. Some felt they had been victimised and became cynical. Some were angry with the 'arrogant' Americans who had destroyed years of hard work, in a supposedly secret weapons program. Some were over-confident and eager to prove that they could exceed the capabilities of their US counterparts, such as those who altered GE designs for the Lungmen nuclear power plant. Overall, these workers seemed to have suffered from a subtle 'Boxer mentality'.[28] On the other hand, with the increased difficulty of recruiting fresh talent, jobs were left to fewer and less qualified personnel to handle all of the increasingly complicated operations and regulations. Unskilled workers were routinely assigned to power plant maintenance and waste management, as in the Orchid Island low-level waste repackaging program. Once projects were contracted out, Taipower showed little interest in understanding the workers' qualifications, workplace safety, or working conditions. All nuclear-related checkups were left to a handful of AEC inspectors.

---

28 'Boxer mentality' describes under-equipped people or groups of people who resort to irrational and rudimentary behaviour against long-term exploitation, humiliation, or suppression. It is a widely used phrase in Chinese-speaking society, derived from the Boxer Rebellion in China (1898–1901), which opposed Western imperialism and Christianity. The Boxers were barely armed but claimed to possess supernatural protection from firearms. They were manipulated by the Qing dynasty empress to declare war on foreign powers.

Taiwan's regulator, the AEC, was established in 1955 basically as a liaison office for international communications. On paper, the AEC is in charge of everything related to nuclear energy except operations, from promoting nuclear research and development, reactors, and nuclear fuel assessment, to radiation monitoring and licensing. In reality, the military ran the CSSI, taking orders directly from the top, and making key decisions on nuclear research and development. Even after the INER moved from the CSSI to the AEC, the latter remained in a supporting role.

Military involvement was severed after the defection of the INER's deputy director Chang, the death of President Chiang Ching-Kuo, and the total destruction of Taiwan's nuclear weapons program in early 1988. Taipower took over the leading role of promoting nuclear energy. The rhetoric of 'only nuclear energy can save Taiwan from electricity shortage' has been used repeatedly in the past, such as in the Lungmen nuclear power plant revival, during nuclear debates (Lin 2011), and in the months leading up to the general election (Huang 2015). In 2015, Taipower issued stern warnings on potential electricity shortages in early summer (Chen 2015). These warnings were undercut when it was found that power plants providing as much as 2.9 GW were scheduled for routine maintenance between May and September, a time of peak demand (Wang 2015).[29]

Since the DPP gained control of the government in 2000, earlier hopes of introducing eight more nuclear reactors faded away. For nuclear proponents, having both the Lungmen nuclear power plant operational and the lifetime extension of existing reactors was their second-best option. President Ma's reluctant decision to terminate the Lungmen project in April 2014 left them with only one option for 'nuclear growth'—the lifetime extension of existing reactors. Even the AEC chair stated that reactor lifetime extensions are necessary to ensure enough electricity supply (Tang 2014).

President Tsai's promise of going nuclear free by 2025 means the end of reactor lifetime extension (C.-W. Huang 2014).[30] But even with spent fuel pools full, Taipower is unwilling to shut reactors down before the expiration dates, as this might be seen as a total retreat for the nuclear

29 For real time information on power production, see Taiwan Power Company (n.d.).
30 Mr Chang Jia-Ju, Minister of Economic Affairs, wondered whether the public would accept blackouts in the event that Taiwan goes nuclear free.

industry. Perhaps the situation is so unbearable for nuclear proponents that they are willing to put the whole society at risk to prevent it from happening.

Secrecy is another problem. It was extremely important to keep the nuclear weapons program moving and free from foreign meddling. Nothing should be revealed to outsiders. To conceal the true intentions, or even to mislead foreign counterparts, a misinformation campaign was initiated. Habitual denial became a reflex. Public health, safety, credibility, and other societal issues became a distant second priority. Some level of sacrifice was deemed necessary. The AEC did whatever policy required of it.

Although the nuclear weapons program was axed nearly three decades ago, the culture of secrecy, denial, and deceit still prevails. As can be seen from the examples provided, most AEC executives still maintain old habits—willing to support prevailing behaviour rather than independently assert the AEC's regulatory powers. The AEC thus provided outdated information in the Taiwan stress test national report, chose to be ignorant of the Orchid Island repackaging conditions, understated the condition of the Lungmen containments, and concealed the Ming Shan Villa records and others. The welfare of ordinary citizens did not and does not enjoy high priority; maintaining the status quo comes first. But, as a result, distrust in government grows, creating further barriers to policy implementation.

In practice, this type of insular organisational culture encourages negligence in work places, weakens safety regulations, invites accidents, and puts the entire society at risk. The many details of dangerous events came to light owing to hundreds of insiders and volunteers who provided key evidence of misconduct to prevent the situation from getting worse, perhaps just in time. But dependence only on whistleblowers and volunteers to correct wrongs is not healthy for Taiwan politically or socially. Moreover, it is unreliable as a check against misinformation and negligence, being far too risky as a guarantor of public safety.

Now is a critical time for the nuclear industry. On the one hand, some industrialised countries have decided to go nuclear free after the Fukushima disaster. On the other hand, many less technically advanced and less transparent countries plan to introduce or expand civil nuclear programs. Managing nuclear power is much more complicated than other types of energy utilities. It requires constant care and vigilance in every respect and can only be sustained by a large number of well-qualified human resources. Many established nuclear countries are already facing a rapidly shrinking supply of well-trained personnel to maintain routine operations, as well as to handle nuclear waste and decommissioning. For the new entrants to nuclear power, there are increasing worries that they may not have sufficient technical capabilities to oversee the construction and safe management of nuclear power. The more important question is whether or not a system of checks and balances is well-established in their national framework. Some of the newcomers have a political system similar to that of Taiwan's a few decades ago. If the regulators are not given enough authority and the operators merely want to get by, Taiwan's past experiences will likely be repeated, or worse.

In order to keep all the parties who participate in nuclear power energy projects vigilant and responsible, a frequent suggestion is that all governments, including Taiwan's, should build transparency into their decision-making mechanisms, open up documents to all, and invite outsiders to scrutinise their work. An international framework for transparency in nuclear power will not only encourage collaboration and cross-national information exchange, but will also help contain potential nuclear weapons ambitions.

References

AEC (Atomic Energy Council), 2002. Brief on quality control of manufacturing reactor base level 2 to 5 of the KMNP unit one. Yonghe City: AEC, 10 June.

AEC (Atomic Energy Council), 2007. Regulatory Committee Report. Yonghe City: AEC, April.

AEC (Atomic Energy Council), 2008a. Regulatory Committee Report. Yonghe City: AEC, February.

AEC (Atomic Energy Council), 2008b. Regulatory Committee Report. Yonghe City: AEC, April.

AEC (Atomic Energy Council), 2008c. Regulatory Committee Report. Yonghe City: AEC, January.

AEC (Atomic Energy Council), 2009. Regulatory Committee Report. Yonghe City: AEC, April.

AEC (Atomic Energy Council), 2010. Regulatory Committee Report. Yonghe City: AEC, February.

AEC (Atomic Energy Council), 2011. LMNP unit number one prefueling joint preparatory inspection with USNRC. AEC Report No. NRD-LM-100-05. Yonghe City: AEC, 18 July.

AEC (Atomic Energy Council), 2012. Investigation report on misconducts in Orchid Island LLW repackaging. Atomic Energy Council Report. Yonghe City: AEC, 7 November.

AEC (Atomic Energy Council), 2013. Taiwan stress test national report. 28 May.

AEC (Atomic Energy Council), 2014. Containment strength OK. Press release, 30 April.

AEC (Atomic Energy Council), 2016. Dry storage control: Regulatory dynamics. www.aec.gov.tw/核物料管制/管制動態/核電廠放射性廢棄物--6_48_169.html (accessed 7 September 2017).

Albright, David, and Corey Gay, 1998. Taiwan: Nuclear nightmare averted. *Bulletin of the Atomic Scientists* 54 (November–December): 54–60. doi.org/10.1080/00963402.1998.11456811

Burr, William, ed., 1999. *New Archival Evidence on Taiwanese 'Nuclear Intentions', 1966–1976*. National Security Archive Electronic Briefing Book No. 20. Washington, DC: George Washington University. nsarchive.gwu.edu/NSAEBB/NSAEBB20 (accessed 2 February 2017).

Central News Agency, 1991a. Both CSNP and KSNP tripped, China Petroleum was forced to reduce consumption. 23 April.

Central News Agency, 1991b. Rolling blackout tomorrow due to malfunction in MSNP. 25 July.

Central News Agency, 1992a. Power shortage looming for the coming summer. 15 April.

Central News Agency, 1992b. Rolling blackout in Kaohsiung this afternoon, caused by Taichung Power Plant malfunction. 4 August.

Central News Agency, 1992c. AEC: Slightly above background radiation level found in part of Ming Shan Villa. 22 August.

Central News Agency, 1992d. AEC: Radiation level in Ming Shan Villa is high. 28 August.

Central News Agency, 1994. Legislator Lu requests an investigation on whether Taipower purposely lower electricity supply. 1 June.

Central News Agency, 1995. Legislator Hsieh accuses Taipower creating electricity shortage. 12 July.

Central News Agency, 2014. NPP4 test run expected end of 2013. 27 November.

CEPD (Council of Economic Planning and Development), 1979. Assessment report of the ten major infrastructure projects. Council of Economic Planning and Development Publication. Taipei: CEPD.

Chang, Wushou, Chang-chuan Chan, and Jungder Wang, 1997. Co-contamination in recycled steel resulting in elevated civilian radiation doses: Causes and challenges. *Health Physics* 73(3): 465–72. doi.org/10.1097/00004032-199709000-00004

Chen, Wei-Ting, 2015. Economic Ministry: Electricity supply will be tight in coming May. Central News Agency, 16 April.

Chiu, I-jun, Hsiu-ting Wang, and Ly-zen Liu, 2013. Orchid Island has the highest cancer death rate in Taiwan. *Liberty Times*, 7 June.

Chung, Yun Shuan, 2009. First time ever, application for CSNP lifetime extension submitted. Central News Agency, 20 October.

Control Yuan, 1994. AEC misled public and authorities on Ming Shan Villa incident, seriously damaging government reputation. Control Yuan Publications no. 5667. 21 June.

Control Yuan, 2012. Evaluation on the electricity reserve margins set by Taiwan Power Company. Investigation Report no. 101000001. 4 January.

CTI, 2012. Embarrassing! Legislator uses photos taken three years ago. CTI Television, 26 October.

Democratic Progressive Party Principles and Guidelines, n.d. Article 64. www.dpp.org.tw/upload/history/20160728102222_link.pdf (accessed 6 March 2017).

ENSREG (European Nuclear Safety Regulators Group), 2013. EU peer review report of the Taiwanese stress tests. 13 November.

Environmental Protection Administration, Taiwan, 2008. Analysis of environmental variations on the spent fuel interim storage project of the first nuclear power plant. Document number 0971283A.

Environmental Protection Administration, Taiwan, 2009. Analysis of environmental variations on the spent fuel interim storage project of the second nuclear power plant. Document number 0980243A.

Gi, GinLing, 1998. Additional nuclear reactors may be part of Taipower's future electricity plan. Central News Agency, 9 July.

Ho, HungRo, 2008. Adding reactors to the existing nuclear plants should be considered. Central News Agency, 5 April.

Hsu, Kuang-Jung, 2013. Comments on Taiwan stress test report by Taiwan environmental protection union. 25 September.

Huang, Chiaw-Wen, 2014. Without LMNP, lifetime extensions of existing reactors must proceed – Minister said. Central News Agency, 17 April.

Huang, Chiaw-Wen, 2015. Economic Ministry warns possible blackouts next year if go nuclear free. Central News Agency, 26 January.

Huang, Chiaw-Wen, 2016. Chair Huang denied Taipower changed its power shortage assessment under nuclear free condition. Central News Agency, 23 March.

Huang, Min-Shi, 2014. Mothball and terminating reactors construction of LMNP. Central News Agency, 27 April.

INER (Institute of Nuclear Energy Research), 2015. News (in Chinese). Press Release. www.iner.gov.tw/index.php/2015-11-17-03-21-34/3400/40-核研所承接台電委託計畫，涉及球員兼裁判.html (accessed 7 September 2017).

Lai, Hsiaotung, 2013. Land and water conservation license not in sight, CSNP dry storage plan is put off. *Liberty Times*, 11 September.

LanYu BiWeekly, 1996. No docking for ElectricLight I–Taipower ship. 28 April.

Lee, Ming-Yang, 2008. Current status and challenges of Taiwan nuclear energy. *Scientific American* (Taiwan ed.) June: 88–9.

Lee, Tsong-You, 1984. Possible origin of radioactive steel. *China Daily Evening News*, 27 December.

Lee, Tsong-You, 2011. More NPP4 design alternations, Taipower executives were sent to court. *China Daily News*, 20 June.

Legislative Yuan, Atomic Energy Council, 1985. Meeting report of the 75th Session. 11th Meeting, 9 April.

Lin, Mon-Ruh, 2016a. Under nuclear free condition, power shortages are rare, Taipower says. Central News Agency, 18 January.

Lin, Mon-Ruh, 2016b. Warning! Low on electricity reserve. Central News Agency, 30 May.

Lin, Shaw-yu, 1984. Army has nothing to do with radioactive contaminated steel. *China Daily Evening News*, 27 December.

Lin, Su-Yuan, 2011. Economic Ministry: Blackouts are expected in eight years of nuclear free. Central News Agency, 25 March.

Liu Ly-Zen, 2002. Where Taiwan's nuclear waste will end up. *Liberty Times*, 7 July.

Lo, Ru-Lan, 2002. Legislators suggest government purchase lanyu for permanent LLW site. *China Daily News*, 22 November.

Now News, 2011. Taiwanese nuclear plants are as safe as Buddha sitting on platform? 15 March.

Shih, Hsiu-Chuan, 2013. Nuclear referendum on ice: KMT caucus. *Taipei Times*, 27 September.

Small and Medium Enterprise Administration, Ministry of Economic Affairs, n.d. Taiwan's economic development. www.moeasmea.gov.tw/ct.asp?xItem=72&CtNode=263&mp=2 (accessed 7 September 2017).

Taipower, n.d.a. Peak load and standby capacity rate of Taipower. data. gov.tw/node/8307 (accessed 7 February 2017).

Taipower, n.d.b. What are feedbacks to local communities after the low level nuclear waste site established. www.taipower.com.tw/content/ new_info/new_info-e47.aspx?LinkID=18 (accessed 5 March 2017).

Taipower, 2005. NPP1 interim spent fuel storage plan: Report of environmental changes and countermeasures: Environmental Impact Assessment No. 0960067020.

Taiwan Power Company, n.d. Household electricity consumption guide. www.taipower.com.tw/content/new_info/new_info_in.aspx? LinkID=27 (accessed 2 February 2017).

Tang, Jian Ling, 2014. AEC: U-turn for not extending reactor lifetime policy. *Liberty Times*, 4 November.

Tang Jia-lin, and Yi-lo Chien, 2014. NPP4 containment leaks. *Liberty Times*, 12 May.

Tsai Chuen-horng, 2013. Atomic Energy Council special report on overall geological survey of operating NPPs and the fourth NPP to the economic committee, Legislative Yuan. Report to the Economic Committee, Legislative Yuan, 17 April.

Tung Sen News, 2011. Worry about radiation? AEC: Koreans do not worry even though they are closer. 14 March.

United Daily News, 1976. Executive Yuan solemnly declares no intention to develop nuclear weapon. 17 September.

Wang, Fong, 2010. Israeli nuclear expert secretly assisted Chiang Kai-Shek developing nuclear weapon. *Yazhou zhoukan* [Asia Weekly], 18 April.

Wang, To-Far, 1991. Does LMNP have to be built? *China Times*, 7 January.

Wang, To-Far, 2015. The curious illness of power shortage. *Taiwan People News*, 6 June.

Wang, Yu-Lin, 1996. Radioactive Formosa: Unearth the radioactive waste scandals. Radiation Safety Improvement Organisation.

Wang, Yu-Su, and Bin Wei, 2008. NPP4 hidden dangers: Taipower altered nuclear power plant design without permission. *Apple Daily News*, 8 February.

Wei, Su, 2005. Nuclear power plant lifetime extension, Atomic Energy Council will thoroughly check its safety. Central News Agency, 10 January.

Weiner, Tim, 2007. *Legacy of Ashes: History of the CIA*. New York: Doubleday.

Yen, Zo-jin, Hsiao-Kuang Shih, Wen-Hua Hsieh, Jia-Chi Lin, and Ji-Hsien Fang, 2011. Radioactive dusts make turns? Scholars question AEC instrument sensitivity. *Liberty Times*, 31 March.

7

Enhancing nuclear energy cooperation in ASEAN: Regional norms and challenges

Mely Caballero-Anthony and Julius Cesar I. Trajano

Abstract

The Fukushima nuclear disaster in 2011 did not dampen plans by Southeast Asian countries to develop nuclear power plants, despite safety concerns. The strong interest in nuclear power development is being driven by strategic considerations as states view nuclear power as an alternative energy source that can help address the dual objectives of energy security and mitigation of climate change effects. Our chapter examines the prospects for the Association of Southeast Asian Nations (ASEAN) to build a stronger regional normative framework to promote nuclear safety and security and prevent the proliferation of nuclear weapons. In light of ASEAN's vision to establish a political and security community, we argue that member states that plan to use nuclear energy need to address critical issues such as legislative and regulatory frameworks, human resources development, radioactive waste management, nuclear safety, emergency planning, and security and physical protection. With the establishment of the ASEAN Community in 2015, we explore the prospects for strengthening the regional framework for nuclear energy in ASEAN post-2015, spearheaded by the ASEAN Network of Regulatory Bodies on Atomic Energy (ASEANTOM).

Introduction

The Fukushima nuclear crisis of March 2011 took place when the nuclear power industry in Asia was on the cusp of a period of growth (IAEA PRIS 2014).[1] However, after an initial wait-and-see period, nuclear energy development plans in Southeast Asia remain mostly in place, despite safety concerns. Some countries in the 10-member Association of Southeast Asian Nations (ASEAN) plan to integrate nuclear power into their long-term energy plans, reflecting their governments' view of nuclear power as an alternative energy source that can help address the dual objectives of energy security and mitigation of climate change effects (Nian and Chou 2014).

To ensure that their energy supplies are secure, affordable, and environmentally sustainable, ASEAN members are moving toward diversifying their energy mix, reducing their overdependence on fossil energy, and gradually integrating nuclear power into their long-term energy plans (see Table 7.1).

Table 7.1 ASEAN electricity generation by source

Fuel	Share	
	2013 (%)	2040 (%)
Coal	32	50
Oil	6	1
Gas	44	26
Nuclear	0	1
Renewables (hydro, geothermal, bioenergy, and others)	18	22
Total	100	100

Source: IEA (2015: 39).

Nuclear energy plans in ASEAN

Several countries in Southeast Asia have been articulating their interest in using nuclear power, as they intend to strengthen their energy security though diversification of their energy mix. Vietnam used to be the lead

---

1 In 2014, there were 439 nuclear reactors operating in 31 countries. Two-thirds of the 69 nuclear reactors under construction are in Asia, led by China, India, and South Korea (IAEA PRIS 2014).

driver of nuclear power development in ASEAN, from 2009 when it decided to build its first nuclear power plant, until November 2016 when its government decided to scrap its plan primarily due to the soaring cost of the project. Prior to the cancellation, Vietnam's 2,400 megawatt (MW) Ninh Thuan 1 nuclear power plant was scheduled to be operative by 2028/29 (after several delays), while the 2,000 MW Ninh Thuan 2 nuclear power plant was set to be commissioned by 2030. Russia's state-owned nuclear firm Rosatom was tipped to build Ninh Thuan 1, while a consortium of Japanese nuclear firms led by Japan Atomic Power was considered for the construction of Ninh Thuan 2 (Pascaline 2016). Nonetheless, although the government already decided to scrap its nuclear power plant project, it will still continue 'promoting' nuclear power (Kyodo News 2016). In this regard, Vietnam plans to build a new research reactor, also known as the Centre for Nuclear Energy Science and Technology, to further enhance the skills and technical know-how of its nuclear professionals and students.

Indonesia has long been preparing for the possible utilisation of nuclear energy with the establishment of three nuclear research reactors: Reactor Triga 2000 in Bandung, established in 1965; Reactor Kartini 250-kilowatt (kW) in Yogyakarta (1979); and RSG-GAS 30 MW in Serpong (1987). In 2006, the International Atomic Energy Agency (IAEA) declared that Indonesia was ready to make a knowledgeable commitment to a nuclear power program, although no government decision has yet been made as to whether Indonesia will proceed to build the nuclear power plants. The Ministry of Energy and Mineral Resources issued a 'White Paper of Indonesia NPPs [nuclear power plants] 5000 MWe in Bangka Belitung 2014–2024', which called for the introduction of nuclear power in order to address Indonesia's rapidly growing energy consumption. Indonesia's electricity demand is projected to increase to 150 gigawatts (GW) by 2025. The contribution of this new energy source is seen as a major energy alternative that can boost the country's power supply (Taryo 2015).

The National Nuclear Energy Agency of Indonesia (BATAN) has recommended that a nuclear power plant be established by 2027. BATAN conducted feasibility studies for possible nuclear power plant sites in Bangka-Belitung Island, West Kalimantan, and Muria and Banten in Java. Bangka-Belitung Island, near Sumatra, has been identified as the site of the country's first nuclear power plant since the island is not within the country's earthquake and volcanic zones. While no official decision has been made on the use of nuclear power, a nationwide public survey commissioned by BATAN in 2014 reported that 72 per cent of Indonesians agree that nuclear power plants should be setup in the country.[2]

In Malaysia, increasing energy needs are cited to justify development of nuclear power. In 2009, Prime Minister Najib Razak announced a plan for a small-scale nuclear reactor. In 2010, the energy mix in peninsular Malaysia consisted of gas (54.2 per cent), coal (40.2 per cent), hydro (5.2 per cent), and oil (0.4 per cent) (Ramli 2013). Nuclear energy development is mentioned in the Eleventh Malaysia Plan 2016–2020, but without a projected percentage of its total energy mix (see Economic Planning Unit 2015).

Nuclear energy has always engendered strong public opposition in Malaysia.[3] Civil society groups have expressed their objections to the nuclear option in a number of forums, including some organised by the government.[4] Despite this opposition, the Malaysian government does not completely rule out the nuclear option. In July 2014, Dato' Mah Siew Keong, minister in the prime minister's department, stated that the government would conduct a feasibility study aimed at building a nuclear power plant in 10 years' time and would carry out a comprehensive study including assessing public acceptance, with input from experts and non-governmental organisations (NGOs) (Bernama 2014). The government has already begun conducting the comprehensive feasibility study, although there is no certainty as to when it will be concluded and publicly released.

In November 2016, Philippines President Rodrigo Duterte gave his go-ahead to the Department of Energy to proceed with a feasibility study to reactivate Bataan nuclear power plant to generate 621 MW of electricity, a turnaround from an earlier stand rejecting the use of nuclear energy during his presidency. But he gave clear instructions to pay special attention to the safety and security aspects of operating the 30-year-old power plant (Lucas 2016).

---

2   Interview with BATAN official, Singapore, 30 October 2015.
3   For further details, see Consumers Association of Penang (n.d.), and Care2 Petitions (n.d.), which is supported by the Malaysian Coalition Against Nuclear.
4   Email interview with a Consumers Association of Penang worker, 10 September 2014.

In May 2016, Cambodia and Russia signed two deals to set up a nuclear energy information centre and a joint working group on peaceful uses of atomic energy. According to Rosatom, the nuclear energy information centre will help Cambodians, especially students, better understand nuclear energy principles and important developments in nuclear energy and industry. The memorandum on a Cambodia–Russia joint working group on the peaceful uses of atomic energy stipulates that the parties will hold regular meetings between experts from the two countries to define and implement joint projects. Russia will provide expertise, research, and training under the terms of the agreement. Although Cambodian Prime Minister Hun Sen had previously stated that his country would not go nuclear, there are repeated calls from within the government to consider nuclear power, prompted by similar interest and moves among its Southeast Asian neighbours. The memorandum and the information centre may lay the groundwork for a nuclear power project in the future if Cambodia chooses to proceed (Tan 2016).

Against these developments, this chapter examines the prospects for building a stronger regional normative framework in promoting nuclear safety and security and preventing proliferation of nuclear weapons in the region. We argue that while ASEAN has already established regional cooperative norms on nuclear safety, security, and safeguards (3S), the extent to which this normative framework is upheld and enhanced in the region still mainly depends on how member states interested in utilising nuclear energy address critical infrastructure issues during the preparatory stages of their respective nuclear power programs. The existing nuclear infrastructure issues, if they remain unaddressed, can pose challenges to these ASEAN norms. We elucidate some of the major nuclear infrastructure issues specifically in three ASEAN members—Vietnam, Indonesia, and Malaysia. These issues are legislative and regulatory frameworks, human resources development, radioactive waste management, nuclear safety, emergency planning, and security and physical protection. With the establishment of the ASEAN Community in 2015, we explore the prospects for strengthening the regional framework for nuclear energy in ASEAN post-2015, spearheaded by the ASEAN Network of Regulatory Bodies on Atomic Energy (ASEANTOM).

Enhancing ASEAN's framework on the safe and peaceful use of nuclear energy

What are ASEAN's norms on the peaceful use of nuclear energy that must be observed by member states? ASEAN first articulated regional norms on nuclear safety and non-proliferation in the 1995 Treaty on the Southeast Asia Nuclear Weapon-Free Zone (the Bangkok Treaty). While this treaty is primarily intended to prohibit member states from developing, manufacturing, and possessing nuclear weapons, it contains several provisions that recognise each state's right to use nuclear energy for peaceful purposes, in particular for economic development and social progress. As such, it establishes the regional normative framework that guides member states should they decide to pursue nuclear energy.

The major regional norms established by the Bangkok Treaty are the following: a state pursuing nuclear energy must (1) use nuclear material and facilities within its territory exclusively for peaceful purposes; (2) subject its nuclear program to rigorous safety assessment, conforming to guidelines and standards recommended by the IAEA for the protection of health and minimisation of danger to life and property; (3) inform fellow members, if requested, of the outcome of the safety assessment; (4) uphold the international non-proliferation system through strict adherence to the Nuclear Non-Proliferation Treaty and the IAEA safeguard system; and (5) dispose of radioactive wastes and other radioactive material in accordance with IAEA standards and procedures (ASEAN 1995).

ASEAN members have underscored adherence to these norms at annual ASEAN leaders' summit meetings and particularly at the ASEAN ministers of energy meetings, in which ministers accentuate the importance of enhancing capacity-building activities on civilian nuclear energy and pursuing regional nuclear safety cooperation. In the Phnom Penh Declaration on ASEAN: One Community, One Destiny (ASEAN 2012b), ASEAN leaders agreed to:

> develop a coordinated ASEAN approach that would contribute to global undertakings to improve nuclear safety, in cooperation with the IAEA and other relevant partners, as well as promote and uphold IAEA standards of safety and security in the development of nuclear energy for peaceful use (ASEAN 2012a).

The ASEAN Political-Security Community Blueprint 2025 (Section B.5.2) also endorses the development of a regional approach to strengthen nuclear safety, in coordination with the IAEA and other relevant international organisations. But, more importantly, the blueprint promotes the strengthening of ASEANTOM so that it can effectively lead the development of the ASEAN regional approach to nuclear safety (see ASEAN Secretariat 2016).

The emerging role of ASEANTOM in strengthening regional cooperation

ASEAN leaders likewise encourage the development of a:

> network amongst nuclear regulatory bodies in Southeast Asia which would enable regulators to exchange nuclear-related information and experiences on best practices, enhance cooperation and develop capacities on nuclear safety, security and safeguards (ASEAN 2012a).

In this regard, in 2011, Thailand's Office of Atoms for Peace (OAP) proposed the creation of ASEANTOM in an informal consultation and received positive comments from ASEAN regulatory bodies. The proposal was later presented to the ASEAN Summit 2011 and received a warm welcome from member states (ASEANTOM 2014a). ASEANTOM was designated in 2015 as an ASEAN sectoral body under the ASEAN Political-Security Community Pillar in Annex 1 of the ASEAN Charter (ASEAN 2015). Its activities are now reported to foreign ministries of ASEAN members and have been recognised in the ASEAN Summit chairman's statement (Biramontri 2016). Realising that they cannot uphold nuclear safety individually and in isolation in the aftermath of the Fukushima disaster, ASEAN members acknowledged that they need to cooperate and share information through ASEANTOM as part of the building blocks of regional frameworks to institutionalise the culture of nuclear safety and security.

ASEANTOM focuses on four issues of mutual interest: emergency preparedness and response, environmental radiation monitoring, nuclear security, and nuclear safety (Biramontri 2016). In 2012, a preliminary meeting among ASEAN regulators was hosted by Thailand's OAP to finalise ASEANTOM's Terms of Reference to enhance the growth of knowledge and resources to ensure the 3S of peaceful nuclear energy applications. ASEANTOM has already conducted three annual meetings. The first meeting (Phuket, 2013) facilitated information exchanges and cooperation in the area of nuclear 3S among the member states, and set up the network's work plan. The second meeting (Chiang Mai, 2014) allowed for a review of activities conducted during the past year and discussed further activities under the work plan for 2015–16. These activities included a number of regional workshops and training courses on emergency preparedness and response as well as on nuclear security culture, safety, and management (ASEANTOM 2014a). The third meeting (Kedah, 2015) designated ASEANTOM as a sectoral body under the ASEAN Political-Security Community and also assigned it to be the key point of contact with the IAEA to facilitate greater cooperation and collaboration on issues related to nuclear 3S (Biramontri 2016).

In 2014, Thailand and Vietnam co-hosted a regional meeting on radiation environment monitoring cooperation aimed at facilitating information exchange and seeking opportunities to establish a regional early warning network and a regional data centre; Malaysia hosted a workshop on nuclear regulation establishment and the current and future national regulatory and legal framework; and Singapore hosted a meeting between Euratom and ASEANTOM in which the former shared its relevant experiences (ASEANTOM 2014a). Under the ASEANTOM framework, Malaysia and Thailand since 2015 have been co-hosting annual nuclear security border exercises, including tabletop and field exercises in their shared borders, and have been involving nuclear regulatory bodies, customs, police, and emergency response teams. All ASEAN members are invited to participate in these exercises that test their capability to jointly interdict illicit trafficking of radioactive materials. Furthermore, ASEANTOM has two ongoing projects, with assistance from the IAEA and the European Union (EU), to strengthen joint nuclear emergency preparedness and response cooperation in ASEAN: a Regional Cooperation Project Concept in South East Asia to Support Regional Environmental Radioactivity Database & Nuclear Emergency Preparedness and Response assisted by the IAEA; and Enhancing Emergency Preparedness and Response in ASEAN: Technical Support for Decision Making assisted by the EU (Biramontri 2016).

ASEANTOM has likewise facilitated inter-regional cooperation between ASEAN and the EU, particularly in regard to nuclear safety and emergency preparedness. In early 2016, the European Commission completed its feasibility study on enhancing regional cooperation within ASEAN on radiological and nuclear emergency preparedness and response. The study was undertaken by the Joint Research Centre of the European Commission, in cooperation with nuclear safety regulatory authorities in six ASEAN members.[5] The study is part of the EU's Instrument for Nuclear Safety Cooperation. It contains a regional strategy, including an action plan for its implementation. It recommends that ASEANTOM implement the action plan with support, inter alia, from the EU and the IAEA. The need for joint emergency preparedness and response in the region has become more urgent due to recent developments, in particular in the context of nuclear power plants being constructed near ASEAN's border in neighbouring countries[6] and plans to use nuclear energy in some ASEAN member states. However, as expressed by some regulatory officials in the region at the International Dissemination Workshop on Enhanced Regional Cooperation on Nuclear Emergency Preparedness and Response, held in Kuala Lumpur in February 2016, one major challenge for ASEANTOM in implementing the regional strategy is the lack of adequate logistical and financial resources of national regulatory bodies.

ASEAN's nuclear 3S regional frameworks and initiatives outlined above clearly demonstrate that its members recognise the importance of upholding the regional norms on the peaceful use of nuclear energy through regional cooperation. However, a regional normative framework on the use of nuclear energy is clearly not sufficient if there are still structural challenges to the preparatory nuclear infrastructure in each of the member countries that are considering utilising nuclear energy. The particular challenges are the availability of human resources, adequate regulatory and legislative frameworks, and institutionalised national radioactive waste management strategies. For instance, if a member state is unable to institutionalise a comprehensive regulatory framework, ASEAN's norms on nuclear safety and security, as well as adherence to global non-proliferation obligations, may not be implemented. A competent nuclear regulatory body typically addresses proliferation concerns by inspecting and verifying that licensees are meeting all applicable safety and security requirements related to material control and accounting, information security, waste management, emergency preparedness, fire safety, radiation safety, and

---

5   The six ASEAN regulatory bodies were the Nuclear Energy Regulatory Agency of Indonesia (BAPETEN), the Atomic Energy Licensing Board (AELB, Malaysia), the Philippine Nuclear Research Institute, the National Environment Agency (Singapore), the OAP (Thailand), and the Vietnam Agency for Radiation and Nuclear Safety (VARANS).

6   China has begun operating three nuclear power stations near its border with Vietnam.

physical protection (US NRC 2015). If the regulatory body cannot fully monitor nuclear facilities, misuse of nuclear materials may occur, posing not only safety and security challenges, but also nuclear proliferation risks. The next section briefly outlines nuclear energy plans in ASEAN and highlights some of the important issues, particularly regarding nuclear safety and security.

Legislative and regulatory frameworks

Nuclear industry players, including exporters of nuclear technology, have claimed that necessary improvements have been made in nuclear safety all over the world since the Fukushima accident of 2011. But nuclear newcomers in Southeast Asia can still derive valuable lessons from states with nuclear power programs when it comes to ensuring safe commissioning and operation of nuclear power plants.

In Southeast Asia, Vietnam, which had the most advanced nuclear power plant plan in the region until the plan was scrapped, has yet to legislate a framework on regulatory independence. The Vietnam Agency for Radiation and Nuclear Safety (VARANS) currently serves as a nuclear regulatory body. Since 2012, Vietnam has been taking steps to develop a more independent regulatory agency. The Ministry of Science and Technology and Japan's Ministry of Economy, Trade, and Industry signed an agreement to enhance the technical and safety competence of Vietnam's nuclear regulatory body. One of the proposed amendments to Vietnam's Atomic Energy Law is to make VARANS an effectively independent regulatory body, since it is only 'partly independent' under the Ministry of Science and Technology, which is the leading agency promoting nuclear energy in Vietnam.[7] VARANS's independence is limited to regulating radioactive sources and materials, mostly for industrial, educational, and medical applications; it cannot regulate nuclear power plant safety and security aspects. The government has yet to act on proposed amendments

---

7　The IAEA's director-general, Amano Yukiya, has repeatedly emphasised that regulatory independence leads to greater transparency and improves public acceptance (Amano 2015). One of the key lessons from Fukushima has been the need to have an independent nuclear safety regulatory body. The 1994 Convention on Nuclear Safety and the IAEA General Safety Requirements call for the establishment of a regulatory body and the need for its separation, or independence, from the promoters of nuclear technology, such as government ministries (IAEA 1994). The primary reason for having an independent regulatory body is to ensure that judgements are made and enforced without pressure from interests that may conflict with safety and security.

to VARANS and it remains uncertain whether a Vietnamese regulatory agency fully independent of ministries promoting nuclear energy can be established.[8] The Vietnamese government was not keen to make VARANS a fully independent regulatory body as it believed that inter-agency cooperation, involving all concerned ministries and VARANS, was far more important at this stage to make the first nuclear power plant project successful.[9]

A national steering committee was set up by the Vietnamese government to oversee the project management of its nuclear power plant program. The committee is composed of the Ministry of Trade and Industry with Electricity of Vietnam as the attached agency; the Ministry of Science and Technology with VARANS, the Vietnam Atomic Energy Institute (VINATOM), and the Vietnam Atomic Energy Agency as attached agencies; and the Ministry of Education and Training. The management of the committee, however, is not efficient and members do not meet regularly as the deputy prime minister (the chair) has been extremely busy with other tasks.

Contrary to the IAEA's prescription, there is no Nuclear Energy Program Implementing Organisation (NEPIO) in Indonesia to lead and manage the effort to consider and develop a nuclear power plant program.[10] Instead, several institutions such as BATAN, the Nuclear Energy Regulatory Agency of Indonesia (BAPETEN), the Ministry of Energy and Mineral Resources, the Ministry of Environment, and the Ministry of Research and Technology carry out separate functions to prepare for the establishment of nuclear power plants (IAEA 2013). This arrangement may compromise the regulatory impartiality of BAPETEN since it is part of the multi-agency nuclear power plant preparatory program and may be involved in activities leading to possible establishment of a nuclear power plant. In an ideal situation, BAPETEN should have objective regulatory oversight of these preparatory activities.

Another issue in Indonesia is the delegation of nuclear power plant–related responsibilities to different agencies, which requires coordination. The absence of a dedicated steering committee signifies a lack of

---

8   Interview with a VARANS official, Hanoi, 8 August 2014.
9   Interview with a VINATOM official, Singapore, 30 October 2015.
10   NEPIO implements 'a nuclear power programme, which may be preparing for a decision to implement, coordinating the implementation among other entities or carrying out the implementation itself' (IAEA 2008: 1).

commitment in pursuing nuclear power plants because although BATAN is the primary institution working on nuclear power as it reports directly to the president, it does not have any authority over other agencies. Inter-agency cooperation to further advance the nuclear power plant program remains weak. While it would be highly recommended to ensure that the regulatory body is completely independent from any agency promoting nuclear power, BAPETEN still needs to have robust cooperation with BATAN to fully develop its regulatory capability (Taryo 2015). More importantly, BAPETEN itself admitted that Indonesia's legislative framework is not yet in full compliance with IAEA standards (Sunaryo 2015). And while existing legal frameworks govern the potential use of nuclear power, they require amendments to incorporate some international conventions, such as the Joint Convention on the Safety of Spent Fuel Management and on the Safety of Radioactive Waste (1997).

In Malaysia, the main legislation relating to nuclear power plants is the *Atomic Energy Licensing Act 1984* (Act 304), which includes detailed provisions on radioactive materials (Bidin 2013). In 2011, the Malaysia Nuclear Power Cooperation (MNPC) was established as a NEPIO. The MNPC is under the supervision of the prime minister's department and assumed the functions of the Nuclear Power Development Steering Committee (Markandu 2013).

The Atomic Energy Licensing Board (AELB) is the assigned nuclear regulatory body in Malaysia. However, it is also part of the Nuclear Power Development Steering Committee chaired by the MNPC, which may compromise its regulatory independence since it is involved in the preparatory initiative to set up a nuclear power plant in the country. The AELB is also an agency attached to the Ministry of Science, Technology, and Innovation, which actively promotes the use of nuclear energy, and a member of the MNPC board of directors. In accordance with IAEA recommendations, a regulatory body should be completely independent of any governmental ministry that has an interest in the establishment and operation of nuclear power plants.

Nuclear safety and security measures

In terms of institutionalising nuclear safety and security measures, Vietnam, Indonesia, and Malaysia have managed to introduce several initiatives that may strengthen their commitment to upholding the nuclear 3S, although some challenges have been identified by observers and even by state agencies. For instance, Vietnam has no prior experience in utilising nuclear energy in terms of scientific and technical knowledge as well as nuclear emergency management. The concept of safety culture even within the regulatory body is not explicitly defined since public awareness of safety culture remains low. A deep understanding of issues related to the safety of nuclear power projects among Vietnamese stakeholders—such as government agencies, scientists, and communities—is still very limited (Vuong 2015). Several Vietnamese nuclear experts have voiced concerns over nuclear safety and the absence of an independent regulatory body, coupled with widespread corruption and transparency issues, and a record of poor safety standards (Ninh 2013).

The Fukushima nuclear disaster raised concerns over Vietnam's capacity to administer and regulate nuclear energy. Based on climate modelling exercises, Vietnam is often listed as one of the most vulnerable countries to the impacts of climate change, such as rising sea levels and stronger typhoons, particularly around the location of the Ninh Thuan nuclear power plant. Ninh Thuan is identified as a disaster-prone coastal province (Mulder 2006) whose coastline is vulnerable to tsunamis potentially originating from a strong tremor in the South China Sea. Despite these risks, the government remains determined to set up its nuclear power plants in Ninh Thuan.

Vietnam works closely with the IAEA to meet all international safety standards and regulatory practices. An IAEA Emergency Preparedness Review was conducted in 2012 to assess Vietnam's radiation emergency preparedness and response capabilities and to provide recommendations (Thiep 2013). Vietnam's national emergency preparedness and response plan was crafted and issued after the conclusion of the review. VARANS has begun to work with relevant national and local government agencies to elucidate a concrete emergency response and evacuation plan. However, there are still implementation challenges for the remaining IAEA recommendations. The director-general of the Vietnam Atomic Energy Agency claimed that staff in key organisations directly working on nuclear infrastructure development have not been trained systematically (Hoang 2013).

Following the IAEA recommendations, Vietnam started devising and implementing nuclear security measures, including a licensing system under VARANS for the transhipment of nuclear material and radioactive sources. The IAEA also provided most of the 12 radiation portal monitors and related systems that have been installed at three ports of Cai Mep, southeast of Ho Chi Minh City: Thi Vai, Ba Ria, and Vung Tau (Vi 2014).

In Indonesia, the plans for nuclear power plant development draw concern both domestically and internationally due to the frequent occurrence of natural disasters such as volcanic eruptions, earthquakes, tsunamis, floods, and landslides (National Agency for Disaster Response 2012). Realising the implications of such geological vulnerability, BATAN has conducted site selection processes based on IAEA guidelines (BCR no. 5/2007 on the Safety Provision for Site Evaluation for a Nuclear Reactor) and best practices from other countries (Suntoko and Ismail 2013). Several proposed sites for the nuclear power plant, such as Muria (Central Java Province) and Banten (West Java Province), have been found to be located in seismically active zones. Bangka-Belitung Island, east of Sumatra Island, is now the preferred site for the first nuclear power plant. It sits outside the Pacific Ring of Fire and has a low risk of natural disasters.[11] BAPETEN has not received any formal application from BATAN, however, suggesting that the plan to construct a nuclear power plant on Bangka-Belitung Island is still at the feasibility study stage.

In order to prepare for nuclear accidents, Indonesia has held a number of nuclear emergency exercises and drills, and Fatmawati Hospital in South Jakarta is a designated referral hospital for nuclear emergencies. Reflecting on recent natural disaster responses performed by the Indonesian National Board for Disaster Management, challenges in inter-agency coordination including division of authority, chain of command and control, and mobilisation of resources remain the source of sub-standard responses. In anticipation of such challenges, BAPETEN formed the Indonesia Center of Excellence on Nuclear Security and Emergency Preparedness in August 2014 (Hadi 2014), a special platform where BAPETEN, BATAN, police, customs, the foreign ministry, and intelligence services communicate and coordinate their efforts for nuclear security and emergency responses (Haditjahyano 2014). To strengthen nuclear security and reduce nuclear proliferation risks, Indonesia has radiation portal monitors at several ports of the archipelago (Sinaga 2012).

---

11   Interview with BAPETEN officials, Jakarta, 14 August 2014.

Since Malaysia is located outside the Pacific Ring of Fire and typhoon belt, it is less susceptible to hazards such as earthquakes, volcanic eruptions, and typhoons (Disaster Management Division of Prime Minister's Department 2011). Floods and landslides are among the few natural disasters that typically hit Malaysia (Asian Disaster Reduction Center 2011). In 2009, Malaysia completed nuclear power plant siting guidelines and, in 2011, five candidate sites were identified. The possible construction of a Malaysian nuclear power plant is still at a very early planning stage, as site selection was made based on digital mapping and no fieldwork has been carried out to date (*Malaysian Insider* 2012; AELB 2010). To boost emergency response and preparedness, the AELB established a nuclear emergency team, and first responders are located at the northern, southern, eastern, and Sabah–Sarawak parts of Malaysia (Teng 2014). The AELB has regularly conducted national radiological emergency response drills, such as the National Radiological Emergency Drill, in the event of a transport accident. It also conducted a National Field Exercise on Research Reactor Emergency Response and a tabletop exercise on Research Reactor Emergency Response in 2007.

To protect its nuclear facilities and adhere to non-proliferation norms, Malaysia is forging a close partnership with the United States through the Global Threat Reduction Initiative.[12] In February 2012, four Radioactive Sources Category 1 Facilities in Malaysia were assessed under the Initiative framework (Nuclear Security Summit 2014). Malaysia also takes part in the Global Initiative to Combat Nuclear Terrorism.[13] As part of its commitment, Malaysia hosted a tabletop exercise with Australia, New Zealand, and the United States in 2014 (European Leadership Network 2014).

Malaysia is finalising its amendments to the *Atomic Energy Licensing Act 1984* (Act 304), which would incorporate the provisions of the IAEA Convention on Physical Protection of Nuclear Material, and its 2005 Amendment Protocol; the International Convention for the Suppression of Acts of Nuclear Terrorism; and the Additional Protocol to the IAEA Comprehensive Safeguards Agreement (Nuclear Security Summit 2016).

---

12 This is a US-led initiative that aims to protect against and reduce excessive civilian nuclear and radiological materials worldwide, particularly highly enriched uranium.

13 The Global Initiative to Combat Nuclear Terrorism is an international partnership to strengthen collective capacity to prevent, detect, and respond to nuclear terrorism. Eighty-five countries take part, including Malaysia, Singapore, and Vietnam. The EU, the IAEA, INTERPOL, and the United Nations Office on Drugs and Crimes are observers.

Human resources development

There is still a tremendous need to educate young people and enhance the skills of older professionals in the nuclear field, particularly in nuclear safety and security. It was emphasised that, as some ASEAN members plan to pursue nuclear power, they need to create and maintain a pool of local nuclear professionals with actual relevant experience in the nuclear industry. Furthermore, well-trained and experienced nuclear professionals are also crucial in institutionalising competent and independent regulatory bodies. The region currently does not have enough human resources that can safely operate its future nuclear power plants.

In Vietnam, the largest challenge is human resources development, particularly in terms of specialists and experts in nuclear engineering. Since it takes years and even decades for a country to master nuclear power technology, depending on a country's existing infrastructure and technical skill base, it is not surprising that Vietnam decided to cancel the construction of its first nuclear power plant. When the project was still ongoing until 2016, several government initiatives were introduced to bolster human resources training in the nuclear field. Following IAEA recommendations made in its first Integrated Nuclear Infrastructure Review mission in 2009, Vietnam established the National Steering Committee on Human Resource Development in the Field of Atomic Energy. After the second review mission in 2012, Vietnam cooperated with the IAEA to organise an expert mission to support its efforts to develop the National Integrated Plan on Human Resource Development for its nuclear power program (Hoang 2014). In 2010, Prime Minister Nguyen Tan Dung approved the National Project for the Training and Development of Human Resources for Atomic Energy, otherwise known as Program No. 1558, with a budget of US$150 million to be spent between 2013 and 2020 (Dung 2010; Thiep 2013). The cancelled project was supposed to train 3,000 undergraduate students, 500 master's degree and doctoral students, and 1,000 teachers in atomic energy. Under this mothballed project, Vietnam had sent students overseas for nuclear energy studies (World Nuclear Association 2015).

Vietnam's Ministry of Education and Training aims to train 1,000 students from 2015 to 2020, while those studying overseas are being trained for three to five years in Russia and Japan.[14] However, a challenge for nuclear programs is the shortage of trained professionals in the construction and operation of nuclear power plants. Although Vietnam is now investing in human resources training and capacity building, criticism has been voiced about an emphasis on theory rather than practice (Ninh 2013). One major concern is the immediate impact on manpower development of the cancellation of the construction of the Ninh Thuan nuclear power plant. Thirty students trained in Russia are expected to have returned home in 2016 and, by 2018–19, additional students will be returning from abroad. Three hundred students are currently being trained in Russia, while 15 are studying in Japan, all of whom are expected to work at the Electricity of Vietnam, which was tasked to operate the country's nuclear power plants. But the Ninh Thuan nuclear power plant will not be constructed, resulting in a lost opportunity for these students to apply what they have learned overseas in operating a nuclear power plant.[15]

Vietnam's education system is not yet fully ready to produce young nuclear professionals. Nuclear engineering is offered as a new course in selected Vietnamese universities in Hanoi (Vietnam National University, Polytechnic University, and Electric Power University), Ho Chi Minh City (University of Science–VNU), and Da Lat (Dalat University). However, these universities do not have experienced professors in the field of nuclear engineering. The education system has focused mainly on nuclear physics, nuclear technique, and radiation technology rather than on the much-needed nuclear engineering (Tran 2015).

Overseas training programs on nuclear power mainly consist of short courses offered by the IAEA, Japan, Russia, South Korea, and other nuclear-powered countries. Vietnam's research and development is not yet fully developed. Although Vietnam has had many years of nuclear research, this research has not been properly focused or organised. There are no local leaders in nuclear research and application as Vietnam lacks leading nuclear scientists and engineers. Research and development infrastructure is also insufficient to facilitate nuclear energy research (Tran 2015).

---

14   Interview with a Ministry of Education and Training official, Hanoi, 8 August 2014.
15   Ibid.

The Vietnamese government views human resources development as key to the success of its nuclear power program. International cooperation will play an essential role in human resources development until Vietnam fully develops its local nuclear expertise. Russia assisted Vietnam in establishing the Centre for Nuclear Energy Science and Technology in 2011. The nuclear energy specialist training program has been recently introduced to train young leaders for Vietnam's nuclear power program. It aims to train 40 top specialists and experts in strategic areas such as nuclear power plant design and construction, nuclear power plant operation and finance, reactor safety, nuclear economics, and nuclear fuel cycles, among others. Trainees undergo nine months training in Vietnam by studying nuclear-related courses. They then receive rigorous training overseas, particularly in the US, Europe, Japan, and South Korea, and on their return to Vietnam they work at nuclear-related agencies such as VINATOM, the Centre for Nuclear Energy Science and Technology, Electricity of Vietnam, and VARANS (Tran 2015).

Meanwhile, Indonesia has a pool of nuclear experts who have worked for over 30 years at BATAN and other nuclear research facilities (Ministry of Energy and Mineral Resources, Republic of Indonesia 2008). Long-serving nuclear experts will soon retire and Indonesia needs to recruit and develop human resources to replace them (*Antara News* 2013). BATAN established four-year bachelor programs in nuclear techno-chemistry and nuclear techno-physics at the College of Nuclear Technology in 2001, but this program is not designed to produce nuclear engineers and technicians needed to operate a nuclear power plant.

BATAN invests in engineering and science graduate recruits to develop specialised expertise in nuclear energy through placement in nuclear power plant companies in South Korea, Japan, and Russia (IAEA 2013). Indonesia has also established a national team of human resources development and drafted a plan of action that includes the establishment of a nuclear training centre (National Team of HRD for NPP 2013). The formation of the centre began in 2010 and remains a work in progress, however.

Due to the growing need to further enhance the country's human resources development program and expertise in operating a nuclear reactor, BATAN plans to construct the Indonesia Experimental Power Reactor, which is scheduled to be operational by 2021/2022. The primary objectives of this project are to demonstrate the safe operation of a small-scale nuclear power plant; to improve the ability of Indonesia's nuclear professionals to master the nuclear power application and technology in preparation for the commissioning of nuclear power plants in the future; to develop research and development for future nuclear power plants and its supporting facilities as well as for human resources development; and to enhance public acceptance of nuclear power plant operation. BATAN also organises site visits for community leaders to experimental reactors as well as public discussions with communities to reassure them that nuclear power plants are safe (Taryo 2015).

In Malaysia, human resources development in nuclear science begins in universities. While the National University of Malaysia is the only tertiary institution with a nuclear science department (Adnan, Ngadiron, and Ali 2012), other universities also offer nuclear-related subjects. The focus of nuclear knowledge and expertise, however, is primarily on non-power applications such as medicine, health, agriculture, industry, and manufacturing (Khair and Hayati 2009).

To operate nuclear power plants, more specialised subjects, such as nuclear reactor design, nuclear safety engineering, and nuclear fuels and materials are needed. However, at present, there are insufficient experienced personnel to teach nuclear engineering courses. Malaysia does not have a dedicated human resources development program for nuclear power plants, and it remains unclear whether Malaysia would have the necessary qualified people by the time it constructed its first nuclear power plant (Khair and Hayati 2009).

From a long-term perspective, ASEAN members may emulate the French and US capacity-building programs in maintaining a local pool of highly qualified nuclear engineers and technicians. Those education and training programs ensure knowledge transfer from an ageing nuclear workforce to the next generation of workers.

Nuclear waste management policy

The failure of states with nuclear power plants to address the disposal of high-level nuclear waste (i.e. spent/used reactor fuel) from the day they started exploring nuclear energy should serve as a crucial takeaway for newcomers in Southeast Asia. Presently, there is no final repository site for high-level waste accumulated globally over six decades. Nevertheless, significant progress has been made in France, Sweden, and Finland in developing deep geological disposal sites that are tentatively to be made available after 2020 (Amano 2015).

But the IAEA has strongly advised newcomers in Asia to first address the waste issue by developing national policy and infrastructure for radioactive waste management, even before commissioning nuclear power plants (Amano 2015). Vietnam has yet to come up with a permanent disposal strategy. As part of its nuclear deal with Moscow, its future spent fuel will be reprocessed in Russia, but the treated wastes will still be returned to Vietnam, where a disposal facility will be required. The lack of a comprehensive plan on the disposal of spent fuel was one key challenge that was supposed to be addressed by Vietnam (Vi 2014). But the cancellation of the project will free up Vietnam from carrying out the difficult task of managing the spent fuel.

Indonesia has ratified the Joint Convention on the Safety of Spent Fuel Management and on the Safety of Radioactive Waste Management. Its nuclear research facilities are capable of managing and disposing of low- and intermediate-level radioactive waste produced from educational, medical, and industrial activities. But no comprehensive plan has yet emerged on the final disposal of high-level waste should Indonesia decide to commission nuclear power plants.

In Malaysia, significant capacity concerns exist around the safe disposal of nuclear waste. The implications for nuclear power plant development and the future safe disposal of nuclear waste are significant. Malaysia has not yet ratified the Joint Convention on the Safety of Spent Fuel Management and on the Safety of Radioactive Waste Management. The controversy over the Lynas rare earth mining site in Pahang is often cited by local critics of nuclear energy to demonstrate the lack of capacity of Malaysian authorities to deal with radioactive waste. The radioactive waste facility at the Lynas site allegedly lacks a sustainable plan for the long-term disposal of radioactive wastes under acceptable conditions, with possible leakage of harmful waste into the environment, according to a report by a German environmental research group, the Öko-Institut. It argues that deficiencies should have already been detected in the licensing process, when application documents were being checked by the nuclear regulator (Schmidt 2013).

Policy pathways to enhance nuclear safety and security in ASEAN

Our discussion has assessed the plans for nuclear power plants of three ASEAN members against the multiple and overlapping security, safety, and safeguards challenges they all face. These three areas point to institutional and human resources capacity and the need to develop a coordinated approach to nuclear energy safeguards, safety, and security. To be sure, nuclear capabilities engender a certain level of apprehension among neighbouring countries that can trigger escalating tensions. We therefore argue that it is imperative for ASEAN members to work together to ensure effective governance of nuclear facilities, materials, and wastes, and to adopt a regional disaster preparedness mechanism.

Upholding nuclear safety and security is extremely important to minimise the possibility of mishaps. Nuclear safety and security are indeed a regional issue, particularly because nuclear incidents can range from accidents with localised radiological impact to large-scale nuclear terrorist attacks, or even nuclear disasters that can cause transnational spill-overs (Heinonen 2016).

One important lesson from the Fukushima accident is the need to have broad perspectives on (and preparedness for) 'unthinkable' events and unforeseen circumstances (Suzuki 2016). In this regard, nuclear emergency preparedness is extremely important, the goal of which is to ensure that an adequate capability is in place within the operating organisation as well as at the local, national, and international levels. Such is necessary for an effective response in a nuclear emergency. Response should also consider crises related to transportation of nuclear and radioactive materials through (or near) the territories and possible terrorist acts. It is crucial to be adequately prepared to prevent and quickly respond to new types of events, for instance, cyber-attacks (Heinonen 2016).

Another important lesson from the Fukushima accident is the need to establish clear responsibility in crisis management. As observed, vague or overlapping responsibilities among stakeholders (operators, local governments, national government, and regulators, among others) are ineffective in crisis management. Regular nuclear emergency drills would help improve cooperation and coordination during an emergency response. Drills should involve the nuclear industry, the regulatory bodies, local and national emergency teams, police, military, customs, the coast guard, local governments, communities, NGOs, and media, among others. Emergency drills should be designed to test the existing response procedures and capabilities of all sectors for various unforeseen scenarios (Heinonen 2016).

Despite criticisms of ASEAN that it is slow and ineffective in tackling regional issues, it remains among the most relevant platform for developing policies and frameworks at the regional level. ASEAN can facilitate regional cooperation on capacity building, information dissemination, and emergency preparedness and response. As there is a risk of radioactive contamination spreading across borders, ASEAN governments must clearly and transparently manage nuclear activities and waste and explore channels for communication with neighbours to address cross-border impacts. As ASEAN members work to establish an ASEAN Community, fostering an ASEAN consensus on nuclear energy–related issues becomes possible.

One key impediment to cooperation, however, is ASEAN's guiding principle of non-intervention in another state's domestic affairs. Many states still perceive energy security as a national security issue and are reluctant to discuss their nuclear energy programs at the regional level. Finding the right balance between national sovereignty and regional cooperation is often challenging since nuclear security always entails confidentiality, as it is considered a national security issue. ASEAN can leverage its strength as an avenue for regional cooperation to address non-traditional security issues such as humanitarian assistance and disaster response in case a nuclear accident occurs. Currently, ASEAN has two sub-groups that promote regional cooperation on nuclear energy: ASEANTOM and a Nuclear Energy Cooperation Sub-Sector Network (NEC-SSN), which is composed of senior officials involved in energy policy and trade. The efficacy of their activities could be boosted by a number of national and regional initiatives.

To complement the normative framework on nuclear energy embodied in the 1995 Bangkok Treaty, ASEAN could explore drafting a blueprint on nuclear 3S. The objective would be the establishment of a robust nuclear governance regime in ASEAN to ensure that nuclear 3S processes are in place in good time before any ASEAN member's nuclear power plans are realised (probably starting with Vietnam in 2026). The blueprint could contain practical and feasible mechanisms, informed by evidence on best practices in other regions, which can facilitate regional cooperation on capacity building, information sharing/dissemination, enhancement of regulatory frameworks, and emergency preparedness and response frameworks. All these subjects would be within the bounds of ASEAN's principle of non-interference in domestic affairs. The important elements of this blueprint might include a regional framework on spent fuel management, cooperation on human resources development, and a feasibility study on a regional nuclear crisis centre and joint nuclear emergency drills.

Concerning the drafting of a possible regional framework on spent fuel management, ASEAN can draw on relevant experiences of Euratom's regional legislative framework. In 2011, the EU ratified binding legislation on spent fuel and radioactive waste management, requiring its members to adopt national programs for handling radioactive waste and to develop specific plans for building waste disposal facilities (European Commission 2014). An ASEAN framework could spell out how the member states can cooperate to contribute to global efforts to find a sustainable approach to disposing of nuclear waste, as well as encourage members interested in pursuing nuclear power to craft their respective comprehensive national plans for management of high-level radioactive waste.

Considering the need to strengthen responses to nuclear crises for the protection of people and the environment, ASEAN could set up a regional nuclear crisis centre in which its first responders, health care practitioners, customs officers, law enforcement, and disaster centre personnel can come together and participate in workshops, training, and joint drills. This cooperative effort would facilitate information and knowledge exchange, and increase response coordination in case member states are affected by radiation plumes. In times of crisis, the centre would act as a special coordinating body for regional and inter-ministerial disaster response. This centre may be formed as a specialised unit within the ASEAN Coordinating Centre for Humanitarian Assistance on disaster management. The specialised unit can help improve regional mechanisms (such as the ASEAN Coordinating Centre) that have so far concentrated on natural disasters to expand their mandate to cover technological disasters including nuclear emergencies, especially since responding to such disasters would require similar efforts.

Relatedly, ASEAN defence ministers can pursue the incorporation of joint nuclear emergency drills into the ASEAN Defence Ministers Meeting-Plus Humanitarian Assistance and Disaster Response/Military Medicine

Exercise. To this end, ASEAN could establish a regional contingent of specially trained nuclear disaster emergency responders, similar to the ASEAN–Emergency Rapid Assessment Team found in the ASEAN Coordinating Centre.

Finally, since human resources development is a key nuclear infrastructure issue that needs to be addressed by member states interested in nuclear power, regional cooperation on this issue can be part of the ASEAN framework on nuclear 3S. They can derive valuable lessons from Euratom's initiatives such as its regional human resources training program. Under the Euratom Fusion Training Schemes, various training actions have been launched since 2006 to ensure that adequate human resources will be available in the future in terms of numbers, range of skills, and high-level training and experience (European Commission 2013).

It must also be noted that there exists a double oversight for nuclear energy cooperation in ASEAN, with the existence of two specialised bodies— ASEANTOM and the NEC-SSN. While ASEANTOM is under the purview of the ASEAN Political-Security Community, the NEC-SNN falls under the ASEAN Ministers of Energy Meeting (Hashim 2016). For 2016, the NEC-SSN meeting's main objectives were to enhance capacity-building activities on civilian nuclear energy and to pursue regional nuclear safety cooperation with ASEAN dialogue partners (ASEAN Centre for Energy 2016). The NEC-SSN likewise facilitates information-sharing among member states with regard to nuclear safety and security. The double oversight signifies the strong commitment of the region to uphold nuclear 3S and foster regional cooperation on nuclear energy governance (Hashim 2016).

The NEC-SSN was tasked by ASEAN energy ministers in 2012 to continue to promote and intensify capacity-building efforts, in collaboration with the IAEA and other relevant partners, so that the region will be more informed and kept updated on the latest nuclear safety standards, developments, and technologies (ASEAN 2012a). Hence, the NEC-SSN needs to accelerate and strengthen its programs under the ASEAN Action Plan on Public Education on Nuclear Energy and Nuclear as the Clean Energy Alternative Option with a view to enhancing public awareness and acceptance of the use of nuclear energy for power generation.

With the assistance of the IAEA, ASEAN members can organise joint training workshops for the region's nuclear security professionals in evaluation methodology, helping them conduct site evaluations and interpret the results. ASEAN members need to ensure that they will be able to conduct the activities already identified during the 2014 meeting of ASEANTOM. These activities include a number of regional workshops and training courses on emergency preparedness and response as well as on nuclear security culture and management (ASEANTOM 2014b)

In conclusion, we reiterate that any nuclear energy program is a long-term commitment that should be expected to take decades, from planning and construction to operation, waste management, and capacity building. It is a sophisticated, uniquely hazardous, and proliferation-prone technology that requires rigorous planning. Vietnam, Malaysia, and Indonesia have already institutionalised several measures that adhere to the region's normative framework on the peaceful use of nuclear energy. Yet the safe development of nuclear power in Southeast Asia faces hurdles to ensure adherence to nuclear 3S norms, including on non-proliferation. Regional cooperation is the key to achieving adherence and, now with the establishment of an ASEAN Community, consensus on nuclear energy-related issues is possible. Member states will, however, have to work around concerns about non-interference in domestic affairs, giving priority to shared concern and interest in a nuclear-safe and nuclear weapons–free ASEAN.

Acknowledgements

This chapter is an updated version of Caballero-Anthony and Trajano (2015).

References

Adnan, Habibah, Norzehan Ngadiron, and Iberahim Ali, 2012. Knowledge management in Malaysian nuclear agency: The first 40 years. Presentation to the Knowledge Management International Conference (KMICe), Johor Bahru, 4–6 July. www.kmice.cms.net.my/ProcKMICe/KMICe2012/PDF/CR89.pdf (accessed 17 January 2017).

AELB (Atomic Energy Licensing Board), 2010. *Nuclear Regulatory Newsletter: Marking AELB's 25th Anniversary (1985–2010)*. December. portal.aelb.gov.my/sites/aelb/nuclearnewsletter/MNRNewsletter Dec2010.pdf (accessed 17 January 2017).

Amano, Yukiya, 2015. Atoms for peace in the 21st century. Transcript of speech delivered at the Energy Market Authority Distinguished Speaker Programme, Singapore, 26 January. www.iaea.org/newscenter/statements/atoms-peace-21st-century-1 (accessed 17 January 2017).

Antara News, 2013. Half of Indonesian nuclear experts will enter retirement age. 14 November (in Indonesian).

ASEAN (Association of Southeast Asian Nations), 1995. Treaty on the Southeast Asia Nuclear Weapon-Free Zone. 15 December.

ASEAN (Association of Southeast Asian Nations), 2012a. Joint Ministerial Statement of the 30th ASEAN Ministers of Energy Meeting (AMEM). Phnom Penh, 12 September. asean.org/joint-ministerial-statement-of-the-30thasean-ministers-of-energy-meeting-amem/ (accessed 17 January 2017).

ASEAN (Association of Southeast Asian Nations), 2012b. Phnom Penh Declaration on ASEAN: One Community, One Destiny. Phnom Penh, 3–4 April.

ASEAN (Association of Southeast Asian Nations), 2015. Chairman's Statement of the 27th ASEAN Summit. Kuala Lumpur, 27 November.

ASEAN (Association of Southeast Asian Nations) Centre for Energy, 2016. 6th Nuclear Energy Cooperation Sub-sector Network's Annual Meeting: Increasing ASEAN's Capacity in Civilian Nuclear Energy. Putrajaya, 12–13 April.

ASEAN (Association of Southeast Asian Nations) Secretariat, 2016. ASEAN Political-Security Community Blueprint 2012. Jakarta: ASEAN Secretariat.

ASEANTOM (ASEAN Network of Regulatory Bodies on Atomic Energy), 2014a. Background information. 1 December.

ASEANTOM (ASEAN Network of Regulatory Bodies on Atomic Energy), 2014b. 2nd Meeting of ASEAN Network of Regulatory Bodies on Atomic Energy, Chiang Mai, 25–27 August.

Asian Disaster Reduction Center, 2011. Information on disaster risk reduction of the member countries: Malaysia. www.adrc.asia/ nationinformation.php?NationCode=458&Lang=en&Nation Num=16 (accessed 18 January 2017).

Bernama, 2014. Govt to conduct feasibility study on building nuclear plant: Mah. *New Straits Times*, 7 July.

Bidin, Aishah, 2013. Nuclear law and Malaysian legal framework on nuclear security. Presentation to the Singapore International Energy Week, Singapore, 31 October. www.esi.nus.edu.sg/docs/default-source/event/presentation-3_aishah-bidin_nuclear-law.pdf?sfvrsn=2 (accessed 17 January 2017).

Biramontri, Siriratana, 2016. Presentation to the RSIS Roundtable at Singapore International Energy Week 2016: Nuclear Safety and Cooperation in ASEAN. Singapore, 28 October.

Caballero-Anthony, Mely, and Julius Cesar I. Trajano, 2015. The state of nuclear energy in ASEAN: Regional norms and challenges. *Asian Perspective* 39(4): 695–724.

Care2 Petitions, n.d. Stop nuclear power plants in Malaysia. www.thepetitionsite.com/745/599/785/public-petition-to-stop-nuclear-power-plants-in-malaysia/ (accessed 18 January 2017).

Consumers Association of Penang, n.d. www.consumer.org.my/ (accessed 18 January 2017).

Disaster Management Division of Prime Minister's Department, 2011. Brief note on the roles of the National Security Council, Prime Minister's Department as National Disaster Management Organisation (NDMO). Presentation to the 3rd ASEAN Inter-Parliamentary Assembly Caucus, Manila, 1 July.

Dung, Nguyen Tan, 2010. Approving the scheme on training and development of human resources in the field of atomic energy. Hanoi, 18 August. www.nti.org/media/pdfs/VietnamHRDevelopment Plan2020.pdf?_=1333145926 (accessed 17 January 2017).

Economic Planning Unit, 2015. *Eleventh Malaysia Plan 2016–2020: Anchoring Growth on People*. Kuala Lumpur: Economic Planning Unit, Prime Minister's Department.

European Commission, 2013. Education and training in Euratom research in EP7 and future perspectives. Luxembourg: Publications Office of the European Union.

European Commission, 2014. Euratom. ec.europa.eu/programmes/horizon2020/en/h2020-section/euratom (accessed 18 January 2017).

European Leadership Network, 2014. Joint statement on the contributions of the Global Initiative to Combat Nuclear Terrorism (GICNT) to enhancing nuclear security. 20 March. www.europeanleadershipnetwork.org/joint-statement-on-the-contributions-of-the-global-initiative-to-combat-nuclear-terrorism-gicnt-to-nuclear-security_1308.html (accessed 11 September 2017).

Hadi, Bambang Sutopo, 2014. Bapeten Bentuk Pusat Unggulan Keamanan Nuklir [BAPETEN creates center of excellence for nuclear security]. *Antara News*, 19 August.

Haditjahyano, Hendriyanto, 2014. Indonesian HRD in nuclear security—Batan's perspective. Presentation to the Workshop on the Asian Centers of Excellence in Nuclear Nonproliferation and Nuclear Security, Washington, DC, 18 July. csis.org/files/attachments/140718_CoEWorkshop_Haditjahyono_Indonesia.pdf (accessed 17 January 2017).

Hashim, Sabar Mohd, 2016. Presentation to the RSIS Roundtable at Singapore International Energy Week 2016: Nuclear Safety and Cooperation in ASEAN. Singapore, 28 October.

Heinonen, Olli, 2016. Presentation to the RSIS Roundtable at Singapore International Energy Week 2016: Nuclear Safety and Cooperation in ASEAN. Singapore, 28 October.

Hoang, Anh Tuan, 2013. Vietnam experience in the IAEA integrated nuclear infrastructure review missions. Presentation to the Support to Nuclear Power Programmes: INIR Missions and Agency Assistance, Vienna, 18 September. www.iaea.org/NuclearPower/Downloadable/News/2013-09-19-inig/INIR_SideEvent_Vietnam.pdf (accessed 17 January 2017).

Hoang, Anh Tuan, 2014. Updates on nuclear power infrastructure development in Vietnam. Presentation to the IAEA Annual Infrastructure Workshop, Vienna, 4–7 February. www.iaea.org/Nuclear Power/Downloadable/Meetings/2014/2014-02-04-02-07-TM-INIG/ Presentations/08_S2_Vietnam_Hoang.pdf (accessed 17 January 2017).

IAEA (International Atomic Energy Agency), 1994. Convention on Nuclear Safety. Vienna, 5 July.

IAEA (International Atomic Energy Agency), 2008. Responsibilities and competencies of a Nuclear Energy Programme Implementing Organization (NEPIO) for a national nuclear power programme. Draft. Vienna: IAEA.

IAEA (International Atomic Energy Agency), 2013. Country nuclear power profiles 2013 edition: Indonesia. www-pub.iaea. org/MTCD/Publications/PDF/CNPP2013_CD/countryprofiles/ Indonesia/Indonesia.htm (accessed 18 January 2017).

IAEA (International Atomic Energy Agency) PRIS (Power Reactor Information System), 2014. The database on nuclear power reactors. www.iaea.org/pris/home.aspx (accessed 17 January 2017).

IEA (International Energy Agency), 2015. *Southeast Asia Energy Outlook 2015: World Energy Outlook Special Report*. Paris: International Energy Agency.

Khair, Nahrul, and Ainul Hayati, 2009. Prospect for nuclear engineering education in Malaysia. Presentation to the 2009 International Conference on Engineering Education (ICEED 2009), Kuala Lumpur, 7–8 December. irep.iium.edu.my/9735/1/Prospect_for_nuclear_ engineering_education_in_Malaysia.pdf (accessed 17 January 2017).

Kyodo News, 2016. Vietnam to scrap nuclear plant construction plans. *Bangkok Post*, 9 November.

Lucas, Daxim L., 2016. Duterte gives nuke plant green light. *Philippine Daily Inquirer*, 12 November.

Malaysian Insider, 2012. Nuclear power project still in infancy, says Najib. 2 October.

Markandu, Dhana Raj, 2013. Roles and organisation of the NEPIO in Malaysia: Case study. Presentation to the Technical Meeting on Topical Issues on Infrastructure Development & Management of National Capacity for Nuclear Power Program, Vienna, 11–14 February.

Ministry of Energy and Mineral Resources, Republic of Indonesia, 2008. Ahli Nuklir Indonesia Berpengalaman 30 Tahun Lebih Operasikan Reaktor Nuklir [Indonesian nuclear experts have more than 30 years of experience operating nuclear reactors]. 26 January.

Mulder, Els, 2006. Preparedness for disasters due to climate change. www.climatecentre.org/downloads/files/articles/preparedness%20 for%20disasters%20related%20to%20climate%20change%20 els%20mulder.pdf (accessed 17 January 2017).

National Agency for Disaster Response, 2012. Potensi dan Ancaman Bencana [The disaster threats outlook]. www.bnpb.go.id/home/ potensi (accessed 7 September 2017).

National Team of HRD for NPP, 2013. Indonesia's update policy and HRD preparation for NPP. Presentation to the Fukui International Meeting on Human Resources Development for Nuclear Energy in Asia, Fukui, 26–27 March. fihrdc.werc.or.jp/achievement/13-pdf/ Session3/3%20Indonesia.pdf (accessed 17 January 2017).

Nian, Victor, and S. K. Chou, 2014. The state of nuclear power two years after Fukushima: The ASEAN perspective. *Applied Energy* 136: 838–48. doi.org/10.1016/j.apenergy.2014.04.030

Ninh, T. N. T., 2013. Human resources and capacity-building: Issues and challenges for Vietnam. Presentation to the ESI-RSIS International Conference on Nuclear Governance Post-Fukushima, Singapore, 31 October.

Nuclear Security Summit, 2014. National progress report: Malaysia. The Hague, 31 March. pgstest.files.wordpress.com/2014/04/malaysia_ pr_2014.pdf (accessed 7 September 2011).

Nuclear Security Summit, 2016. National progress report: Malaysia. Washington, DC, 31 March. www.nss2016.org/document-center-docs/2016/3/31/national-progress-report-malaysia (accessed 17 January 2017).

Pascaline, Mary, 2016. Vietnam nuclear power program: National Assembly scraps atomic energy project with Russia, Japan. *International Business Times*, 23 November.

Ramli, S., 2013. National nuclear power programme in Malaysia— An update. Presentation to the Technical Meeting on Technology Assessment of SMR for Near Term Deployment, Chengdu, 2–4 September. www.iaea.org/NuclearPower/Downloadable/Meetings/2013/2013-09-02-09-04-TM-NPTD/9_malaysia_ramli.pdf (accessed 17 January 2017).

Schmidt, Gerhard, 2013. Description and critical environmental evaluation of the REE refining plant LAMP near Kuantan/Malaysia. Berlin: Öko-Institut.

Sinaga, M., 2012. Development of nuclear security in Indonesia. Presentation to the NRC's International Conference on Nuclear Security, Washington, DC, 4–6 December.

Sunaryo, Geni Rina, 2015. Development of nuclear power programme in Indonesia. Presentation to the 5th Nuclear Power Asia Summit, Kuala Lumpur, 27 January.

Suntoko, H., and Ismail, 2013. Current status of siting for NPP in Indonesia. Presentation to the Second Workshop on Practical Nuclear Power Plants Construction Experience, Beijing, 24 October.

Suzuki, Tatsujiro, 2016. Presentation to the RSIS Roundtable at Singapore International Energy Week 2016: Nuclear Safety and Cooperation in ASEAN. Singapore, 28 October.

Tan, Hui Yee, 2016. Cambodia and Thailand edging closer to nuclear power. *Straits Times*, 30 May.

Taryo, Taswanda, 2015. Development of nuclear power plant in Indonesia. Presentation to the RSIS Roundtable at the Singapore International Energy Week 2015: Is Southeast Asia Ready for Nuclear Power? Singapore, 29 October.

Teng, I. L., 2014. Post Fukushima: Environmental survey and public acceptance on nuclear program in Malaysia. Presentation to the International Experts' Meeting on Radiation Protection after the Fukushima Daiichi Accident: Promoting Confidence and Understanding, Vienna, 21 February.

Thiep, Nguyen, 2013. Safety infrastructure development for Vietnam's nuclear power programme. Presentation to the Second ASEM Seminar on Nuclear Safety: International Instruments for Ensuring Nuclear Safety, Vilnius, 4–5 November.

Tran Chi Thanh, 2015. Implementation of the nuclear power program: Challenges and difficulties. Presentation to the RSIS Roundtable at the Singapore International Energy Week 2015: Is Southeast Asia Ready for Nuclear Power? Singapore, 29 October.

US NRC (Nuclear Regulatory Commission), 2015. NRC's support of US nonproliferation objectives in the licensing of enrichment and reprocessing facilities. *Backgrounder*, February. www.nrc.gov/reading-rm/doc-collections/fact-sheets/nonproliferation.pdf (accessed 17 January 2017).

Vi, Nguyen Nu Hoai, 2014. Viet Nam's experience in the area of nuclear security. Presentation to the International Cooperation to Enhance a Worldwide Nuclear Security Culture, Amsterdam, 20 March.

Vuong, Huu Tan, 2015. Safety culture in Vietnam. Presentation to the IAEA Technical Meeting on Topical Issues in the Development of Nuclear Power Infrastructure, Vienna, 2–6 February.

World Nuclear Association, 2015. Nuclear power in Vietnam. www.world-nuclear.org/info/Country-Profiles/Countries-T-Z/Vietnam/ (accessed 17 January 2017).

Part III
The real costs of going nuclear

8

Health implications of ionising radiation

Tilman A. Ruff

Abstract

The biological effects of ionising radiation is one of the most disputed and politicised topics relevant to analysis of nuclear technologies, including nuclear power. The first part of this chapter addresses the nature and sources of ionising radiation and its effects on biology and human health, including important new evidence that points to these effects being greater than previously estimated. Because the field of radiation and health has such a politicised history and is beset with contestation and interference by vested interests, the second part draws a brief historical landscape of who's who in radiation and health, in the hope of assisting readers to navigate and consider some critical questions of history, motivation, and interests behind who is saying what and why. The third part considers potential sources of human origin of large-scale radiation exposures.

Nature, sources, and effects of ionising radiation

What is ionising radiation?

Ionising radiation includes various kinds of transmitted energy. Some types, like X and gamma rays, are part of the electromagnetic spectrum that spans from long wavelength, low energy radio, micro- and infrared waves, through visible light and ultraviolet radiation, to short wavelength, high energy X and gamma rays. Other types of ionising radiation consist of subatomic particles and fragments of atoms. Both electromagnetic and particulate radiation of high energy are called 'ionising' because the various types (see Table 8.1) are of sufficient energy to eject one or more electrons from atoms (ionisation) and break chemical bonds. Whenever radiation is referred to in this chapter, unless otherwise specified, ionising radiation is meant.

Table 8.1 Common types of ionising radiation

Ionising radiation type		Radiation weighting factor (biological effect, compared with photons)*	Stopped by
Electromagnetic radiation (photons)	Gamma rays (similar to X-rays)	1	Penetrating (providing the basis for the use of X-rays for imaging), stopped by dense materials (e.g. lead or concrete), not by clothing
Subatomic particles	Alpha (helium nucleus) and other heavy fission fragments like atomic nuclei	20	Outer layer of skin, a sheet of paper (harm derives from internal exposure, e.g. when inhaled or ingested)
	Beta (electron)	1	A layer of clothing; some can penetrate to basal layer of human skin
	Neutron	5–20 depending on neutron energy	Penetrating; concrete or earth most effective protection

* A weighting factor of 1 means that for this amount of radiation energy, the particular type of radiation causes the same amount of biological damage as X or gamma rays; a factor of 2 means that type of radiation is twice as biologically damaging as gamma rays of the same energy, and so on.

Source: Centers for Disease Control and Prevention (2015); European Nuclear Society (n.d.).

Ionising radiation can be produced by the spontaneous decay of radioactive elements, which fragment into smaller atoms, often repeatedly, each step usually also releasing energy in the form of ionising radiation. These decay chains often have multiple intermediaries that are also radioactive, until they eventually end up as elements that are stable. The decay chain, the rate at which each step occurs, and the type of radiation emitted, are fixed physical properties of each radioactive element. The rate of radioactive decay is described by the half-life—the time it takes for half the amount of a radioactive element to disintegrate. Elements with long half-lives persist for long periods, but their radioactivity is less intense than for elements that decay rapidly. Many chemical elements exist in different atomic forms (known as isotopes, with the same number of protons but varying numbers of neutrons in their nucleus), some or all of which may be radioactive, in which case they are called radioisotopes.

Ionising radiation can also be produced, such as in an X-ray machine, when rapidly moving charged particles collide with a substance. Neutron radiation emitted by a nuclear explosion can induce radioactivity in materials that are not normally radioactive, such as by converting nitrogen in the air to carbon-14, which is radioactive with a half-life of 5,730 years.

Different types of radiation have widely differing capacities to penetrate tissues. Alpha particles are stopped by a thin layer of paper or clothing and do not penetrate through the dead upper layer of normal human skin. However, their high energy makes them highly damaging if taken internally, into the lungs or gut. Some beta particles (electrons) penetrate to the basal layer of skin where new cells are produced. Gamma and X-rays are highly penetrating, which is the basis for the value of X-rays in medical imaging. Thick concrete or lead are used to stop them.

Radiation sources and exposure pathways

People are exposed to radiation via different pathways—essentially internal or external. Penetrating radiation from cosmic sources travelling down through the atmosphere, from radioactive materials in or on the ground, in building materials, or in the air, irradiate people externally. Direct contact with a source of penetrating radiation causes greatest exposure to the part in closest proximity. Inhalation, ingestion (via food, water, or environmental sources like soil and dust, particularly in children), or contamination of wounds or other skin breaks can cause exposure to radioactivity from radioactive materials that get inside the body. In this case, even radiation that penetrates only very short distances can be harmful, especially if radioactive particles are retained in the body for long periods. Internal contamination, especially with substances that do not emit highly penetrating radiation (like plutonium, strontium, and tritium) are more difficult to measure and may be neglected in assessing radiation exposure.

Many biologically important radioisotopes behave chemically, and are therefore handled by our bodies, like other elements that we need (see Table 8.2). Some of these are concentrated in living things and also up the food chain, and may be recycled in the biosphere. For example, the concentration of cesium-137 in fish in freshwater lakes may be 10,000 times higher than the concentration in the water in which the fish live.

Table 8.2 Selected radioactive isotopes from nuclear power plants significant in human health impact

Radioisotope	Main type(s) of radioactivity emitted	Half-life	Health significance and predominant means of exposure
Iodine-131	Beta, gamma	8 days	Ingestion, concentrated up food chain (especially milk); concentrated in thyroid gland; causes thyroid disease including cancer—children most vulnerable
Cesium-137	Beta, gamma	30 years	External and ingestion, body handles like potassium, the main positively charged ion inside cells; bio-concentrated; associated with many cancers; dominant cause of radiation exposure to world's people from atmospheric nuclear test explosions to date, and nuclear power plant accidents
Strontium-90	Beta	28 years	Ingestion, handled by the body like calcium, concentrates in bones and teeth; bioconcentrated; retained; causes leukaemia, bone cancer
Plutonium-239	Alpha	24,400 years	Inhalation, retained; internal hazard; especially when inhaled, causes lung cancer. Plutonium isotopes are inevitably produced inside nuclear reactors by the absorption of neutrons by uranium atoms
Tritium (hydrogen-3)	Beta	12.3 years	Ingestion, internal hazard; becomes incorporated in water molecules, does not bioaccumulate

Source: Based on UNSCEAR (1993: Annex B, 128–9).

Radioactive materials may be solid, liquid, or gaseous. Materials, objects, and organisms that contain radioactive materials or become surface-contaminated by them emit radioactivity; whereas objects and organisms exposed to radiation but not containing radioactive materials are not a source of potential exposure or hazard to others. For example, people injected with a radioactive chemical for medical purposes become radioactive for a period; whereas someone undergoing diagnostic X-rays or external radiotherapy for cancer treatment is not rendered radioactive.

The bulk of our natural background radiation exposure is derived from radon gas, the heaviest of the noble gases, a ubiquitous carcinogen produced by the decay of primordial uranium-235 and 238 and thorium-232 present in the Earth's crust. Radon decays via a number of intermediaries (polonium, bismuth, and tellurium), which are more reactive and attach to aerosols and dust in the atmosphere, which may lodge in the lungs when inhaled. These radon progeny deliver most of the radiation dose associated with radon. Radon is the second-most important cause of lung cancer worldwide, exceeded only by tobacco smoking.

In many parts of the world, medical radiation exposure has increased markedly in recent decades, and in some countries now accounts for similar (for example, Australia, the US) or greater (for example, Japan) levels of radiation exposure across the population than does naturally occurring radiation. This is particularly due to increasing use of CT scans (computed tomography, a sophisticated type of X-ray examination).

Most modern nuclear weapons contain both highly enriched uranium and plutonium. Uranium or plutonium can fission (split) in about 40 different ways, producing altogether some 300 different radioactive nuclides, with half-lives varying from a fraction of a second to many millions of years. Inside a nuclear reactor, uranium and plutonium are also fissioned, but the neutrons that propagate the controlled chain reaction in a reactor are slower than the 'fast' neutrons involved in a nuclear weapon exploding. In addition, because the initial fission fragments, which are mostly short-lived, are not dispersed in a reactor, their longer-lived decay products accumulate. Thus, a nuclear reactor accumulates proportionately more long-lived radioisotopes than are produced by a nuclear bomb.

As well as creating hundreds of new radioactive elements that did not exist before (see Table 8.3), a nuclear explosion and the operation of a nuclear reactor both increase the total radioactivity from that of the starting material a million times or more.

Table 8.3 Common radioisotopes in routine releases from nuclear plants

Airborne	
Type	Common isotopes released (atomic number)
Fission and activation gases	Krypton (85, 85m, 87, 88) Xenon (131, 131m, 133, 133m, 135, 135m, 138) Argon (41)
Halogens	Iodine (131, 132, 133, 134, 135) Bromine (82)
Particulates	Cobalt (58, 60) Cesium (134, 137) Chromium (51) Manganese (54) Niobium (95)
Tritium	Hydrogen (3)
Liquid	
Mixed fission and activation products	Iron (55) Cobalt (58, 60) Cesium (134, 137) Chromium (51) Manganese (54) Zirconium (95) Niobium (95) Iodine (131, 133, 135)
Tritium	Hydrogen (3)
Dissolved and entrained noble gases	Krypton (85, 85m, 87, 88) Xenon (131, 133, 133m, 135, 135m)

Source: Adapted from National Research Council (2012: 37–8).

Why is ionising radiation of biological importance?

Ionising radiation is intensely biologically injurious, not because it contains extraordinarily large amounts of energy, but because its energy is bundled and delivered to cells in large packets. The energy of a diagnostic X-ray, for example, is typically around 15,000 times as great as the energy of a chemical bond. A whole-body dose of ionising radiation of 4 gray (Gy) causes potentially lethal acute radiation sickness in humans. Yet the energy delivered to a 70 kg adult human body by that dose of radiation amounts to only 280 joules—the same amount of energy as the heat absorbed by drinking one 3 mL sip of hot (60 degrees Celsius) tea or coffee.

Large complex molecular chains, especially of DNA, define who we are, regulate many biological processes, and are both our most precious inheritance and the most vital legacy we pass on to our children. One of the strands of the double DNA helixes inside each of our cells is derived from our mother, the other from our father. These large molecules are particularly vulnerable to disruption by ionising radiation. Radiation may cause direct damage to DNA, or cause indirect damage through the production of highly reactive chemicals, like free radical ions, which then react with DNA. A variety of types of damage may result—single and double-strand DNA breaks, oxidative changes to the nucleotide bases that make up DNA, deletions of sections of DNA, and resulting gene and chromosomal damage. The frequency of chromosomal aberrations, particularly dicentric forms, in blood lymphocytes can be used within weeks of whole-body radiation to estimate the dose received. Stable and persistent chromosomal changes that do not kill affected cells, like translocations (rearrangements of segments of chromosomes), have been demonstrated at increased frequencies even more than 50 years after exposure in Japanese *hibakusha* (nuclear bombing survivors) and New Zealand nuclear test veterans (Wahab et al. 2008).

DNA damage from radiation can have various outcomes, including effective repair, cell death (especially at high doses), impaired function, induction of cancer, or result in DNA changes transmissable to subsequent generations. Cells have mechanisms to repair DNA damage, but these are not complete or error-free. DNA is most susceptible to radiation damage when cells are dividing, so rapidly dividing and growing tissues are most vulnerable, such as blood-forming cells in the bone marrow, germ cells in the ovary and testis, cells lining the gastro-intestinal tract, and hair follicles. Radiation exposure to a foetus in the womb can lead to foetal damage (such as mental retardation) and malformations. Young children and foetuses are especially sensitive to radiation effects, and a cancer-prone mutation occurring early in prenatal life is likely to transmit to a larger number of daughter cells than a mutation occurring later, when a cell undergoing a mutation will produce fewer daughter cells.

The science of radiation and health is still evolving. It has often been considered that the same dose of radiation delivered quickly is one-and-a-half to two times more injurious than the same total dose delivered over a longer time. However, as discussed later, recent evidence suggests that this is not the case. *Bystander effects* are a feature of many types of radiation, whereby radiation damage to one cell damages nearby cells, even without initial DNA damage occurring. Inflammatory responses are thought to be involved. *Genomic instability* describes radiation-related gene damage causing increased susceptibility to further damage, and can be transmitted from parent to daughter cells. Both bystander effects and genomic instability can be delayed.

Radiation levels and effects

Radiation is measured in different ways. The most basic unit of radioactivity measures the frequency of atomic disintegration—1 becquerel (Bq) is 1 radioactive decay per second. The absorbed dose of radioactivity is measured by the gray—1 Gy is 1 joule of energy deposited per kilogram of mass (usually tissue).

The *equivalent dose* measures the biological effect of the energy absorbed for a particular organ or tissue—it is the absorbed dose multiplied by the relevant tissue weighting factor, which reflects how sensitive a tissue is to radiation effects. There are five groups of tissue weighting factors spanning a 40-fold difference in radiation sensitivity. Most sensitive are the gonads (ovary and testis); the next most sensitive group includes red bone marrow (where blood cells are made), stomach, colon, and lung. The *effective dose* is a summation of the equivalent doses to tissues and organs exposed, adjusting for the varying radiosensitivity of different tissues. It gives an indication of overall risk. Such summations are not an exact science, and are less meaningful where doses are divergent for different tissues. For example, a brain CT scan may involve a 40–50 milliGray (mGy) dose to the brain, which particularly increases the risk of brain cancer; however, this translates into an effective whole-body dose of around 4.5 milliSievert (mSv) (Mathews et al. 2013). Both equivalent dose and effective dose are measured in sievert (Sv). For penetrating radiation like X and gamma rays, the dose in Gy and Sv is the same.

The average global background level of radiation we are all exposed to from inhalation of radon gas produced by the decay of uranium in the Earth's crust, cosmic sources, soil and rocks, and ingestion of low levels of naturally occurring radioactive substances, is about 2.4 mSv per year. A single back-to-front chest X-ray typically involves a dose of 0.01 mSv;

a CT scan typically involves doses of 3–12 mSv or more. Acute exposures over 100 mSv produce effects on chromosomes measurable by laboratory testing. Doses below 100 mSv are generally categorised as 'low dose'.

Ionising radiation in doses over 100–250 mSv causes acute effects detectable by commonly available blood tests, and symptoms of acute radiation sickness develop at higher doses. Doses over 100 mSv cause both reversible and persistent effects to various organs. Acute symptoms are increasingly likely at acute doses above a few hundred mSv; without intensive medical care, doses around 4 Sv (4,000 mSv) will be fatal for many of those exposed. Much higher doses targeted to particular body sites are used in treatment of various cancers (the cells of which usually divide faster than normal cells) when the purpose is to kill cancer cells. Some acute effects of radiation, such as skin burns, hair loss, sterility, and acute symptoms such as headache, nausea, and vomiting occur only above certain thresholds of radiation dose received. Recovery may occur from many of the effects of acute radiation sickness, especially with good medical care.

In contrast, any and all levels of ionising radiation exposure, including doses too low to cause any short-term effects or symptoms, are associated with increased risks of long-term genetic damage, chronic disease, and increases in almost all types of cancer,[1] proportional to the dose. Radiation both increases the chance of developing cancer and brings its onset earlier. These excess risks persist for the lifetime of those exposed. It has been conclusively established that there is no dose of radiation below which there is no incremental health risk—all radiation exposure adds to long-term health risks. This applies also to natural background radiation. Higher background exposures in some geographic areas have been shown to be linked to adverse effects such as increased mutation rates, immunological changes, physical body changes, and increased cancer rates in many diverse plant and animal species, including humans (Møller and Mousseau 2013). Even before the availability of recent data on unexpectedly high and early cancer rates in children following CT scans, it was estimated that 2.5 mSv of natural background radiation to the bone marrow each year would be the cause of up to 30 per cent of childhood leukaemia (Wakeford, Kendall, and Little 2009).

---

1   Cancers are often considered in two broad types—cancers of blood-forming organs (leukaemia), and cancers of solid organs.

Arguably the most authoritative and rigorous periodic assessments of radiation health risks are the Biological Effects of Ionizing Radiation (BEIR) reports produced by the US National Academy of Sciences. However, the most recent report, BEIR VII (Committee to Assess Health Risks from Exposure to Low Levels of Ionizing Radiation 2006), was published in 2006, and substantial new evidence has accumulated since then. The BEIR VII report estimates that the overall increase in risk of solid cancer incidence (occurrence) across a population is about one in 10,000 for each 1 mSv of additional radiation exposure. The increased risk for leukaemia is about 10 per cent of this. As about half of all cancers are fatal, the estimated increased risk of death from cancer is about half that—about 1 in 20,000 per mSv.

The maximum permitted dose limit recommended by the International Commission on Radiological Protection (ICRP) and most national radiation protection agencies for any additional non-medical exposures for members of the public is 1 mSv per year (corresponding to about 0.11 microSv per hour, a common unit of radiation exposure measurement) (ICRP 2009). Some authorities apply more protective criteria, for example, the US Environmental Protection Agency specified level for clean-up of radioactively contaminated sites is 0.12 mSv per year (US EPA 2014).

There has been a consistent trend over time that the more we know about radiation effects, the greater the evidence indicates those effects to be. Maximum permitted radiation dose limits have never been raised over time; they have always been lowered. For example, from 1950 to 1991, the maximum recommended whole-body radiation annual dose limits for radiation industry workers declined from approximately 250 to 20 mSv. These recommended dose limits are not doses below which there is no health risk. Rather, they represent the most recent compromise between safety and optimally protecting people on the one hand, and commercial and other vested interests and cost considerations on the other.

Ionising radiation also increases the risk of occurrence and death from some non-cancer diseases, including cardiovascular and respiratory disease. This has been clearly demonstrated at moderate and high doses, and recent evidence indicates that circulatory disease mortality also increases at low total doses and dose rates, such as occur in many nuclear industry workers. The increased risk of death from heart and other circulatory diseases is estimated to be comparable in magnitude to the radiation-related cancer risk, meaning that the total extra risk of dying because of exposure to radiation is likely to be around double the increased risk of death from cancer alone (Little et al. 2012).

One important factor in understanding radiation health effects is that small doses received by a large number of people may cause significant consequences. For example, a UK study showed that the great majority (over 85 per cent) of radon-related lung cancer deaths in the UK occurred among people living in homes where the level of radon was less than 100 becquerel per cubic metre (Bq/m³) of air, well below the level recommended as warranting remedial action of 200 Bq/m³ (Gray et al. 2009). An additional average radiation dose across a population of 100 million people of only 1 mSv, using the traditional risk estimates above, can be expected to result in an additional 10,000 cancer cases.

Populations with increased vulnerability to radiation health harm

Radiation risk is not uniform across a population. It is highest in very young children and declines gradually with age. Infants are overall about four times as sensitive to radiation's cancer-inducing effects as middle-aged adults (Committee to Assess Health Risks from Exposure to Low Levels of Ionizing Radiation 2006). A single X-ray to the abdomen of a pregnant woman, involving a radiation dose to the foetus of about 10 mSv, was shown in pioneering studies by Alice Stewart in the UK in the 1950s to increase the risk of cancer during childhood in her offspring by 40 per cent (Doll and Wakeford 1997).

In the BEIR VII assessment, following uniform whole-body exposure to the same level of radiation, women and girls are 52 per cent more likely to develop cancer than men and boys, and 38 per cent more likely to die of cancer than males. The difference is greatest at younger ages of exposure—for the same exposures occurring between zero and five years of age, girls are 86 per cent more likely to develop cancer than boys (Makhijani, Smith, and Thorne 2006: 35–40).

The greater vulnerability of children than adults to radiation damage is substantial—exposures in infancy (below one year of age) for boys are 3.7 times more likely to lead to cancer than the same exposure for a 30-year-old man; for infant girls compared with 30-year-old women, that risk is 4.5 times greater (Committee to Assess Health Risks from

Exposure to Low Levels of Ionizing Radiation 2006: 470–99). These differences relate to both increased sensitivity of the young and the usually longer remaining years of life for effects to become manifest. The relationship between overall cancer risk and age is depicted in Figure 8.1.

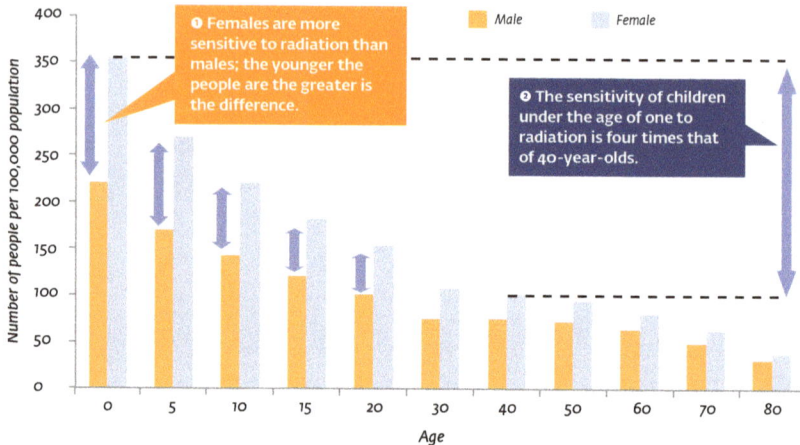

Figure 8.1 Increased lifetime cancer risk by age and gender associated with an extra radiation dose of 10 mSv

Source: NAIIC (2012), based on data from Committee to Assess Health Risks from Exposure to Low Levels of Ionizing Radiation (2006).

Increased vulnerability of young people also applies to the non-cancer health risks of radiation exposure. In recent research for the British population, by way of example, a similar increase in risk occurs associated with younger age of exposure for cardiovascular disease. It was estimated that the increased lifetime risk of death from circulatory disease is about 10 times higher for a child exposed to radiation before 10 years of age compared with exposure after age 70. Similarly, an exposed child's risk of death from solid cancer was estimated to be more than 20 times higher than for exposure occurring over the age of 70, and about double that associated with the same exposure at age 30–39 years (Little et al. 2012).

The combined effects of exposure during early childhood and greater female susceptibility can be dramatic. For example, for intake of fluid containing the radioactive isotope strontium-90, infant girls exposed to the same level of contamination are assessed to have a 20.6-fold higher risk of breast cancer than women aged 30 years. For the same level of contamination of ingested iodine-131, the risk for infant girls compared with 30-year-old women is 32.8 times higher. This means that for the same level of radioactive contamination, the cumulative breast or thyroid cancer risk by ingestion over the first five years of life for girls is greater than that accumulated by women over their entire adult lives (Makhijani, Smith, and Thorne 2006: 40).

These differential vulnerabilities are obscured by averaging risks across populations.

The evolving evidence of radiation-related health harm

The long-term follow-up studies of Hiroshima and Nagasaki *hibakusha* have provided the bulk of historic data on which radiation health risks have been estimated—and based on which the recommended dose limits for nuclear industry and other radiation-exposed workers and the public have been set. The most recent published data from ongoing studies of the Hiroshima and Nagasaki survivors confirm a linear dose–response relationship between radiation dose and overall cancer risk, with no threshold (Ozasa et al. 2012). They also suggest that the risk per unit dose is greater at lower radiation doses. They do not show a reduction in cancer risk for the same total exposure when radiation is delivered over a longer time, as some bodies such as the ICRP still assume. They certainly do not show an absence of evidence of increased cancer risk at doses less than 100 mSv, as Japanese official bodies and educational materials for students and teachers still claim (Cabinet Office et al. 2016). Even within the context of Japan, the extreme and unfounded nature of this official mantra is contradicted by the Japanese government recently awarding workers' compensation to a man who developed leukaemia after working for 14 months as a clean-up worker at the damaged Fukushima Daiichi nuclear plant (Soble 2015). The level of radiation exposure required to claim workers' compensation in Japan is 5 mSv per year.

The Japanese survivor studies, however, have a range of methodological flaws that lean towards underestimation of radiation risk (see Table 8.4) (Richardson, Wing, and Cole 2013; Mathews et al. 2013). The most significant gaps are that the *hibakusha* studies reflect a hardy survivor population, misclassify some exposures, include relatively few people who received low radiation doses, and miss cancer deaths occurring in the first five years and cancer cases over the first 13 years.

Table 8.4 Features of the Japanese *hibakusha* Life Span Study that tend to underestimate radiation health risks

Selected hardy population	Those who survived the atomic bombings were a selected population who had already survived the hardship, poor nutrition, and deprivation of war. The added stresses of the atomic bombings would favour the survival of those who were hardier, better nourished, less likely to be injured or suffering pre-existing health problems, and thus less vulnerable to the adverse health effects of the bomb.
Early cancer deaths missed	Cancer deaths were not recorded until 1950—cancer deaths between 1945 and 1950 were not counted.
Early leukaemia and cancer cases missed	Leukaemia cases were only counted from 1950, and other cancer cases only from 1958, 13 years after the atomic bombings; whereas increases can be expected to have begun by 1947–48.
Some counted as unexposed to radiation were in fact exposed	Those who were between 2.5 km and 10 km from the epicentre were considered 'unexposed'; many of these will have had some exposure. Those who entered the bombed cities in the week after the bombings were exposed to neutron-induced radioactivity (estimated at 82 mSv colon (average internal organ) dose in Hiroshima on day two). People in the vicinity of Hiroshima who were exposed to radioactive fallout as black rain had high mortality in 1950–62, yet were counted in the control group. All these types of exposures were ignored, which would tend to reduce the observed effects of radiation exposure on survivors. Further, in various analyses, doses of <10mGy were counted as unexposed.
Dose estimates are uncertain	While repeated careful dose estimates for *hibakusha* have been made, considerable uncertainty about individual doses remains.
Relatively few *hibakusha* received low radiation doses	There were about 70,000 survivors estimated to have received low doses (<100 mSv). This is considerably fewer than in recent worker and CT scan studies, which have greater statistical power to detect an effect, and with greater precision.
Missing radiation doses were disproportionately among the more highly exposed	In both Hiroshima and Nagasaki, about 3,500 survivors with 'missing doses', who were close to the bombs and therefore highly exposed, generally young and commonly migrated out of the bombed cities, were disproportionately excluded from analyses.
Immortal person time	Survivors were enrolled into the Life Span Study in 1950, but estimation of radiation doses and counting of cancer cases among those enrolled were not complete until 1965. Thus some people were included in the study for up to 15 years during which time any cancers they developed were not included.
Radiation-related cancers are still occurring	Particularly for those exposed at a young age for whom the risks are highest, excess cancer cases will continue to occur as long as they live.
Stigma and discrimination	Many *hibakusha* suffered ostracism, discrimination, and social isolation with reduced opportunities for marriage, providing motivation for concealment of their status as *hibakusha*.

Sources: Author's original work; Richardson, Wing and Cole (2013); Mathews et al. (2013).

Powerful new epidemiological studies over the past decade have provided estimates both more accurate and demonstrating greater radiation-related health risks than previously estimated (Kitahara et al. 2015). These studies are made possible by electronically linking data on radiation exposure, especially at low doses, and health outcomes for large numbers of people, such as for children who underwent a CT scan funded by national health insurance, who subsequently developed a cancer reported to their local cancer registry. The most important of these new studies are outlined below.

Childhood leukaemia near nuclear power plants

Apparent excesses of leukaemia occurring in children living near nuclear power plants have caused concern and controversy over decades. Perhaps the most prominent was an excess of leukaemia and lymphoma cases around the Sellafield nuclear plant in England in the 1980s, which was the location of the Windscale accident and fire in 1957, and, before the 1986 Chernobyl disaster, the most radioactively polluting nuclear facility in Europe. An investigation recommended by a government-commissioned committee unexpectedly found that the risks for leukaemia and lymphoma were higher in children born within 5 km of Sellafield, and in children with fathers employed at the plant, particularly those recording high radiation doses before their child's conception (Gardner et al. 1990). A 2007 meta-analysis supported by the US Department of Energy examined all of the reliable data available worldwide, confirming a statistically significant increase in leukaemia for children living near nuclear power plants (Baker and Hoel 2007).

The most definitive findings come from a large national German study, which examined leukaemia among children living near any of Germany's 16 operating nuclear plants over a 25-year period. It showed that the risk of leukaemia more than doubled for children living within 5 km of a nuclear plant, with elevated risk extending beyond 50 km from a plant (Kaatsch et al. 2008). This finding was highly statistically significant. A subsequent but less powerful study in France found a similar increase. While these findings have been challenged on grounds that they are not explicable on the basis of prevailing estimates of the radiation exposures involved and their predicted effects, this in no way changes the strength of the association, whatever its cause, and no possible cause other than radiation has been identified. Actual data on real-world outcomes always trumps any theoretical model.

Childhood cancer following CT scans

A major part of growing medical radiation exposure worldwide is due to CT scans. These use X-rays to take spiralled images to show closely spaced cross-sections of the body, and involve effective whole-body exposures of 1 to 10 or more (up to 20+) mSv. A number of studies have now documented cancer risks following CT scans in children that are much greater than previously estimated. The largest to date is an Australian study of cancer risk after CT scans in 680,000 young people (aged less than 20 years), compared with the 10.3 million young Australians who did not have CT scans, over the same 20-year period (Mathews et al. 2013). The study involved 10 times as many people exposed and four times the total radiation dose as the Japanese survivor data for low doses of radiation (approximately 70,000 people who received less than 100 mSv).

The CT study demonstrated a 24 per cent increase in cancer in the decade following one CT scan delivering an average effective dose of only 4.5 mSv, and 16 per cent greater cancer risk for each additional scan (Mathews et al. 2013). Cancers occurred as early as two years after exposure. The average length of follow-up after the first CT scan was close to a decade, so new cancers will continue to occur through the life of exposed individuals. For similar ages of exposure and lengths of follow-up, the risk for leukaemia related to CT radiation was similar to that among *hibakusha*; however, the risk of solid cancer in the more powerful CT study was 12.5 times higher for brain cancer (Smoll et al. 2016) and nine times higher for solid cancers overall (Mathews et al. 2013) than in the *hibakusha* studies. The findings for leukaemia and brain cancer are quite similar in the Australian study and a smaller British study (which did not include other solid cancers) (Pearce et al. 2012).

The Australian study is now the largest population-based study of low-dose radiation ever conducted, in children who are the group most susceptible to radiation, giving its results great importance. These studies fill important gaps in the *hibakusha* studies regarding low doses, early onset cancers, and children. Longer term follow-up of these children and examining the risks associated with nuclear medicine procedures are underway and can be expected to yield important new findings in coming years. Already the results of these studies warrant upward revision of radiation risk estimates and reduction of recommended radiation dose limits in order to effectively protect the most vulnerable. One strong likelihood is that the dose–response curve for radiation-related cancer risk is not linear as generally assumed, but steeper at low doses, with a greater effect per mSv at low doses than at higher doses, particularly for children (Smoll et al. 2016). It is also likely that the greatest increase in cancers related to radiation is in those occurring early after exposure, among people most susceptible.

Cancer risks for nuclear industry workers

Updated results of large long-term studies of hundreds of thousands of nuclear industry workers coordinated by the International Agency for Research on Cancer on risks for leukaemia (Leuraud et al. 2015) and cancer (Richardson et al. 2015) were reported in 2015. The studies included 308,000 workers from France, the UK, and the US, some of them followed up since 1944, with a mean follow-up period of 26 years, to an average age of 58 years, and involving total measured colon radiation dose (a common measure of internal organ exposure) more than five times the collective dose received by *hibakusha* who received low doses. The mean dose rate for the workers involved was only 1.1 mGy per year, less than background radiation in most places, with cumulative doses well within the current most widely recommended dose limit for nuclear industry workers of an average of no more than 20 mSv per year (the average total dose received by each worker in the study during their average 12-year employment in the industry was close to 20 mSv).

The solid cancer risk was statistically compatible with, but 50 per cent higher than, that in 20–60-year-old male *hibakusha*, and will continue to rise as the subjects age. The leukaemia risk identified was similar to that in 20–60-year-old male *hibukusha*. It is important to note that at the average age of workers in the study of 58 years, the incidence of cancer and chronic diseases is beginning to accelerate.

These large and powerful studies show risks even at very low-dose rates and doses well within recommended occupational limits. They do not support a reduction of risk for the same total dose if the dose is delivered over a longer time (low-dose rates compared with high-dose rates). BEIR VII assumes a dose rate reduction factor for low-dose rates of 1.5 and a number of radiation protection bodies such as the ICRP use a factor of 2, whereas such reduction factors were abandoned by the World Health Organization (WHO) in its 2013 report on health consequences of the Fukushima nuclear disaster (WHO 2013), and by the United Nations Scientific Committee on the Effects of Atomic Radiation (UNSCEAR) in 2013 (UNSCEAR 2014). Together, the CT scan and worker studies conclusively demonstrate the absence of a threshold for ionising radiation-related cancer risk.

Cancer and other health effects in people exposed to the Chernobyl and Fukushima Daiichi nuclear disasters

The effects of the 1986 Chernobyl nuclear disaster have recently been independently reviewed (Fairlie 2016). Major findings include:

- an estimated 40,000 excess fatal cancers in Europe by 2065;

- 6,000 additional cases of thyroid cancer have already occurred. An additional 10,000 are expected by 2065. Initially these were almost exclusively in children; more recently, cases are also occurring at older ages. Increases in thyroid cancer have also been found in a number of other countries, such as Austria, Slovakia, the Czech Republic, and Poland. It is likely that at least some of this increase is due to Chernobyl;

- increasing rates of leukaemia and thyroid cancer among the estimated total 600,000–800,000 clean-up workers, as well as increased risk of cataracts at a lower threshold dose than previously thought (100–250 mGy);

- despite international agencies assuming that no increases in congenital malformations will be detectable in Chernobyl-contaminated areas, increases in nervous system birth defects have been found in the highly contaminated Rivne-Polissia region of Ukraine. These include neural tube defects like spina bifida, anencephaly, microcephaly, and small or missing eyes (Dancause et al. 2010);

- increasing rates of breast cancer in the most contaminated regions of Belarus and Ukraine; and

- dislocation of lives due to radioactive contamination and long-term worry about radiation risks can also have adverse health consequences—among clean-up workers, depression and post-traumatic stress disorder rates are elevated even decades later, and mothers of exposed young children are at high risk of depression, anxiety, and other mental disorders.

While various Japanese and international agencies stated that no radiation-related adverse health consequences were likely to be detected as a result of the Fukushima nuclear disaster, this implausible assessment has already been shown to be in error. The Japanese government's Reconstruction

Agency estimated 3,407 nuclear-disaster related deaths to early 2016 in Fukushima prefecture (including due to inadequate evacuations and continuing care of chronically ill patients in contaminated regions, and suicides). A lack of comprehensive health screening and follow-up for the exposed population and inadequate cancer registries in many of the relevant areas of Japan mean that the capacity to detect and respond to health problems is constrained.

The one area where more effective screening is taking place in Fukushima prefecture (but not in other fallout-affected areas) is in periodic ultrasound examinations of the thyroid glands of children aged less than 18 years at the time of the disaster. Even though 24–29 per cent of the eligible population have not participated (to September 2016), such an active search for thyroid abnormalities can be expected to find more cysts and nodules than would come to medical attention in the absence of an active screening program, the findings to date suggest that despite thyroid radiation doses being estimated to be much lower in Fukushima than following the Chernobyl disaster, early evidence of an epidemic of thyroid cancer appears to be emerging. This evidence is summarised on the basis of 113 thyroid cancers in children in Fukushima diagnosed to the end of 2015, including 51 diagnosed in the second round of ultrasound screening (Tsuda et al. 2016a, 2016b):

- the rates of thyroid cancer detected initially in Fukushima were between 20 and 50 times higher than the Japanese national average;
- among the cancers diagnosed on the second ultrasound screening, two years after the first, the rate is still 20 to 38 times the national average, likely too great a difference to be explained by active screening alone;
- within Fukushima prefecture, the rate in the most contaminated district was 2.6 times higher than in the least contaminated areas; and
- the cancers diagnosed were not disproportionately benign—92 per cent of the operated cases had spread outside the thyroid gland, to lymph nodes, or to distant organs.

To September 2016, the number of reported thyroid cancer cases among children screened in Fukushima had increased to 145.

Radiation effects for non-human species living in contaminated areas

Timothy Mousseau and Anders Møller's chapter in this volume outlines the extraordinary body of evidence—much of it gathered by them—on the effects of radiation on non-human biota living in contaminated areas of Chernobyl and Fukushima. In virtually every species and ecological community studied—from soil bacteria and fungi through trees, to various insects, spiders, diverse birds, and large and small mammals—adverse biological effects have been found, in direct proportion to the degree of radioactive contamination, without any apparent threshold, and with most effects apparent across the range of 1–10 mGy per year.

It is biologically implausible that humans will be immune from similar effects. As for recent human data, there is a similar trend for non-human species of larger biological impacts of radiation to be identified than previously recognised. The work of Mousseau and Møller demonstrates the importance of assessing effects on organisms in the real ecological context and conditions under which they live, and of including the most vulnerable developmental stages, tissues and organs, including gametes and embryos.

A brief history of radiation and health

Delays in translating evidence into policy and practice, undue influence of vested interests, and corruption of science and medicine

Historically, it has often taken decades, in some cases many decades, for evidence about public health risks to be translated into policy and action, particularly when the risks are associated with long lag periods between exposures and consequences, and when there are powerful vested interests at stake. Examples include smoking, asbestos, alcohol, unhealthy 'junk' food, fossil fuels, and low-dose radiation. For radiation and fossil fuels especially, an additional insidious factor fuelling denial, inattention, and inertia is that most victims are not able to be identified with a face and a name. There is generally no way to distinguish a cancer (or a heart attack) caused by radiation from one caused by smoking or chemical exposures or other factors, and most cancers have a number of interacting causes.

So many radiation victims cannot be personally identified. They melt into the crowd. Radiation cannot be seen or felt or smelt or tasted. Except for acute radiation sickness occurring following high doses, you cannot feel it doing you harm. Long-term genetic damage and cancer typically manifest years, often decades, later. These factors mean that effects of radiation are often inadequately recognised or downplayed. However, this does not make the people and families affected any less real, or less deserving of efforts to prevent suffering and premature death. An example is the global fallout from atmospheric nuclear weapons test explosions, which every human being carries in their body, and over time will cause over 2 million cancer deaths and as many non-fatal cancers (Ruff 2015). Yet apart from some test veterans and some members of downwind communities, most of the individuals suffering and dying from these excess cancers will not be personally identifiable.

A wider and historical context is crucial to understanding the dynamics, institutions, and conflict that enmeshes the field and evidence of radiation health effects. Nuclear weapons are by far the most destructive of all weapons, constituting the paramount existential threat to life on Earth. Yet they are central to the politics and practices of some states, including the largest. Almost half a century after nuclear-armed states made legally binding commitments in the Nuclear Non-Proliferation Treaty to abolish their nuclear weapons, there are still 14,930 nuclear weapons (Kristensen and Norris 2017), no disarmament negotiations underway, and all nine nuclear-armed states are investing massively (a projected US$1 trillion investment over the next 30 years in the US alone) in the modernisation and deployment of their nuclear weapons for the indefinite future (Kristensen and McKinzie 2015). Treaties to prohibit and eliminate chemical and biological weapons, landmines, and cluster munitions are largely being implemented. A historic treaty that for the first time comprehensively prohibits nuclear weapons was finally adopted at the United Nations (UN) on 7 July 2017 by 122 nations. The treaty is based on the clear evidence that the most indiscriminate and inhumane weapons of all cannot be used for any legitimate purpose and effectively constitute global suicide bombs. However, the treaty has been opposed by all nuclear-dependent and nuclear-armed states, which continue to refuse to eliminate their nuclear arsenals.

Over the past 70 years, the perceived potency of nuclear weapons as instruments of state power has been a driver not only for their proliferation, but for the uranium enrichment plants, nuclear reactors, and spent fuel reprocessing plants that provide the capacity and materials to build them. Following US President Dwight Eisenhower's 'Atoms for Peace' speech at the UN in 1953, nuclear research and isotope production reactors and nuclear power plants were actively promoted globally. US Department of Defense consultant Stefan Possony advocated to the Psychology Strategy Board in 1953: '[T]he atomic bomb will be accepted far more readily if at the same time atomic energy is being used for constructive ends' (quoted in Kuznick 2011). US Atomic Energy Commissioner Thomas Murray promoted nuclear power specifically in Japan:

> Now, while the memory of Hiroshima and Nagasaki remain so vivid, construction of such a power plant in a country like Japan would be a dramatic and Christian gesture which could lift all of us far above recollection of the carnage of those cities (quoted in Kuznick 2011).

Beginning with the Manhattan Project, accompanying the massive investment in burgeoning Cold War nuclear arsenals, donation of research reactors by both the US and the USSR to over 100 countries, and promotion and subsidisation of nuclear power, there has been a concerted and ongoing effort to manipulate, distort, downplay, and sideline evidence of the extent of radiation-related health risks. This has involved the large institutions, both government and private, with strong funding, career, institutional, political, and commercial vested interests in either or both nuclear weapons and reactors. Unwelcome research has frequently been suppressed and shut down; independent researchers collecting and gathering unwelcome evidence have been undermined, de-funded, dismissed, and discredited. While a detailed history of radiation health is not feasible here, it is important to recognise the courage and salute the scientific contributions of independent scientists and physicians like Alice Stewart, George Kneale, Thomas Mancuso, Edward Martell, and Carl Johnson, to name but a few who have suffered because their seminal scientific work on the extent of radiation health risks was unwelcome for powerful vested nuclear interests (Quigley, Lowman, and Wing 2012).

A tobacco industry executive in 1969 explained: 'Doubt is our product since it is the best means of competing with the "body of fact" that exists in the mind of the general public. It is also the means of establishing a controversy' (Brown and Williamson, Minnesota Lawsuit 1969). Similar denial and minimisation of risks and promotion of a perception of controversy continue to be widespread in the field of radiation health risks. A brief discussion illustrating some concerns about key institutions in the field follows.

The World Health Organization

As the world's lead technical agency in health, with its assembly including all nations, WHO has great authority and convening power. Its reports, recommendations, technical standards, and guidelines command attention and respect. However, WHO's capacity is constrained by chronic budgetary shortages. In radiation and health there has been a lack of leadership as well as capacity, exacerbated by what is widely perceived as excessive deference and inadequate independence of WHO in relation to the International Atomic Energy Agency (IAEA). A 1959 agreement between the agencies stipulates: 'Whenever either organization proposes to initiate a program or activity on a subject in which the other organization has or may have a substantial interest, the first party shall consult the other with a view to adjusting the matter by mutual agreement' (IAEA 1959: Article 1.3).

The IAEA is structurally conflicted, as both global regulator of nuclear industry standards and safeguards, while at the same time promoting nuclear technology including power—in effect promoting the means for nuclear weapons proliferation, which it is also mandated to discourage (IAEA 1957). On some significant occasions, such as the 2006 multi-agency UN report on the consequences of the Chernobyl nuclear disaster, the IAEA took the lead role in the report's conclusions, dissemination, and public comments while inappropriately downplaying the health impacts of the disaster (IAEA 2006). The report claimed only 4,000 excess deaths could be expected from the Chernobyl disaster; whereas the International Agency for Research on Cancer—WHO-linked but more research-active and independent—in the same year estimated 41,000 cancer cases and 16,000 (6,700–38,000) deaths to 2065 (Cardis et al. 2006).

WHO produced two landmark reports on the health effects of nuclear war (WHO 1983, 1987), the greatest potential source of radioactive exposure and contamination. The World Health Assembly recognised 'that nuclear weapons constitute the greatest immediate threat to the health and welfare of [hu]mankind' (WHO 1983). In 1987, the assembly decided that investigation of the health effects of nuclear war should be continued, and requested the WHO director-general to report periodically to the assembly on progress in this field. However no such follow-up has occurred.

In relation to the Fukushima nuclear disaster, WHO's role has been essentially limited to compiling a report on radiation doses, and a report assessing the health risks of those exposures (WHO 2012, 2013). It has had no active role in international input into the ongoing public health management of the disaster and protecting the health of the affected population. Raising questions about the Japanese government and about WHO's independence are reports of pressure by Vice-Minister Shinji Asonuma directly on the WHO director-general to reduce the estimated radiation doses to the thyroid glands of Japanese children as a result of the Fukushima nuclear disaster. The reported initial WHO estimates of 300–1,000 mSv in more contaminated areas and 10–100 mSv in Tokyo and Osaka were reduced to 100–200 mSv and 1–10 mSv respectively in the final WHO report, with the Japanese government reported to have sought further downward revision until moments before the report's release (*Asahi Shimbun GLOBE* 2014).

WHO's chronic budget crisis weakens its capacity, and its dependence on the goodwill of governments for a large proportion of its income undermines its independence. The WHO budget has increased from US$1.4 billion in 1990–91 to US$4.4 billion projected in 2016–17, yet over this time the assessed (regular) contributions of governments have remained stagnant at less than US$1 billion annually. WHO thus now depends on voluntary contributions—many of them earmarked for specified purposes—from governments and charities for 79 per cent of its budget (WHO 2015).

The International Commission on Radiological Protection

The ICRP was established in 1928 and has had a major role in recommending radiation protection standards. It claims to be an independent organisation, committed to public benefit, impartial, transparent, and accountable. According to its 2014 annual report, it consists of 232 individuals from over 30 countries, nominated through an open process and invited to join as independent experts on a voluntary basis. Formally an independent registered charity in the UK, with a small secretariat in Canada, the ICRP operates as something of a club. Even a brief review of its website in July 2016 raises serious questions about its independence. Its assistant scientific secretary, Dr Haruyuki Ogino, is provided cost-free by the Central Research Institute of Electric Power Industry in Japan, that is, the commercial Japanese nuclear reactor operators. The ICRP membership includes a large number of employees of governments active in nuclear power and/or weapons, and employees of companies with clear vested nuclear interests, including Cameco, a large uranium mining company, Areva Resources Canada Inc., and nuclear reactor operators in the US, Europe, and Japan and their associations. A number of nuclear corporations also provide funding. These multiple layers of close involvement in a scientific body by vested interests is clearly improper by any usual standard.

Damning evidence of corruption and undue influence of vested interests in the ICRP was unearthed by the Fukushima Nuclear Accident Independent Investigation Commission established by the Japanese Diet (parliament) (NAIIC 2012). The commission found internal documents showing that the Federation of Electric Power Companies (FEPC) in Japan successfully lobbied radiation specialists, including members of the ICRP, so as to relax radiation protection standards, and minimise the radiation protection standards adopted in Japan. It emerged that the FEPC covered international travel costs for the ICRP's members from Japan under the guise of paying expenses for another group. The FEPC documents stated '[d]ose constraints on occupational exposure should not be covered by regulation ... Special dose limits for women, special medical checkups for workers ... and legislated dose limits in case of emergency exposure should be abolished'. The outcome was that '[a]ll the views of the operators concerning the ICRP 2007 recommendations were reflected' (NAIIC 2012: Chapter 5, section 5.2.3). The ICRP has to date given no indication that it has addressed such corrupt and improper practices.

The Diet Commission also described reactor operators as seeking 'to steer research concerning the health effects of radiation in a direction that would find less damage and to steer the views of experts in Japan and elsewhere concerning radiation protection in a direction that would relax protection and control'. The FEPC documents stated, '[i]f it can be proven scientifically that the effects of dosage does not accumulate, significant relaxation including the review of dose limits can be expected in the future'. Mr Sakae Muto, former vice president of the Tokyo Electric Power Company, or TEPCO (owner and operator of the Fukushima nuclear plants), advised, '[k]eep an eye on the research trends so they won't be hijacked by bad researchers and pushed in a bad direction' (NAIIC 2012: Chapter 5, section 5.2.3).

The United Nations Scientific Committee on the Effects of Atomic Radiation

In the face of growing global concern and protest about atmospheric nuclear test explosions, their global radioactive fallout, and increasing levels of strontium-90 in the teeth of babies worldwide, UNSCEAR was established by the UN General Assembly in 1955. According to its website, 'purportedly with the intention to deflect a proposal calling for an immediate end to all nuclear explosions, it was proposed to the General Assembly of the United Nations that it establish a Committee to collect and evaluate information on the levels and effects of ionizing radiation' (UNSCEAR 2016). The creation of UNSCEAR was also related to the suspicion held by nuclear-armed states of an international movement of scientists through the United Nations Educational, Scientific and Cultural Organization and the International Council of Scientific Unions to sponsor scientific study of nuclear test fallout independently of governments (Herran 2014). Nestor Herran describes the scientific hegemony of the US and the UK in the early years of UNSCEAR, and the establishment of a pattern of downplaying the hazards of radioactive fallout—for example in the first major UNSCEAR report of 1958, omitting reference to carbon-14, which is responsible for the bulk of long-term human radiation exposure caused by nuclear tests (Ruff 2015).

Fifteen member states were initially designated as members; additional states were added in 1973 and 2011, to a current total of 27. Except for Sudan, which in 2016 signed a framework agreement with China for construction of nuclear power reactors, all the member states currently have nuclear weapons, nuclear power plants, and/or nuclear research reactors. Their UNSCEAR representatives and experts are government-appointed, and are generally staff of their nuclear power or nuclear regulatory agencies. They therefore cannot be regarded as independent experts appointed on their scientific and medical merits. Some have close links with the nuclear industry, and there is overlap in membership with the ICRP. A notable example is Dr Douglas R. Boreham, who was the Canadian delegate to UNSCEAR in 2012 and a member of the Canadian delegation in other years, and who works for Bruce Power, a nuclear power company in Ontario. Dr Boreham's views on radiation risks do not reflect available evidence or the views of national or international radiation protection bodies in his repeated assertions that low-dose radiation does more good than harm, for example: 'If anything, there is health benefit instead of risk at low levels of exposure' (Higson et al. 2007: 259), and that CT scans may reduce rather than increase the risk of cancer (Scott et al. 2008). Dr Boreham has made at least three visits to Australia on behalf of uranium mining companies Toro Energy, Uranium One, and Heathgate Resources to undertake 'employee radiation training' and 'community consultation on radiation and health'—a euphemism for events designed to downplay and foment confusion about radiation risks (Toro Energy Limited 2008; MAPW 2012).

While many would hope and assume that international bodies like UNSCEAR and the ICRP would be pillars of independent rigour and scientific integrity protecting global health, like the extensive peer-reviewed processes of the Intergovernmental Panel on Climate Change, in contrast these bodies are excessively dominated by vested government and commercial interests, lacking in transparency, inadequate in their ethical practices related to conflicts of interest of their members, and with a recurring pattern of selective interpretation of evidence and downplaying radiation risks. For example, UNSCEAR reports for years have dismissed as inconsistent with their preferred models the growing body of empiric evidence of significant effects of Chernobyl and Fukushima contamination on plants and animals discussed in Mousseau and Møller's chapter in this volume, and have ignored much of the evidence of health effects of the Chernobyl disaster.

Potential future large-scale population radiation exposures

While life on Earth has evolved with the constant evolutionary pressure and biological risks associated with background radiation, the advent of nuclear reactors and weapons has created unprecedented potential for radioactive releases enormous in size as well as extent in time and space.

Nuclear weapons

The nuclear bombings of Hiroshima and Nagasaki, leaks and waste from nuclear weapons production, and a total of 2,056 nuclear test explosions are together responsible for the largest environmental radioactive contamination by human hands, and will continue to exact a toll in human health for many millennia (Ruff 2015). However, these pale in comparison to the potential population radiation exposure following a nuclear war (WHO 1987; IOM 1986), even one using a small fraction of the current global arsenal of 14,930 nuclear weapons (Kristensen and Norris 2017). The consequences of any given dose of radiation will be far worse after nuclear war because they will invariably co-exist with multiple other injuries, stresses, and health risks—and effectively functioning health services cannot be expected.

The risk of nuclear war is an ever-present danger as long as nuclear weapons exist, and is generally assessed by those most knowledgeable to be rising as a result of failure to implement disarmament; extensive nuclear modernisation programs of all the nuclear-armed states; 1,800 weapons still on high alert ready to be launched within a few minutes; growing risks of cyber-attack on nuclear command and control systems; deteriorating relationships and increasingly aggressive posturing of nuclear forces between the US/North Atlantic Treaty Organization (NATO) and Russia, between India and Pakistan, and in the South China Sea and the Korean Peninsula; and policies in many nuclear-armed states to use nuclear weapons first and potentially early in escalation of an armed conflict (Helfand et al. 2016). The hands of the *Bulletin of the Atomic Scientists* Doomsday Clock were moved from five to three minutes to midnight in 2015, kept there in 2016, and moved forward to two and a half minutes to midnight in 2017. Former US Defense Secretary William Perry and former Russian Foreign Minister Igor Ivanov are among those assessing the danger of nuclear war being greater now than it was during the Cold War, and growing (Helfand et al. 2016).

No effective medical response is feasible in the event of even a single nuclear detonation in a city, and the urgent elimination of nuclear weapons is the only means to durably prevent their otherwise inevitable use, and is therefore an absolute global health imperative. The decision by the United Nations General Assembly (2016: 4) to convene negotiations in 2017 on a 'legally binding instrument to prohibit nuclear weapons, leading towards their total elimination' is a historic opportunity to break the logjam in nuclear disarmament. In addition to prohibiting and eliminating nuclear weapons arsenals, achieving and sustaining a world freed from nuclear weapons will require controlling and eliminating the fissile materials—highly enriched uranium and separated plutonium—from which nuclear weapons can be built. This will require ending production and disposing of stockpiles as irreversibly as possible. It is the intrinsic capacity for uranium enrichment plants to be used to produce highly enriched uranium and for the plutonium inevitably created in nuclear reactors to be extracted from spent reactor fuel, that are the greatest planetary health dangers associated with the generation of nuclear power.

Ending production of highly enriched uranium (von Hippel and IPFM 2016), placing all uranium enrichment plants under international control (Diesendorf 2014), and ending reprocessing of spent nuclear fuel to extract plutonium (IPFM 2015b) would be important steps towards a safer world.

Radioactive releases from nuclear facilities

Nuclear facilities release radioactivity during routine operations, and minor accidents are very common. At nuclear power plants, large amounts of highly radioactive and long-lived materials accumulate in the reactor fuel, both in reactors and in the spent fuel pools where the intensely hot and radioactive used fuel must be cooled in circulating water for three to five years before it can be placed in dry storage. These pools do not have the multiple engineered layers of containment that reactors do, but are covered by a simple building. At the Fukushima Daiichi site at the time of the 2011 disaster, 70 per cent of the total onsite radioactivity was in the spent fuel pools (Stohl et al. 2011). Like reactors, these pools are vulnerable to fire and explosion if the continuous water cooling and the power that drives it are interrupted even briefly.

Official and vendor accounts state that the Fukushima Daiichi reactor core meltdowns only occurred as a result of the tsunami and not the earthquake itself. However, there is evidence that radioactive leaks began after the earthquake struck and before the tsunami hit (Stohl et al. 2011; NAIIC 2012). This has implications for every nuclear plant worldwide. Disruption to reactor and spent fuel pools cooling could occur as a result of some combination of poor design, construction faults, and natural disaster like earthquakes and tsunamis, as on 11 March 2011; or deliberately, by physical disruption to power supply or cooling water supply or circulation; or potentially by cyber-attack. The Israeli/ US-developed Stuxnet computer worm exposed in 2010 is perhaps the best-known example of a number of cyber-attacks on nuclear facilities since at least 1992. Stuxnet targeted the Siemens Step 7 SCADA system used by Iran's nuclear facilities, making uranium enrichment centrifuges spin too fast, partially destroying around 1,000 of them (Baylon, with Brunt and Livingstone 2015). An average-sized, 1 gigawatt nuclear power reactor contains more (and longer-lived) radioactivity than is released by a 1 megaton nuclear bomb (67 times the explosive yield of the bomb that destroyed Hiroshima). Thus, each of the 402 nuclear power reactors operating as at 1 July 2016 (Schneider et al. 2016) is in fact a potential huge, pre-positioned, radiological terrorist weapon ('dirty bomb').

There have been about 20 accidents worldwide since the early 1950s known to have resulted in melting of reactor cores (Burns, Ewing, and Navrotsky 2012). These have occurred in military and civilian reactors of various designs in different countries. Not all have resulted in releases of radioactivity into the environment, although all had the potential to do so. There have also been a total of 20 nuclear accidents categorised as Level 4 or higher at the severe end of the International Nuclear and Radiological Event Scale (INES). Level 4 accidents have local consequences, including 'release of significant quantities of radioactive material within an installation with a high probability of significant public exposure' (IAEA n.d.). Such accidents have occurred in Argentina, Canada, France, Japan, Slovakia, Switzerland, the UK, the US, and the USSR/Russia (Lelieveld, Kunkel, and Lawrence 2012). The actual historic frequency of core melt accidents is around one per 800 years of reactor operation (reactor years). For boiling water reactors with Mark 1 or 2 (early design) containments, like the Fukushima Daiichi reactors and many US reactors, the historic frequency of core melt accidents is around one in 630 years (Cochran 2011). Thus, with 400-odd nuclear reactors operating, a core melt accident can be expected every few years.

In Lelieveld and colleagues' analysis of the global frequency and fallout from major nuclear power reactor accidents (Lelieveld, Kunkel, and Lawrence 2012), on average more than 90 per cent of emitted cesium-137 would be transported beyond 50 km and about 50 per cent beyond 1,000 km. They define more than 37 kilobecquerel per square metre (kBq/m^2) of cesium-137 as significant contamination, being associated with a human radiation dose during the first year after a major accident of about 1mSv. Using the Chernobyl disaster as the basis for estimating consequences, and the historic frequency of the worst category of accidents (INES Level 7—Chernobyl and Fukushima), they estimate that there are large parts of North America, East Asia, and Europe with risks of being contaminated by a major nuclear accident of more than 1 per cent per year. The average area contaminated with more than 40 kBq/m^2 of cesium-137 after a catastrophic core melt is 138,000 km^2, affecting an average of 28 million people if such an accident occurred in Western Europe, and 34 million people if in South Asia.

Dispersal of radioactive materials

Dispersal of radioactive material in water or food supplies or with conventional explosives is technically simple, and there is abundant potential for access to radioactive materials. The most hazardous radiological materials available in large quantities are high-level radioactive waste (consisting of spent reactor fuel and wastes left from reprocessing of spent reactor fuel to extract plutonium) and plutonium separated from spent fuel. Around 505 tonnes of separated plutonium exists worldwide, about half of it in civilian facilities (IPFM 2015a). High-level waste is dangerously radioactive for many hundreds of thousands of years and must be strictly isolated from groundwater and the biosphere for up to 1 million years. The average nuclear power reactor produces 30 tonnes of high-level waste each year—around 12,000 tonnes annually worldwide, with a global total stockpile in 2015 of about 390,000 tonnes. No country yet has a functioning permanent repository for such waste, and the possibility of such material being stolen or diverted, particularly from a spent fuel reprocessing plant, and then deliberately dispersed in one or multiple cities is real and will exist over geological time frames. Such events could cause significant localised radioactive contamination.

References

Asahi Shimbun GLOBE, 2014. Revision demanded of the Fukushima radiation exposure report. 7 December.

Baker P. J., and D. G. Hoel, 2007. Meta analysis of standardised incidence and mortality rates of childhood leukaemia in proximity to nuclear facilities. *European Journal of Cancer Care* 16: 355–63. doi. org/10.1111/j.1365-2354.2007.00679.x

Baylon, Caroline, with Roger Brunt, and David Livingstone, 2015. Cyber security at civil nuclear facilities: Understanding the risks. Chatham House Report. London: Royal Institute of International Affairs.

Brown and Williamson, Minnesota Lawsuit, 1969. Smoking and health proposal. Truth Tobacco Industry Documents. www.industrydocumentslibrary.ucsf.edu/tobacco/docs/#id=psdw0147 (accessed 19 January 2017).

Burns, Peter C., Rodney C. Ewing, and Alexandra Navrotsky, 2012. Nuclear fuel in a reactor accident. *Science* 335: 1184–8. doi.org/10.1126/science.1211285

Cabinet Office, Consumer Affairs Agency, Reconstruction Agency, Ministry of Foreign Affairs, Ministry of Education, Culture, Sports, Science and Technology, Ministry of Health, Labour and Welfare, Ministry of Agriculture, Forestry and Fisheries, Ministry of Economy, Trade and Industry, Ministry of the Environment, and Secretariat of the Nuclear Regulation Authority (Japan), 2016. Basic information on radiation risk. February. www.reconstruction.go.jp/english/topics/RR/index.html (accessed 19 January 2017).

Cardis, Elizabeth, Daniel Krewski, Mathieu Boniol, Vladimir Drozdovitch, Sarah C. Darby, Ethel S. Gilbert, Suminori Akiba, Jacques Benichou, Jacques Ferlay, Sarah Gandini, Catherine Hill, Geoffrey Howe, Ausrele Kesminiene, Mirjana Moser, Marie Sanchez, Hans Storm, Laurent Voisin, and Peter Boyle, 2006. Estimates of the cancer burden in Europe from radioactive fallout from the Chernobyl accident. *International Journal of Cancer* 119: 1224–35. doi.org/10.1002/ijc.22037

Centers for Disease Control and Prevention, 2015. Radiation and your health. 7 December. www.cdc.gov/nceh/radiation/health.html (accessed 6 March 2017).

Cochran, Thomas B., 2011. Statement on the Fukushima nuclear disaster and its implications for US nuclear power reactors. Testimony to the US Senate Joint Hearings of the Subcommittee on Clean Air and Nuclear Safety and the Committee on Environment and Public Works, 12 April. www.nrdc.org/sites/default/files/tcochran_110412.pdf (accessed 19 January 2017).

Committee to Assess Health Risks from Exposure to Low Levels of Ionizing Radiation, 2006. *Health Risks from Exposure to Low Levels of Ionizing Radiation: BEIR VII, Phase 2*. Washington, DC: National Academies Press.

Dancause, Kelsey Needham, Lyubov Yevtushok, Serhiy Lapchenko, Ihor Shumlyansky, Genadiy Shevchenko, Wladimir Wertelecki, and Ralph M. Garruto, 2010. Chronic radiation exposure in the Rivne-Polissia region of Ukraine: Implications for birth defects. *American Journal of Human Biology* 22(5): 667–74. doi.org/10.1002/ajhb.21063

Diesendorf, Mark, 2014. *Sustainable Energy Solutions for Climate Change*. Sydney: UNSW Press.

Doll, Richard, and Richard Wakeford, 1997. Risk of childhood cancer from fetal irradiation. *British Journal of Radiology* 70: 130–9. doi.org/10.1259/bjr.70.830.9135438

European Nuclear Society, n.d. Radiation weighting factors. www.euronuclear.org/info/encyclopedia/r/radiation-weight-factor.htm (accessed 6 March 2017).

Fairlie, Ian, 2016. TORCH-2016: An independent scientific evaluation of the health-related effects of the Chernobyl nuclear disaster, Version 1.1. Vienna: GLOBAL 2000/Friends of the Earth Austria/Vienna Office for Environmental Protection, 31 March. www.global2000.at/sites/global/files/GLOBAL_TORCH%202016_rz_WEB_KORR.pdf (accessed 19 January 2017).

Gardner, Martin J., Michael P. Snee, Andrew J. Hall, Caroline A. Powell, Susan Downes, and John D. Terrell, 1990. Results of a case-control study of leukaemia and lymphoma among young people near Sellafield nuclear plant in West Cumbria. *British Medical Journal* 300: 423–9. doi.org/10.1136/bmj.300.6722.423

Gray, Alastair, Simon Read, Paul McGale, and Sarah Darby, 2009. Lung cancer deaths from indoor radon and the cost-effectiveness and potential of policies to reduce them. *British Medical Journal* 338: a3110. doi.org/10.1136/bmj.a3110

Helfand, Ira, Andy Haines, Hans Kristensen, Patricia Lewis, Zia Mian, and Tilman Ruff, 2016. The growing threat of nuclear war and the role of the health community. *World Medical Journal* 62(3): 86–94.

Herran, Nestor, 2014. 'Unscare' and conceal: The United Nations Scientific Committee on the effects of atomic radiation and the origin of international radiation monitoring. In *The Surveillance Imperative: Geosciences during the Cold War and Beyond*, edited by Simone Turchetti and Peder Roberts, 69–84. New York: Palgrave Macmillan. doi.org/10.1057/9781137438744_4

Higson, D. J, D. R. Boreham, A. L. Brooks, Y-C. Luan, R. E. Mitchel, J. Strzelczyk, and P. J. Sykes, 2007. Effects of low doses of radiation: Joint statement from the following participants at the 15th Pacific Basin Nuclear Conference, sessions held in Sydney, Australia, Wednesday 18 October 2006. *Dose-Response* 5(4): 259–62. doi.org/10.2203/dose-response.07-017.Higson

IAEA (International Atomic Energy Agency), 1957. The Statute of the IAEA (ratified 29 July 1957). www.iaea.org/about/statute (accessed 19 January 2017).

IAEA (International Atomic Energy Agency), 1959. Agreement between the International Atomic Energy Agency and the World Health Organization. www.ippnw.de/commonFiles/pdfs/Atomenergie/Agreement_WHO-IAEA.pdf (accessed 19 January 2017).

IAEA (International Atomic Energy Agency), 2006. Chernobyl's legacy: Health, environmental and socio-economic impacts and recommendations to the governments of Belarus, the Russian Federation and Ukraine. The Chernobyl Forum: 2003–2005. 2nd rev. edn. IAEA/PI/A.87 Rev.2/06-09181. Vienna: IAEA.

IAEA (International Atomic Energy Agency), n.d. INES: The international nuclear and radiological event scale. www.iaea.org/sites/default/files/ines.pdf (accessed 19 January 2017).

ICRP (International Commission on Radiological Protection), 2009. Application of the Commission's recommendations to the protection of people living in long-term contaminated areas after a nuclear accident or a radiation emergency. ICRP Publication 111. *Annals of the ICRP* 39(3).

IOM (Institute of Medicine), 1986. *The Medical Implications of Nuclear War*. Washington, DC: National Academy Press.

IPFM (International Panel on Fissile Materials), 2015a. Global fissile material report 2015: Nuclear weapon and fissile material stockpiles and production. fissilematerials.org/library/gfmr15.pdf (accessed 19 January 2017).

IPFM (International Panel on Fissile Materials), 2015b. Plutonium separation in nuclear power programs: Status, problems, and prospects of civilian reprocessing around the world. fissilematerials.org/library/rr14.pdf (accessed 27 June 2017).

Kaatsch, Peter, Claudia Spix, Renate Schulze-Rath, Sven Schmiedel, and Maria Blettner, 2008. Leukaemia in young children living in the vicinity of German nuclear power plants. *International Journal of Cancer* 1220: 721–6. doi.org/10.1002/ijc.23330

Kitahara, Cari M., Martha S. Linet, Preetha Rajaraman, and Estelle Ntowe, 2015. A new era of low-dose radiation epidemiology. *Current Environmental Health Reports* 2: 236–49. doi.org/10.1007/s40572-015-0055-y

Kristensen, Hans M., and Matthew G. McKinzie, 2015. Nuclear arsenals: Current developments, trends and capabilities. *International Review of the Red Cross* 97(899): 563–99. doi.org/10.1017/S1816383116000308

Kristensen, Hans M., and Robert S. Norris, 2017. Status of world nuclear forces (updated 8 July). Federation of American Scientists. fas.org/issues/nuclear-weapons/status-world-nuclear-forces/ (accessed 7 September 2017).

Kuznick, Peter, 2011. Japan's nuclear history in perspective: Eisenhower and atoms for war and peace. *Bulletin of the Atomic Scientists*, 13 April.

Lelieveld J., D. Kunkel, and M. G. Lawrence, 2012. Global risk of radioactive fallout after major nuclear reactor accidents. *Atmospheric Physics and Chemistry* 12: 4245–58. doi.org/10.5194/acp-12-4245-2012

Leuraud, Klervi, David B. Richardson, Elisabeth Cardis, Robert Daniels, Michael Gillies, Jacqueline A. O'Hagan, Ghassan B. Hamra, Richard Haylock, Dominique Laurier, Monika Moissonnier, Mary K. Schubauer-Berrigan, Isabelle Thierry-Chef, and Ausrele Kesminiene, 2015. Ionising radiation and risk of death from leukaemia and lymphoma in radiation-monitored workers (INWORKS): An international cohort study. *Lancet Haematology* 1: e276–81. doi. org/10.1016/S2352-3026(15)00094-0

Little, Mark. P., Tamara V. Azizova, Dimitry Bazyka, Simon D. Bouffler, Elisabeth Cardis, Sergey Chekin, Vadim V. Chumak, Francis A. Cucinotta, Florent de Vathaire, Per Hall, John D. Harrison, Guido Hildebrandt, Victor Ivanov, Valeriy V. Kashcheev, Sergiy V. Klymenko, Michaela Kreuzer, Olivier Laurent, Kotaro Ozasa, Thierry Schneider, Soile Tapio, Andrew M. Taylor, Ioanna Tzoulaki, Wendy L. Vandoolaeghe, Richard Wakeford, Lydia B. Zablotska, Wei Zhang, and Steven E. Lipshultzet, 2012. Systematic review and meta-analysis of circulatory disease from exposure to low-level ionizing radiation and estimates of potential population mortality risks. *Environmental Health Perspectives* 120(11): 1503–11. doi.org/10.1289/ehp.1204982

Makhijani, Arjun, Brice Smith, and Michael C. Thorne, 2006. Science for the vulnerable: Setting radiation and multiple exposure environmental health standards to protect those most at risk. Takoma Park, MD: Institute for Energy and Environmental Research, 19 October.

MAPW (Medical Association for Prevention of War), 2012. Toro Energy promotes radiation junk science. Statement of 45 medical doctors, May. www.mapw.org.au/download/doctors-slam-uranium-miner-over-junk-science-radiation-safety-statement-issued-mapw-1-may-2 (accessed 19 January 2017).

Mathews, John, Anna Forsythe, Zoe Brady, Martin Butler, Stacy Goergen, Graham Byrnes, Graham Giles, Anthony Wallace, Philip Anderson, Tenniel Guiver, Paul McGale, Timothy Cain, James Dowty, Adrian Bickerstaffe, and Sarah Darby, 2013. Cancer risk in 680,000 people exposed to computed tomography scans in childhood or adolescence: Data linkage study of 11 million Australians. *British Medical Journal* 346: f2360. doi.org/10.1136/bmj.f2360

Møller, Anders P., and Timothy A. Mousseau, 2013. The effects of natural variation in background radioactivity on humans, animals and other organisms. *Biological Reviews* 88: 226–54. doi.org/10.1111/j.1469-185X.2012.00249.x

NAIIC (National Diet of Japan Fukushima Nuclear Accident Independent Investigation Commission), 2012. The official report of the Fukushima Nuclear Accident Independent Investigation Commission. Tokyo: National Diet of Japan.

National Research Council, 2012. *Analysis of Cancer Risks in Populations near Nuclear Facilities: Phase 1.* Washington, DC: National Academies Press.

Ozasa, Kotaro, Yukiko Shimizu, Akihiko Suyama, Fumiyoshi Kasagi, Midori Soda, Eric J. Grant, Ritsu Sakata, Hiromi Sugiyama, and Kazunori Kodama, 2012. Studies of the mortality of atomic bomb survivors, report 14, 1950–2003: An overview of cancer and noncancer diseases. *Radiation Research* 177(3): 229–43. doi.org/10.1667/RR2629.1

Pearce, Mark S., Jane A. Salotti, Mark P. Little, Kieran McHugh, Choonsik Lee, Kwang Pyo Kim, Nicola L. Howe, Cecile M. Ronckers, Preetha Rajaraman, Sir Alan W. Craft, Louise Parker, and Amy Berrington de González, 2012. Radiation exposure from CT scans in childhood and subsequent risk of leukaemia and brain tumours: A retrospective cohort study. *The Lancet* 380(9840): 499–505. doi.org/10.1016/S0140-6736(12)60815-0

Quigley, Dianne, Amy Lowman, and Steve Wing, 2012. *Tortured Science: Health Studies, Ethics and Nuclear Weapons in the United States.* Amityville, NY: Baywood Publishing Company Inc.

Richardson, David, Steve Wing, and Stephen R. Cole, 2013. Missing doses in the lifespan study of Japanese atomic bomb survivors. *American Journal of Epidemiology* 177(6): 562–8. doi.org/10.1093/aje/kws362

Richardson, David B., Elisabeth Cardis, Robert D. Daniels, Michael Gillies, Jacqueline A. O'Hagan, Ghassan B. Hamra, Richard Haylock, Dominique Laurier, Klervi Leuraud, Monika Moissonnier, Mary K. Schubauer-Berrigan, Isabelle Thierry-Chef, and Ausrele Kesminiene, 2015. Risk from occupational exposure to ionizing radiation: Retrospective cohort study of workers in France, the United Kingdom, and the United States (INWORKS). *British Medical Journal* 351: h5359. doi.org/10.1136/bmj.h5359

Ruff, Tilman A., 2015. The humanitarian impact and implications of nuclear test explosions in the Pacific region. *International Review of the Red Cross* 97(899): 775–813. doi.org/10.1017/S1816383116000163

Schneider, Mycle, and Antony Froggatt, with Julie Hazemann, Ian Fairlie, Tadahiro Katsuta, Fulcieri Maltini, and M. V. Ramana, 2016. *The World Nuclear Industry Status Report 2016.* Paris: Mycle Schneider Consulting Project.

Scott, Bobby R., Charles L. Sanders, Ron E. J. Mitchel, and Douglas R. Boreham, 2008. CT scans may reduce rather than increase the risk of cancer. *Journal of American Physicians and Surgeons* 13(1): 8–11.

Smoll, Nicholas R., Zoe Brady, Katrina Scurrah, and John D. Mathews, 2016. Exposure to ionizing radiation and brain cancer incidence: The Life Span Study cohort. *Cancer Epidemiology* 42: 60–5. doi.org/10.1016/j.canep.2016.03.006

Soble, Jonathan, 2015. Japan to pay cancer bills for Fukushima worker. *New York Times*, 20 October.

Stohl, A., P. Seibert, G. Wotawa, D. Arnold, J. F. Burkhart, S. Eckhardt, C. Tapia, A. Vargas, and T. J. Yasunari, 2011. Xenon-133 and caesium-137 releases into the atmosphere from the Fukushima Dai-ichi nuclear power plant: Determination of the source term, atmospheric dispersion, and deposition. *Atmospheric Chemistry and Physics Discussions* 11(10): 28319–94. doi.org/10.5194/acpd-11-28319-2011

Toro Energy Limited, 2008. Radiation information seminar. www.ausimm.com.au/Content/wir/doug_boreham_invit.pdf (accessed 19 January 2017).

Tsuda, Toshihide, Akiko Tokinobu, Eiji Yamamota, and Etsuji Suzuki, 2016a. Thyroid cancer detection by ultrasound among residents ages 18 years and younger in Fukushima, Japan: 2011 to 2014. *Epidemiology* 27: 316–22. doi.org/10.1097/EDE.0000000000000385

Tsuda, Toshihide, Akiko Tokinobu, Eiji Yamamota, and Etsuji Suzuki, 2016b. Thyroid cancer under age 19 in Fukushima – As of 57 months after the accident. Presentation to the International Physicians for the Prevention of Nuclear War (IPPNW) Congress – Five Years Living With Fukushima, 30 Years Living With Chernobyl, Berlin, 27 February. www.tschernobylkongress.de (accessed 19 January 2017).

United Nations General Assembly, 2016. Taking forward multilateral nuclear disarmament negotiations. A/C.1/71/L.41, 14 October.

UNSCEAR (United Nations Scientific Committee on the Effects of Atomic Radiation), 1993. *Sources and Effects of Ionizing Radiation: UNSCEAR 1993 Report to the General Assembly with Scientific Annexes.* New York: United Nations.

UNSCEAR (United Nations Scientific Committee on the Effects of Atomic Radiation), 2014. *Sources, Effects and Risks of Ionizing Radiation: UNSCEAR 2013 Report: Volume I. Scientific Annex A: Levels and Effects of Radiation Exposure Due to the Nuclear Accident after the 2011 Great East-Japan Earthquake and Tsunami.* New York: United Nations. www.unscear.org/docs/reports/2013/13-85418_Report_2013_Annex_A.pdf (accessed 19 January 2017).

UNSCEAR (United Nations Scientific Committee on the Effects of Atomic Radiation), 2016. Historical milestones. www.unscear.org/unscear/about_us/history.html (accessed 19 January 2017).

US EPA (Environmental Protection Agency), 2014. Radiation risk assessment at CERCLA sites: Q&A. EPA 540-R-012-13, May.

von Hippel, Frank, and IPFM (International Panel on Fissile Materials), 2016. Banning the production of highly enriched uranium. fissilematerials.org/library/rr15.pdf (accessed 27 June 2017).

Wahab, M. A, E. M. Nickless, R. Najar-M'Kacher, C. Parmentier, J. V. Podd, and R. E. Rowland, 2008. Elevated chromosome translocation frequencies in New Zealand nuclear test veterans. *Cytogenetic and Genome Research* 121: 79–87. doi.org/10.1159/000125832

Wakeford, Richard, G. M. Kendall, and Mark P. Little, 2009. The proportion of childhood leukaemia incidence in Great Britain that may be caused by natural background radiation. *Leukemia* 23: 770–6. doi.org/10.1038/leu.2008.342

WHO (World Health Organization), 1983. *Effects of Nuclear War on Health and Health Services.* Geneva: World Health Organization.

WHO (World Health Organization), 1987. *Effects of Nuclear War on Health and Health Services.* 2nd edn. Geneva: World Health Organization.

WHO (World Health Organization), 2012. Preliminary dose estimation from the nuclear accident after the 2011 Great East Japan earthquake and tsunami. Geneva: World Health Organization.

WHO (World Health Organization), 2013. Health risk assessment from the nuclear accident after the 2011 Great East Japan earthquake and tsunami, based on a preliminary dose estimation. Geneva: World Health Organization.

WHO (World Health Organization), 2015. Investing in the World's Health Organization: Taking steps towards a fully-funded Programme Budget 2016–17. *Financing Dialogue 2015.* www.who.int/about/finances-accountability/funding/financing-dialogue/Programme-Budget-2016-2017-Prospectus.pdf?ua=1 (accessed 19 January 2017).

9

Nuclear energy and its ecological byproducts: Lessons from Chernobyl and Fukushima

Timothy A. Mousseau and Anders P. Møller

Abstract

Given increasing energy needs related to global development, and the spectre of climate change related to carbon dioxide (CO_2) emissions from fossil fuels, there is an urgent need for large-scale energy production that does not involve the production of greenhouse gases. Nuclear energy is one possible solution that has been embraced by developing and developed countries alike (for example, China and the US). But the accidents at Three Mile Island, Chernobyl, and most recently Fukushima have demonstrated the vulnerability of this technology to human error, design flaws, and natural disasters, and these accidents have resulted in enormous health, environmental, and economic costs that must be factored into any energy policy that includes nuclear as an option. In the past, such analyses have largely ignored the potential costs of accidents for ecological systems in affected regions. Studies of natural systems are essential since they provide a bellwether for the potential long-term consequences for human populations that by necessity and government policy continue to inhabit contaminated regions. In this chapter, we discuss studies of the non-human biota living in Chernobyl and Fukushima. Extensive research on birds, insects, rodents, microbes, and trees has demonstrated potentially significant injury to individuals, species, and ecosystem functioning related to radiation exposure that has previously been underappreciated. We present an overview of the effects of radiation on DNA, birth defects, infertility, cancer, and longevity, and its consequences for the health and long-term prospects of wildlife living in radioactive regions of the world.

Introduction

The nuclear disasters at Chernobyl and Fukushima have had enormous direct economic impacts, estimated in the hundreds of billions of US dollars (Samet and Seo 2016), with much of these costs associated with the decommissioning of damaged reactors and clean-up of affected regions. And yet very little investment has been made into the ecological consequences of the radionuclides that were dispersed at continental scales. In large part, the lack of investment in basic scientific research has stemmed from the perception, often perpetuated by nuclear regulatory bodies, that the direct effects of these contaminants have been minimal. There have even been suggestions that the wildlife are thriving as a result of these disasters because of reduced hunting pressures in the exclusion zones, leaving the public with the notion that radioactive contaminants are of little concern. In this chapter, we review some of the recent scientific studies that have been conducted over the past decade aimed at assessing the health and population success of wildlife in Chernobyl and Fukushima in relation to radioactivity in these regions. It is proposed that studies of these accident sites can provide valuable insights to possible consequences for biota, and perhaps even humans, exposed to radioactivity from accidents and other sources.

The hazards related to nuclear energy extend far beyond catastrophic accidents

No matter how one personally feels about nuclear energy, the truth is that it is here now and will continue to be a significant component of the world's energy portfolio for many years to come. There are currently on the order of 438 nuclear reactors in operation around the world, and 65 more are under construction (for example, 22 in China and four in the US), with 165 more on order or planned, and 325 additional reactors proposed. Most people do not realise that *every* nuclear reactor generates large amounts of radioactive effluents as a normal part of day-to-day operations, although total emissions have dropped significantly over the past decades. For example, a typical boiling water reactor in the US generates between 1,000 and 100,000 gigabecquerels of radioactive noble gases each year, while pressurised water reactors generate 10 to 100 times less, on average (Burris et al. 2012). In addition, a potpourri of other, potentially more concerning radionuclides (for example, iodine-131, cesium-137, and strontium-90) are also produced and released by normal operations at nuclear power stations. Although there have been a few studies of the possible effects of such releases for humans (for example, childhood leukaemia; Fairlie 2014), studies of non-human biota have largely been limited to the effects of thermal pollution rather than any effects of radioactivity per se. Given the prevalence of nuclear power for the foreseeable future, basic studies of their impacts on ecological systems seem warranted.

In addition to lawful, regulated emissions from nuclear power plants, it is now evident that many of these facilities are leaking unregulated quantities of radioactive effluents into the environment. A notable recent discovery of this sort was at the Vermont Yankee nuclear power plant in the US, where in 2010 large unregulated leaks of tritium (radioactive hydrogen) were discovered and linked to faulty cooling pipes. The discoveries of these leaks undoubtedly played a significant role in the closure of this nuclear power plant in 2014. Since then, it has been reported that more than three-quarters of the US's commercial power plant sites have had some kind of radioactive leak. In part, such leaks are the inevitable consequence of an ageing nuclear fleet and it seems likely that many more leaks will be discovered in the future. Despite these obvious issues, very little is known about the potential ecological impacts of such emissions. A report commissioned by the US Government Accountability Office indicated that little was known about the hazards related to tritium leaks and that further research was warranted (US GAO 2011).

In fact, all parts of the nuclear fuel cycle release vast quantities of radioactive contaminants, from the tailings generated by mining operations, to the processing and packaging of nuclear fuels. At the end of the cycle, enormous spent fuel stockpiles have amassed at nuclear power stations over the course of their operations. All of these sources represent potential hazards to the surrounding human population and ecological landscape, especially in the event of accidents or natural disasters. However, 'events' at nuclear facilities are not well-documented and there have been few studies of their ecological consequences.

The potential risks associated with nuclear energy have recently been comprehensively explored by Wheatley, Sovacool, and Sornette (2016), where they estimated a 50 per cent chance of a Fukushima-scale event or larger in the next 50 years, a Chernobyl event (or larger) in the next 27 years, and a Three Mile Island event (or larger) in the next 10 years.

Based on this analysis, future accidents appear inevitable, and yet there is still relatively little investment in the basic research needed to accurately assess the likely hazards to ecological systems from such accidents.

A research program to assess ecological consequences of nuclear accidents

In an attempt to at least partially fill the void in current knowledge concerning the hazards of radioactive contamination for natural systems, Timothy Mousseau and Anders Møller initiated a collaboration in 2000 to investigate the consequences of the Chernobyl disaster on bird populations in the region. In 2005, these studies expanded to include a variety of organisms, including insects, spiders, and plants, and research in Fukushima began in July 2011. The organising principles used to direct this research initiative were related to the following questions:

1. Do the radiation levels observed in Chernobyl (and now Fukushima) generate doses sufficient to increase mutation rates and genetic damage in natural populations?
2. Are there phenotypic consequences to elevated mutation rates and genetic damage in these regions?
3. Are there fitness consequences (i.e. changes in survival and/or reproduction) of elevated mutation rates?
4. Is there any evidence that populations are adapting to elevated radiation levels in these regions?
5. Are there consequences of radiation effects for abundance and diversity of natural populations?
6. Are there ecosystem effects that result from radiation effects on populations?

Genetic effects of radiation on non-human biota

Often, the first thought that comes to mind when discussing radioactivity concerns the possible genetic consequences of exposure to this mutagen. There is now an overwhelming body of evidence to suggest that, indeed, genetic systems are directly affected by chronic exposure to low doses of ionising radiation in the environment. The evidence comes from a plethora of single species studies, and more recently from meta-analyses of compilations of these single species studies.

Perhaps the first test for radiation effects on nuclear DNA mutation rates for a Chernobyl population used microsatellite markers (i.e. DNA fingerprints) to examine *de novo* mutation rates in barn swallows (*Hirundo rustica*) by comparing microsatellite DNA fingerprints for parents and their offspring (reviewed in Mousseau and Møller 2014). This study found mutation rates for these markers to be two- to 10-fold higher in Chernobyl when compared to control populations in Ukraine and Italy, a finding that was paralleled by a study of the offspring of Chernobyl accident liquidators (i.e. humans). Surprisingly, there have been no other similar studies to assess *de novo*, heritable genetic mutations related to the Chernobyl accident. Given the plummeting costs of the genomic tools needed to assess changes at the level of individual DNA, much progress could be made towards a fundamental understanding of how the interaction between mutagens and genes are induced and transmitted to subsequent generations. Such studies are greatly needed, not just for issues related to conservation biology, but also to address fundamental questions in evolutionary genetics where the search for direct links between variability at the level of DNA and consequent changes in expression of phenotypic characters has long been a high priority.

However, there have been many other studies that have employed indirect techniques to assess genetic damage (for reviews see Møller and Mousseau 2006, 2015) and, when taken collectively, there is little doubt that the radioactive contaminants associated with the Chernobyl disaster have generated genetic damage and increased mutation rates, with many studies also finding phenotypic effects that were correlated to the levels of genetic damage reported. Surprisingly, the first summary of genetic effects stemming from exposure to Chernobyl-derived radiation was presented in Møller and Mousseau (2006: 205, Table 1), which listed 33 studies that had investigated mutations or cytogenetic effects of increased radiation around Chernobyl compared with control areas in a variety of plant and animal species. Although there was considerable heterogeneity in the results, 25 of the studies showed a significant increase in mutations or cytogenetic abnormalities related to radiation exposure. Several studies showed an increase in mutation rates for some loci, but not for others. However, many studies were based on small sample sizes, with a resulting low statistical power and were thus unable to show differences of 25 per cent as being statistically significant. Only four of these studies investigated germ-line mutations (i.e. mutations that could be passed to the next generation) and these all found significant increases. Of relevance here is the fact that many of these studies were not even considered by the highly influential International Atomic Energy Agency (IAEA) Chernobyl Forum reports, which downplayed the potential injury to natural populations.

Møller and Mousseau (2015) have recently extended their studies of mutation rates in Chernobyl populations and used a meta-analysis to examine the effects of radiation in Chernobyl across 45 published studies, covering 30 species. Meta-analysis is a relatively new statistical technique that permits the combination of datasets from disparate sources to permit global analyses of hypotheses of interest. Based on their meta-analysis, the overall effect size of radiation effects, estimated as Pearson's product-moment correlation coefficient, was very large ($E = 0.67$; 95 per cent confidence intervals (CI) 0.59 to 0.73), accounting for 44.3 per cent of the total variance in an unstructured random-effects model (Møller and Mousseau 2015: 2, Figure 1). In simple terms, this means that radiation effects explained almost half of the total variation observed among studies, which is extraordinary by any standard. By using a 'fail-safe' sensitivity analysis, it was possible to determine just how robust this finding was. Fail-safe calculations reflect the number of unpublished null results that would be needed to eliminate this average effect size. In this study, the fail-safe number was 4135 demonstrating the extreme robustness of this finding (Rosenberg's method: 4135 at $p = 0.05$). Indirect tests did not provide any evidence of publication bias. The effect of radiation on mutations varied among taxa, with plants showing a larger effect than animals. Humans were shown to have intermediate sensitivity of mutations to radiation compared to other species. Effect size did not decrease over time, providing no evidence for an improvement in environmental conditions. The surprisingly high mean effect size suggests a strong impact of radioactive contamination on individual fitness in current and future generations, with potentially significant population-level consequences, even beyond the area contaminated with radioactive material. Overall, this study provides perhaps the strongest evidence so far of the mutagenic consequences of chronic exposure to ionising radiation in natural populations.

To date, there have been relatively few studies of genetic effects related to the Fukushima disaster. Joji Otaki of the University of the Ryukus in Okinawa, Japan, has conducted a series of seminal studies of butterflies exposed to radioactive contaminants associated with the Fukushima disaster and found strong evidence for increased mutation rates as a direct consequence of exposure to radionuclides (reviewed in Mousseau and Møller 2014). These studies were greatly strengthened by laboratory experiments that used both internal and external radiation sources, and these unambiguously validated observations of the elevated mutation rates and phenotypic effects observed in the field (Mousseau and Møller 2014). Later studies by Otaki's group provided additional support for acute and chronic effects of radiation effects, with effects decreasing over time, possibly due to reduced dose rates after several years. Of particular note was the suggestion that acquired mutations were in some cases transmitted to offspring. Collectively, these studies of butterflies provide some of the most rigorous and comprehensive experimental analyses of chronic radiation effects in natural populations.

Additional support for the hypothesis that low-dose rate exposures can lead to elevated mutation rates comes from a recent meta-analysis of the effects of naturally occurring radioactive materials on plant and animal populations around the world (Møller and Mousseau 2013). Natural radiation levels vary greatly across the planet largely in relation to variation in surface deposits of radioactive uranium and thorium. Well-known areas include Ramsar, Iran, Kerala, India, and Guarapara, Brazil, among many others. This study surveyed the results from more than 5,000 publications to arrive at 46 studies conducted with sufficient rigour to be included in the meta-analysis. Although many of the individual effects were small and statistically insignificant on their own, overall there were many more that were greater than zero than expected by chance, with an overall average effect size of 0.093 (95 per cent CI = 0.039–0.171) indicating that exposure to naturally occurring radiation accounted for about 1 per cent of the variance in the traits examined. Albeit a small effect, this could still prove significant on an evolutionary timescale. The principal conclusion from this analysis was that there is extensive evidence for small, but significant negative effects of natural variation in background radiation on immune systems, mutation rates, and disease expression across a range of different animals and plants (Møller and Mousseau 2013). In other words, there was no evidence of any threshold below which effects are not potentially observable given sufficient statistical power. Studies of naturally radioactive areas may also provide opportunities to investigate evolutionary processes of adaptation, although to our knowledge no such studies have yet been conducted. Of more relevance here, perhaps, is the finding that 'natural background levels' of radiation can be sufficient to cause injury to individuals, contrary to frequent statements by nuclear industry and regulatory bodies that, because emissions from nuclear power plants are often of the same order of magnitude as background levels, they do not need to be considered from a public health perspective. This study suggests that there are good reasons to consider radioactive releases, even if they are similar to natural background levels.

Developmental effects: Albinism, asymmetry, brain size, cataracts, sperm, and tumors

There is an increasing array of empirical studies in Chernobyl, and now Fukushima, that document a wide range of physiological, developmental, morphological, and behavioural consequences of exposure to radioactive contaminants. It is presumed that most of these effects have an underlying genetic basis, although in some cases direct toxicity cannot be ruled out. Among the first visible signs of exposure were the appearance of white spots on feathers of birds and perhaps the fur of mammals (i.e. cattle in Fukushima). These 'partial albinos' (also sometimes referred to as partial leucism) have been well-documented for barn swallows in Chernobyl and for a number of other bird species as well (Mousseau and Møller 2014). Barn swallows with aberrant white feathers were first detected in Fukushima in 2012 by amateur bird watchers and were observed in apparently increasing frequencies in 2013. However, such a trend could in part be related to a 'screening effect' due to higher levels of scrutiny for this trait following the disaster and further investigation is needed. Although such partial albinos are believed to have reduced probabilities of survival, there are sufficient data to suggest that this character can be inherited and may at least in part result from a mutation(s) in the germ line, based on parent–offspring resemblance. Although the presence of white feathers in and of itself seems unlikely to directly affect individual performance (i.e. reproduction and survival), it may serve as a useful biomarker for radiation effects on individuals. Further research is needed to determine any links between the expression of this trait and any underlying genetic or physiological mechanism related to radiation exposure.

Analysis of gametes has served as a proxy for estimates of germ line mutation rates for several species of birds in Chernobyl. For example, it has been reported that the frequency of abnormal sperm in barn swallows was up to 10 times higher for Chernobyl birds as compared to sperm from males living in control areas (Mousseau and Møller 2014). It was found that abnormality rates were correlated with reduced levels of antioxidants in the blood, liver, and eggs of these birds, supporting the hypothesis that antioxidants likely play a significant role in protecting DNA from the direct and indirect consequences of exposure to radionuclides. And a more recent analysis of Chernobyl birds found that in nine out of 10 species examined, sperm abnormality rates were much larger for birds living in Chernobyl than those living in control areas across Europe, with the highest damage levels observed for species with longer sperm, suggesting that sperm abnormalities are likely common for birds living in radioactive areas. Similar effects on sperm morphology of small rodents have recently been reported (Kivisaari et al. 2016). It has been found that barn swallow sperm swimming ability is negatively related to radiation levels and that plasma oxidative status could predict sperm performance, further supporting the role antioxidants are known to play in protecting spermatogenesis from the effects of ionising radiation (reviewed in Mousseau and Møller 2014). Overall, these studies provide convincing evidence that spermatogenesis can be significantly impacted by low-dose radiation and the resulting male infertility may in part explain the smaller population sizes of many species that have been documented for the region (see below).

Studies of plant pollen and seed germination may also be informative with respect to radiation effects on reproductive tissue and hence fitness of individuals. A recent study of 111 plant species in Chernobyl found small but significant negative effects of radiation on pollen viability (Møller, Shyu, and Mousseau 2016), which may in part explain reduced germination rates in many of these species (Møller and Mousseau 2017).

Many other cell types and tissues have been shown to be affected by Chernobyl contaminants. For example, it has been demonstrated that the frequency of visible tumors on birds was significantly higher in radioactive areas, presumably reflecting elevated mutation rates in somatic tissues (reviewed in Mousseau and Møller 2014). Visible tumor rates in birds from Chernobyl were in excess of 15 per 1,000 birds, while tumors have never been reported for Danish populations despite extensive surveys

(0 per 35,000 birds observed). Recent surveys of rodents from Chernobyl suggest increased frequency of tumors related to radiation dose as measured by whole body burdens of cesium-137.

Radiation cataract was detected in the eyes of atomic bomb survivors shortly after the end of the Second World War and showed a very significant dose response relationship (Otake and Schull 1991). Similarly, in Chernobyl, the frequency and magnitude of cataract expression in eyes was related to radiation exposure: birds from areas with high background radiation were more likely to display opacities in one or both eyes (Mousseau and Møller 2014). As with radiation-related cataract in humans, there was no relationship with the age of the birds, further supporting the hypothesis that radiation was the underlying cause of cataract expression. Lehmann et al. (2016) recently reported significantly increased rates of cataracts in rodents living in radioactively contaminated regions of Ukraine providing additional support for the use of cataract incidence as a reliable biomarker for exposure to ionising radiation. Although never explicitly tested, it seems likely that impaired vision related to cataracts would have significant fitness consequences for animals (for example, the probability of evading predators or finding food).

Neurological development has long been known to be sensitive to the effects of ionising radiation. Many studies of prenatally exposed survivors of the atomic bombings of Hiroshima and Nagasaki suggested that serious mental retardation and small head sizes are a direct consequence of exposure to ionising radiation (Otake and Schull 1998). Chernobyl birds also show reduced brain size in regions of high radioactivity and smaller brain size was associated with reduced survival prospects (reviewed in Mousseau and Møller 2014). Similar effects have also been observed for rodents living in both Chernobyl and Fukushima (Mappes et al. 2016).

A wide range of other morphological and behavioural abnormalities have been reported for wild organisms living in radioactive regions of Chernobyl. Time, effort, and imagination appear to be the only constraints to the discovery of the biological consequences of the Chernobyl accident. This is exemplified by a recent study of calling song behaviour in the European cuckoo, *Cuculus canorus* (Møller et al. 2016). In this study, the authors documented the number of 'syllables' in the calls of 129 male cuckoos as well as the occurrence of 'aberrant' calls from a number of locations in Ukraine spanning a large range of radioactivity, from 0.01 to 218 microSievert per hour (µSv/h). Overall, males produced fewer and more aberrant syllables in radioactively contaminated areas, and this effect persisted even after correcting for the potential effects of other environmental variables. Although it is not possible to extrapolate from the call of a cuckoo to health effects in humans, radioactive contaminants can very clearly influence natural systems in a wide variety of ways.

Population abundances and biodiversity in regions of high radiation

A key issue for conservation biologists concerns the fitness consequences of mutation accumulation and resulting developmental effects that have been observed for wild populations living in Chernobyl and Fukushima. To this end, we have conducted demographic studies aimed at documenting population sizes, numbers of species (i.e. biodiversity), sex ratios, survival and reproductive rates, and patterns of immigration for animals in both Chernobyl and Fukushima. Because of the highly heterogeneous nature of radionuclide deposition inside the contaminated regions of Chernobyl and Fukushima, it is possible to identify areas that represent the full spectrum of radiation levels, from relatively 'clean' uncontaminated habitats all the way to large areas of very high radiation levels, all within short geographical distances. This heterogeneity makes it possible to conduct highly replicated tests for the effects of radiation on biological populations and communities for a single large-scale event. In effect, the distribution of radioactive contaminants, especially in Chernobyl, is more akin to a mosaic or quiltwork than diffusion from a point, allowing the uncoupling of radiation levels from distance from the source. It is this lack of geographic structure for radiation levels when combined with multiple tests for radiation effects across multiple habitat types that permits a sensitive analysis of radiation effects independent of other biotic and abiotic factors.

Abundance and diversity of birds, butterflies, and other invertebrates

Comprehensive surveys of animal abundance and diversity in Chernobyl were conducted by Møller and Mousseau starting in the mid-2000s. The basic sampling protocol was a 'massively replicated biotic inventory' design whereby point counts of birds and invertebrates (chiefly, butterflies, dragonflies, bees, grasshoppers, and spiders) were conducted at about 300 locations across northern Ukraine and southeastern Belarus in 2006–08. An identical protocol was used to conduct surveys in Fukushima at 400 distinct locations in 2011–16. To date (2016), a total of 1,146 and 1,900 biotic inventories have been generated for the Chernobyl and Fukushima regions, respectively. In addition to quantitative estimates of animal abundances and species diversity at each site, a large number of additional biotic and abiotic factors were measured or estimated, including the type of vegetation, the distance to open water, soil type, ambient meteorological conditions, latitude, longitude, elevation, and time of day. All of these variables were included in a multivariate model and used to generate predictions for expected numbers of organisms of each species or group for each location. This model was then used to provide estimates of the variation in numbers explained by radiation independent of all the other potentially contributing factors; in essence, a partial relationship between abundance and radiation. To our knowledge, this approach has not been used in this way before to assess radioactively contaminated areas, although it has been used for monitoring bird populations in Europe and North America since the 1960s. This approach is perhaps the only solution for complex ecological questions of this type, short of large-scale experimental manipulations, which are generally not possible for testing the effects of nuclear fission products at a landscape scale. This approach has the added advantage of permitting assessment of ecological effects even in the absence of pre-disaster baseline data, as it uses contemporary observations of distribution and abundance from unaffected areas to infer expected patterns in contaminated areas.

Contrary to popular notions, in 2006–09 the abundance and diversity of forest and grassland birds in Chernobyl were dramatically lower in contaminated areas, showing a dose–response-like relationship, with about one-third as many birds and half as many species present in high contamination areas relative to that predicted by the models and abundances found in relatively 'clean' parts of the same general region (reviewed in Mousseau and Møller 2014). Although not every species showed declines with radiation levels, and a few even appeared to be unaffected (Galván et al. 2014) and perhaps showed evolutionary adaptation to radiation, the overall patterns of decline were very apparent and the analyses were statistically robust. Birds of prey also showed patterns of reduced numbers in contaminated regions of Chernobyl, although it was not apparent if reduced numbers were a consequence of direct exposure to radionuclides via ingestion or via indirect effects on behaviour mediated by reduced prey. In addition to population censuses, there are other lines of evidence supporting the observed decline in population sizes of birds in Chernobyl including changes in adult sex ratios (more males than females), and reductions by half in the number of older birds relative to juveniles and one-year-olds. Also, there was evidence from analyses of stable isotopes in feathers that the Chernobyl region is acting like a population sink with a higher proportion of immigrants present than in control areas or when compared to birds in historical museum collections from the same area.

The overall pattern was very similar for birds in Fukushima in July 2011, with the strength of the negative relationship between abundance and radiation significantly stronger in Fukushima when comparing the 14 bird species that were common to both regions (Mousseau and Møller 2014). The observed stronger relationship in Fukushima could reflect the difference between acute and chronic exposures, with Chernobyl bird populations showing a response to 20+ years of selection for resistance, or this could reflect the effects of other radionuclides (for example, iodine-131 and cesium-134) that were present at high levels in Fukushima during the spring of 2011 that are no longer present in Chernobyl.

Field studies in Fukushima were also conducted in 2011–14, and the initial analyses showed a strengthening of the negative relationship between ambient radiation levels and abundance and species richness at a given site over time (Møller et al. 2015; Møller, Nishiumi, and Mousseau 2015). Although no comprehensive surveys of raptors in Japan have yet been conducted, a recent study of goshawk (*Accipiter gentilis Fujiyama*) has reported significant declines in reproduction for this bird of prey in Fukushima following the disaster (Murase et al. 2015), although only three study areas were included in the analysis and thus attribution of the observed effect to radionuclide exposure is preliminary. Surveys of barn swallows showed significant drop-offs in abundance in the more radioactive regions of Fukushima, although preliminary analyses did not indicate any relationship with genetic damage to blood cells in nestlings (Bonisoli-Alquati et al. 2015).

In the first study of its sort, Garnier-Laplace et al. (2015) calculated doses for 57 species of birds (almost 7,000 individuals) living in Fukushima following the nuclear disaster of 11 March 2011 that were surveyed by Møller et al. (2015), and Møller, Nishiumi, and Mousseau (2015). Doses were calculated based on radiological conditions at the point of observation and corrected by including ecological and life history attributes of the species in the model. Dose was used to predict total number of birds while statistically controlling for potentially confounding environmental variables (for example, habitat type, elevation, presence of water bodies, ambient meteorological conditions, and time of day). Total dose was found to be a strong predictor of abundances ($P < 0.0001$), which showed a proportional decline with increasing doses with no indications of a threshold or intermediate optimum. Overall, the $ED_{50\%}$ (the total absorbed dose causing a 50 per cent reduction in the total number of birds) was estimated to only be 0.55 gray.

It is interesting to note that, as a group, butterflies also showed significant declines with radiation levels in both Chernobyl and Fukushima (Mousseau and Møller 2014). We speculate that there is something peculiar about the female ZW sex determination system shared by birds and butterflies (i.e. females are heterogametic) that make these groups particularly vulnerable to mutagenic substances.

Unlike mammals where males are 'XY' and females are 'XX' with respect to sex chromosomes, in both birds and butterflies it is the female that is the equivalent of XY. This 'reversal' of the genetic system underlying sex determinism could greatly enhance the deleterious effects of mutations on reproduction and hence population growth rates.

Our hypothesis is that because in these groups the sex that is responsible for egg production is heterogametic, mutational load effects on reproduction stemming from mutation accumulation on the Z chromosome are likely to be expressed immediately following exposure as opposed to species where females are homogametic, as is the case for most sexually reproducing organisms (for example, mammals). In species where the female is homogametic (i.e. XX in mammals), deleterious effects of mutations on these chromosomes may not be expressed immediately because of redundancies in the genetic material. This might be particularly important given the apparent lack of gene dosage compensation in birds and Lepidoptera. In addition, slightly deleterious mutations may accumulate faster on sex chromosomes than on autosomes, and this could be, at least in part, responsible for the observed greater sensitivity to radiation reported for birds and Lepidoptera.

Put more simply, female birds and butterflies may be more likely to express accumulated mutations arising from ionising radiation than females from other species because of their reversed sex determination system. And, because females are in large part responsible for propagation of the species (i.e. they make the eggs), mutational effects could lead to direct effects on fecundity, which is often the most important determinant of population growth rates.

In most other invertebrate groups examined (for example, grasshoppers, dragonflies, bees, and spiders), population sizes were significantly reduced in areas of high contamination in Chernobyl 20+ years after the disaster, while there was no evidence for similar declines in Fukushima; in fact, spiders showed significant increases in numbers, at least during the first summer following the disaster (Mousseau and Møller 2014). It has been proposed that such differences in the time course for population effects might reflect the consequences of multi-generational mutation accumulation of recessive deleterious mutations in Chernobyl, which is also consistent with the immediate effects on birds and butterflies observed in Fukushima. Alternatively, increases in spider numbers could simply reflect a reduction in predation pressure (for example, birds), a finding similar to that reported for large mammals living in the Chernobyl zone where the lack of hunting pressure has been associated with increased population sizes in some species.

Recent evidence suggests that DNA repair may also be involved in determining sensitivity to the mutagenic properties of radionuclides. A recent analysis found a significant relationship between the strength of population declines of a given species with radiation and historical mitochrondial DNA substitution rates for 32 species of birds in Chernobyl (Møller et al. 2010). Species with higher substitution rates showed the greatest declines with radiation levels, suggesting that variation in DNA repair capability may be influencing population success, although this hypothesis remains to be tested experimentally. In essence, individuals of some species are less able to cope with the increased levels of genetic damage associated with ionising radiation in Chernobyl and Fukushima.

Large mammals: A special case?

Recently, it has been suggested that some of the large mammals of the Chernobyl Exclusion Zone are thriving, and perhaps it would not be surprising if this were indeed the case for animals that normally face significant hunting pressure. Inside the exclusion zones of both Chernobyl and Fukushima, hunting pressures are significantly reduced, if not completely eliminated, thus providing a refuge for game animals. Two recent studies have suggested increased numbers of wolves, deer, elk, and wild boar living in the Chernobyl Exclusion Zone. However, these studies were conducted in a manner that did not permit rigorous analysis of radiation effects on relative abundances or animal health. Prior, finer-scaled analyses of mammal distribution and abundance have demonstrated significantly fewer individuals of all species (except wolves) in the more radioactive areas of the zone (Mousseau and Møller 2014), and more recent studies of rodents have demonstrated significant decreases in abundances and fertility (Mappes et al. 2016), as well as high frequencies of cataracts in the more radioactive areas of the Chernobyl Exclusion Zone (Lehmann et al. 2016). In sum, these studies of mammals suggest that, although there are clear indications of radiation effects, hunting pressure must also be factored into such analyses. Similar effects have recently been reported for fish species in Fukushima that would normally have been under significant fishing pressure.

Adaptation to radiation?

Evolution by natural selection is an inevitable and ubiquitous consequence of simple biological processes: all organisms are capable of reproduction, some more, some less, with some portion of the variation in reproductive success being related to the phenotypic attributes of the individual that are more or less genetically determined. Previous studies of genetic variability within natural populations (for example, Mousseau and Roff 1987) suggest that genetic variation generally exists in most species for most characters, and laboratory studies have repeatedly demonstrated that some organisms can adapt to high radiation levels (for example, tardigrades; Jönsson et al. 2008). However, relatively few studies have attempted to assess adaptive responses to radiation of organisms living under natural conditions. Møller and Mousseau (2016) recently reviewed all purported studies of evolutionary responses of organisms living in Chernobyl and found very little evidence to suggest that adaptation might be a common response to ionising radiation. Of the 14 studies conducted with sufficient rigour to address this question, only one study (of bacteria) showed any signs of an evolved, adaptive response (Ruiz-González et al. 2016). In addition, there are no studies to date that provide any evidence for hormesis in these natural populations. There are several reasons why adaptation might not evolve under radioactive conditions, including the possibility that there is no genetic variation within a population for adaptive responses, or that there may have been insufficient time since the accident to allow for evolutionary responses. Studies of naturally radioactive regions of the world where organisms have had millennia to respond to this type of selection pressure may provide valuable insights to the mechanisms underlying adaptive responses as well as likelihood of such responses in regions affected by nuclear accidents (Mousseau and Møller 2014).

Ecosystem consequences of nuclear accidents

Ecosystems provide many of the basic requirements for life on the planet. Ecosystem services related to humans include the provisioning of water, medicines, foods, and plant productivity, among many other functions. Given the wide range of radiation effects on individuals, populations, and communities of plants, animals, and microbes, it would not be surprising to find consequences at the level of ecosystem function. To date, very few ecosystem-level studies have been performed at either Chernobyl or Fukushima. However, based on limited data, it is very clear that ecosystems are not immune to the impacts of nuclear accidents. Recent studies have found that plant primary productivity is negatively impacted by radiation stemming from the Chernobyl accident, as evidenced by reduced growth rates of trees following 1986, especially during years when there was additional stress induced by drought (Mousseau et al. 2013). Experimental studies of the rate of decomposition of leaf litter at the soil surface have demonstrated dramatically reduced decomposition rates in areas of high ambient radiation that likely reflect effects on the microbial community (especially fungi and bacteria) (Mousseau et al. 2014). It is very likely that such effects on decomposition have large cascading effects on nutrient turnover rates in the soil, potentially impacting plant productivity at landscape scales. In fact, the reduced growth rates of trees observed in Chernobyl could be in part related to the indirect consequences of radiation effects on the soil microbial communities, which are often essential for the acquisition of mineral nutrients by plants. Other studies have documented radiation effects on the interactions between plants, insect pollinators, and fruit productivity that are clearly affecting other components of the ecosystem in negative ways (for example, Møller, Barnier, and Mousseau 2012). And it now seems apparent that stress due to radiation can interact with climate change to influence patterns of ecological succession in ways that may pose an ongoing threat to human populations adjacent to contaminated regions. For example, reduction in decomposition rates in radioactive areas has resulted in the accumulation of dead organic matter (for example, leaf litter) at the soil interface, thus dramatically increasing the fuel available for forest fires, which have increased in frequency and intensity in recent years as a consequence of climate change in the region (Mousseau et al. 2014). Given that this soil litter is itself often highly radioactive, forest fires have the potential to volatilise radionuclides and to disperse these contaminants to populated regions in surrounding countries (Evangeliou et al. 2015), as was demonstrated following several fires in the Chernobyl Exclusion Zone during the summer of 2015 (Evangeliou et al. 2016).

Concluding remarks

In conclusion, the radiological disasters at Chernobyl and Fukushima provide a unique opportunity to investigate genetic, ecological, evolutionary, and ecosystem consequences of acute and chronic exposures to mutagenic sources in natural populations at regional and landscape scales. Recent advances suggest many small and large effects on biological systems, from molecules to ecosystems, that will likely influence ecosystem form and function for decades to centuries to come. Recent surveys of population effects in Chernobyl (Garnier-Laplace et al. 2013) suggest that populations living under the full range of natural stressors (biotic and abiotic) are almost 10 times more sensitive to ionising radiation than predicted by conventional approaches used by some regulatory and governmental agencies, providing some potential insights to the cause of the apparent discrepancy between empirical ecological studies and predictions from conventional radio-ecological models of radiation effects. The opportunity to compare and contrast organisms from both Chernobyl and Fukushima provides for a possible level of scientific rigour (i.e. replication) not previously available for studies of this sort, as well as analysis of the time frame over which responses may occur, and the development of predictive models to aid the management and conservation of biological systems following future nuclear accidents. Given recent advances in molecular genetic technologies, it seems likely that much new knowledge could be gained from a sustained and expansive investment in basic research related to the biological effects of radioactive mutagens within an ecosystem context that could extend far beyond the disasters at Fukushima and Chernobyl. Given the near certainty of additional nuclear accidents small and large in the near future, investment in the basic research needed to characterise the environmental consequences of past accidents would seem prudent.

Acknowledgements

We thank our many friends and colleagues in Ukraine, Belarus, Japan, and elsewhere, without whom this work would never have been conducted. Our studies have been financially supported by a large number of sources including the Centre National de la Recherche Scientifique (CNRS, France), the National Science Foundation, the National Institutes of Health, CRDF Global, the National Geographic Society, the University of South Carolina, the Fulbright Foundation, the Samuel Freeman Charitable Trust, the American Council of Learned Societies, Chubu University, and private citizens of the US and Japan. This essay is dedicated to the memory of Eugene Pisanets, Dmitry Grodzinsky, and Alexey Yablokov.

References

Bonisoli-Alquati, A., K. Koyama, D. J. Tedeschi, W. Kitamura, H. Suzuki, S. Ostermiller, E. Arai, A. P. Møller, and T. A. Mousseau, 2015. Abundance and genetic damage of barn swallows from Fukushima. *Scientific Reports* 5: 9432. doi.org/10.1038/srep09432

Burris, J. E., J. C. Bailar III, H. L. Beck, A. Bouville, P. S. Corso, P. J. Culligan, P. M. Deluca Jr, R. A. Guilmette, G. M. Hornberger, M. Karagas, R. Kasperson, J. E. Klaunig, T. Mousseau, S. B. Murphy, R. E. Shore, D. O. Stram, M. Tirmarche, L. Waller, G. E. Woloschak, and J. J. Wong, 2012. *Analysis of Cancer Risks in Populations Near Nuclear Facilities: Phase I*. Washington, DC: National Academies Press.

Evangeliou N., Y. Balkanski, A. Cozic, W. M. Hao, F. Mouillot, K. Thonicke, R. Paugam, S. Zibtsev, T. A. Mousseau, R. Wang, B. Poulter, A. Petkov, C. Yue, P. Cadule, B. Koffi, J. W. Kaiser, and A. P. Møller, 2015. Fire evolution in the radioactive forests of Ukraine and Belarus: Future risks for the population and the environment. *Ecological Monographs* 85(1): 49–72. doi.org/10.1890/14-1227.1

Evangeliou, N., S. Zibtsev, V. Myroniuk, M. Zhurba, T. Hamburger, A. Stohl, Y. Balkanski, R. Paugam, T. A. Mousseau, A. P. Møller, and S. I. Kireev, 2016. Resuspension and atmospheric transport of radionuclides due to wildfires near the Chernobyl nuclear power plant (CNPP) in 2015: An impact assessment. *Scientific Reports* 6: 26062. doi.org/10.1038/srep26062

Fairlie, I., 2014. A hypothesis to explain childhood cancers near nuclear power plants. *Journal of Environmental Radioactivity* 133: 10–17. doi.org/10.1016/j.jenvrad.2013.07.024

Galván, I., A. Bonisoli-Alquati, S. Jenkinson, G. Ghanem, K. Wakamatsu, T. A. Mousseau, and A. P. Møller, 2014. Chronic exposure to low-dose radiation at Chernobyl favours adaptation to oxidative stress in birds. *Functional Ecology* 28(6): 1387–403. doi.org/10.1111/1365-2435.12283

Garnier-Laplace, J., K. Beaugelin-Seiller, C. Della-Vedova, J. M. Métivier, C. Ritz, T. A. Mousseau, and A. P. Møller, 2015. Radiological dose reconstruction for birds reconciles outcomes of Fukushima with knowledge of dose-effect relationships. *Scientific Reports* 5: 16594. doi.org/10.1038/srep16594

Garnier-Laplace, J., S. Geras'kin, C. Della-Vedova, K. Beaugelin-Seiller, T. G. Hinton, A. Real, and A. Oudalova, 2013. Are radiosensitivity data derived from natural field conditions consistent with data from controlled exposures? A case study of Chernobyl wildlife chronically exposed to low dose rates. *Journal of Environmental Radioactivity* 121: 12–21. doi.org/10.1016/j.jenvrad.2012.01.013

Jönsson, K. I., E. Rabbow, R. O. Schill, M. Harms-Ringdahl, and P. Rettberg, 2008. Tardigrades survive exposure to space in low Earth orbit. *Current Biology* 18(17): R729–31. doi.org/10.1016/j.cub.2008.06.048

Kivisaari, K., Z. Boratynski, S. Calhim, P. Lehmann, T. Mappes, T. A. Mousseau, and A. P. Møller, 2016. Cut to the chase: Radiation effects on sperm structure at Chernobyl. In review.

Lehmann, P., Z. Boratynski, T. Mappes, T. A. Mousseau, and A. P. Møller, 2016. Fitness costs of increased cataract frequency and cumulative radiation dose in natural mammalian populations from Chernobyl. *Scientific Reports* 6: 19974. doi.org/10.1038/srep19974

Mappes, T., Z. Boratynski, K. Kivisaari, G. Milinevski, T. A. Mousseau, A. P. Møller, E. Tukalenko, and P. Watts, 2016. Radiation effects on breeding and population sensitivity in a key forest mammal of Chernobyl. In review.

Møller, A. P., and T. A. Mousseau, 2006. Biological consequences of Chernobyl: 20 years on. *Trends in Ecology and Evolution* 21(4): 200–7. doi.org/10.1016/j.tree.2006.01.008

Møller, A. P., and T. A. Mousseau, 2013. The effects of natural variation in background radioactivity on humans, animals and other organisms. *Biological Reviews of the Cambridge Philosophical Society* 88(1): 226–54. doi.org/10.1111/j.1469-185X.2012.00249.x

Møller, A. P., and T. A. Mousseau, 2015. Strong effects of ionizing radiation from Chernobyl on mutation rates. *Scientific Reports* 5: 8363. doi.org/10.1038/srep08363

Møller, A. P., and T. A. Mousseau, 2016. Are animals and plants adapting to low-dose radiation at Chernobyl? *Trends in Ecology and Evolution* 31(4): 281–9. doi.org/10.1016/j.tree.2016.01.005

Møller, A. P., and T. A. Mousseau, 2017. Radiation levels affect pollen viability and germination among sites and species at Chernobyl. *International Journal of Plant Species* 178(7): 537–45. doi.org/10.1086/692763

Møller, A. P., F. Barnier, and T. A. Mousseau, 2012. Ecosystem effects 25 years after Chernobyl: Pollinators, fruit set, and recruitment. *Oecologia* 170: 1155–65. doi.org/10.1007/s00442-012-2374-0

Møller A. P., J. Erritzøe, F. Karadas, and T. A. Mousseau, 2010. Historical mutation rates predict susceptibility to radiation in Chernobyl birds. *Journal of Evolutionary Biology* 23(10): 2132–42. doi.org/10.1111/j.1420-9101.2010.02074.x

Møller, A. P., F. Morelli, T. A. Mousseau, and P. Tryjanowski, 2016. The number of syllables in Chernobyl cuckoo calls reliably indicate habitat, soil and radiation levels. *Ecological Indicators* 66: 592–7. doi.org/10.1016/j.ecolind.2016.02.037

Møller, A. P., T. A. Mousseau, I. Nishiumi, and K. Ueda, 2015. Ecological differences in response of bird species to radioactivity from Chernobyl and Fukushima. *Journal of Ornithology* 156(S1): 287–96. doi.org/10.1007/s10336-015-1173-x

Møller, A. P., I. Nishiumi, and T. A. Mousseau, 2015. Cumulative effects of radioactivity from Fukushima on the abundance and biodiversity of birds. *Journal of Ornithology* 156(S1): 297–305. doi.org/10.1007/s10336-015-1197-2

Møller, A. P., J. C. Shyu, and T. A. Mousseau, 2016. Ionizing radiation from Chernobyl and the fraction of viable pollen. *International Journal of Plant Sciences* 177(9): 727–35. doi.org/10.1086/688873

Mousseau, T. A., G. Milinevsky, J. Kenney-Hunt, and A. P. Møller, 2014. Highly reduced mass loss rates and increased litter layer in radioactively contaminated areas. *Oecologia* 175(1): 429–37. doi.org/10.1007/s00442-014-2908-8

Mousseau, T. A., and A. P. Møller, 2014. Genetic and ecological studies of animals in Chernobyl and Fukushima. *Journal of Heredity* 105(5): 704–9. doi.org/10.1093/jhered/esu040

Mousseau, T. A., and D. A. Roff, 1987. Natural selection and the heritability of fitness components. *Heredity* 59(Pt 2): 181–97. doi.org/10.1038/hdy.1987.113

Mousseau, T. A., S. M. Welch, I. Chizhevsky, O. Bondarenko, G. Milinevsky, D. Tedeschi, A. Bonisoli-Alquati, and A. P. Møller, 2013. Tree rings reveal extent of exposure to ionizing radiation in Scots pine *Pinus sylvestris*. *Trees: Structure and Function* 27(5): 1443–53. doi.org/10.1007/s00468-013-0891-z

Murase, K., J. Murase, R. Horie, and K. Endo, 2015. Effects of the Fukushima Daiichi accident on goshawk reproduction. *Scientific Reports* 5: 9405. doi.org/10.1038/srep09405

Otake, M., and W. J. Schull, 1991. A review of forty-five years study of Hiroshima and Nagasaki atomic bomb survivors: Radiation cataract. *Journal of Radiation Research* 32: 283–93. doi.org/10.1269/jrr.32.SUPPLEMENT_283

Otake, M., and W. J. Schull, 1998. Radiation-related brain damage and growth retardation among the prenatally exposed atomic bomb survivors. *International Journal of Radiation Biology* 74(2): 159–71. doi.org/10.1080/095530098141555

Ruiz-González, M. X., G. Á. Czirják, P. Genevaux, A. P. Møller, T. A. Mousseau, and P. Heeb, 2016. Resistance of feather-associated bacteria to intermediate levels of ionizing radiation near Chernobyl. *Scientific Reports* 6: 22969. doi.org/10.1038/srep22969

Samet, J. M., and J. Seo, 2016. The financial costs of the Chernobyl nuclear power plant disaster: A review of the literature. Zurich: Green Cross Switzerland.

US GAO (Government Accountability Office), 2011. Nuclear Regulatory Commission: Oversight of underground piping systems commensurate with risk, but proactive measures could help address future leaks. Report GAO-11-563. Washington, DC: US Government Accountability Office.

Wheatley, S., B. K. Sovacool, and D. Sornette, 2016. Reassessing the safety of nuclear power. *Energy Research & Social Science* 15: 96–100. doi.org/10.1016/j.erss.2015.12.026

Part IV
A post-nuclear future

10

Decommissioning nuclear power reactors

Kalman A. Robertson

Abstract

Global demand for decommissioning services is poised to rise rapidly over the next 20 years, creating major technical and administrative challenges for a large number of states and operators that have only limited experience in this field. This chapter explains the radiological risks associated with each step from shutting down a reactor to releasing the former reactor site for a new use. The selection of a strategy for decommissioning a reactor involves competing policy imperatives that may be assessed in light of two key principles related to funding decommissioning and assuring safety, inter-generational equity and the polluter/user pays principle. Based on an assessment of current trends in decommissioning, there are opportunities to improve cost estimates for decommissioning and strengthen international cooperation to meet rising demand. Risk communication and public participation also warrant special attention due to the highly technical nature of the risks associated with decommissioning and remediation of reactor sites.

Introduction

As commercial nuclear power generation enters its seventh decade, the world is set to undergo an unprecedented increase in the number of reactors requiring decommissioning. The combination of ageing reactor fleets and early shutdowns will see a doubling in the number of reactors undergoing decommissioning within the next 20 years, and it is projected to create a global market for decommissioning and waste storage worth over US$100 billion by 2030 (*Nucleonics Week* 2016).

In principle, many of the risks to nuclear safety associated with a reactor site progressively decrease as it is shut down and decommissioned. However, due to the long time periods typically involved in the life cycle of a power reactor, decommissioning poses unique choices and challenges. Worldwide, experience with complete decommissioning of full-scale power reactors is restricted to a handful of cases. Decommissioning costs and requirements vary significantly with the design of the reactor, its operational history, and the state in which it is located. Maintaining continuity of knowledge over the conditions at a site is also difficult— decisions that were made during design, construction, or operation of a reactor, as well as accidents during its operational life, can have important implications many decades later during decommissioning.

The projected upsurge in decommissioning is coming at a time when the issue of disposal of radioactive waste, particularly spent fuel (high-level waste), has not yet been completely resolved in any state. This has important implications for all other decommissioning activities and for the end-state of the former reactor site. Choices involved in scheduling decommissioning activities involve complex trade-offs between different generations' interests in radiation protection for workers and the public, environmental protection, and financial expenses.

This chapter outlines each of the basic steps that are typically involved in decommissioning with reference to examples of power reactors that have reached advanced stages of the process. It explains current challenges in the field of decommissioning, including managing the increasing number of reactor shutdowns, handling unexpected changes in the cost/timing of decommissioning, and achieving unrestricted release of decommissioned sites for safe use by the public. It makes recommendations for ensuring adequate finance for decommissioning, promoting transparency in decommissioning, and developing international cooperation to cope with the emerging demand for decommissioning services.

Defining 'decommissioning'

Power reactors are among the most complicated industrial facilities to decommission. During operation, a nuclear reactor maintains a controlled, self-sustained fission chain reaction (IAEA 2002: paragraph 5.5).[1] A power reactor uses this reaction to generate useful energy, typically electricity. By contrast, a research reactor is not used to produce electricity and therefore does not use turbines and generators. A research reactor tends to have smaller physical dimensions and lower levels of radioactivity because of its relatively low thermal power.

There are six major stages in the lifetime of a reactor: siting, design, construction, commissioning, operation, and decommissioning. In addition to radioactive spent fuel, an operating reactor generates two basic categories of radiation hazards: contamination and activation.[2] Contaminants are radioactive materials that have been deposited on a solid surface or in a liquid or gas. Since the reactor's primary coolant is in contact with radioactive material (chiefly the fuel itself) in the core, it tends to become highly contaminated during reactor operation. All surfaces that come in contact with this coolant (for example, pipes and pumps) also tend to be contaminated. Depending on the chemical form of those contaminants, it may be relatively easy for them to subsequently leave the surface and enter the surrounding environment. Activation primarily occurs when neutrons from the fission process in the reactor core are absorbed by a material that is not fissile (for example, traces of cobalt in the wall of a reactor), causing that material to become radioactive. Since neutrons may travel significant distances in some materials before being absorbed, activation products may be found deep inside the building materials of old reactors.

---

1   For present purposes, a critical assembly can be thought of as a small research reactor with fewer provisions for cooling and shielding. Note that radioisotope thermoelectric generators, which use the heat released by the decay of radioactive material to generate electricity (usually for long-term, low-power applications like spacecraft), are not nuclear reactors because they do not involve a fission chain reaction.

2   The radiation hazards associated with nuclear energy involve ionising radiation—radiation that, by virtue of its type and/or energy, is capable of ionising atoms or molecules in body tissue. For general information on radiation, see Knoll (2010). This chapter only covers nuclear safety and radiation protection aspects of decommissioning. It does not cover the challenges posed by other hazardous substances that may be present at nuclear sites, such as solvents, non-radioactive heavy metals, and asbestos.

Once a reactor ceases operations, it must eventually be decommissioned so that the site can be made safe for other uses.[3] The International Atomic Energy Agency (IAEA) defines decommissioning as: 'Administrative and technical actions taken to allow the removal of some or all of the regulatory controls from a facility' (IAEA Department of Nuclear Safety and Security 2007: 48; IAEA 2016a: 34). Decommissioning usually involves dismantling a facility or decontaminating buildings to reduce radiation risks, ensure the long-term protection of the public and the environment, and free up the site for a new use.

In essence, decommissioning involves two key principles related to assuring safety: inter-generational equity and the polluter/user pays principle. Despite the long timescales involved in constructing, operating, and decommissioning a reactor, and the even longer timescales associated with the decay of radioactive waste, it is generally accepted that reactors should be decommissioned in such a way as to avoid unduly burdening future generations (Bråkenhielm 2005). Decisions on decommissioning should avoid compromising acceptable standards of public safety, environmental sustainability, nuclear security, and resource availability for future generations (see Taebi and Kadak 2010).

The polluter/user pays principle means that those who benefit from nuclear power (i.e. utility companies and end-users) should be responsible for ensuring that decommissioning is completed and should pay the entire cost, rather than passing the cost onto taxpayers as a whole. The IAEA recommends that each state place primary responsibility for decommissioning on facility licensees acting under the supervision of national regulators, including the national decommissioning authority, health and safety regulators, local authorities, and environmental regulators (Stoiber et al. 2010: 73). In this respect, nuclear regulation has improved significantly since the early days of nuclear power, although financing unforeseen costs in decommissioning remains a challenge (see below). Today, most states with nuclear power reactors (or contemplating their construction) require utilities to have a decommissioning plan drafted prior to commissioning of the reactor (see Laraia 2012). By contrast, many of the first-generation reactors were built without detailed

---

3   The term 'decommissioning' is also used to refer to other fuel cycle facilities but this chapter will only consider reactors.

consideration of how they would eventually be decommissioned (Samseth et al. 2013). It is telling that the IAEA's first major guidance document on decommissioning was not published until the mid-1970s (IAEA 1976).

Global status of decommissioning and current outlook

Demand for decommissioning services is likely to rise rapidly over the next 15 years. Worldwide, there were 443 power reactors either operational or in temporary shutdown at the end of 2015.[4] The average age of these reactors is approximately 30 years. Although some reactors are receiving life extensions, most have a projected operating life of 40 years. Germany will complete an early phase-out of its nuclear power plants by 2022 (Schneider et al. 2015; International Energy Agency 2016).

A large proportion of states currently operating power reactors have little, if any, experience with decommissioning them (see Table 10.1). The United States has the most experience, with the Nuclear Regulatory Commission (NRC) releasing much of the land at several decommissioned power plant sites over the past 15 years, usually while continuing to regulate residual spent fuel storage installations. Some states may be able to fall back on their experience in decommissioning research reactors. Worldwide, 33 states (plus Taiwan) have together decommissioned a total of 352 research reactors (IAEA 2016b). Compared with power reactors, the cost and technical complexity of decommissioning research reactors tend to be limited because of their comparatively small physical dimensions and low levels of radioactive contamination/activation. Research reactor pressure vessels may often be removed and buried in one piece without undertaking the arduous process of cutting the vessel, which may expose workers to radioactive dust particulates. Consequently, experience with research reactor decommissioning is not necessarily equivalent to experience

---

4   This chapter does not cover decommissioning of naval propulsion reactors. Quoted values for the cost of decommissioning nuclear submarines generally cover the cost for the entire vessel, rather than the reactor unit, making comparisons with stationary reactors difficult. On the effects of disposal of Soviet naval propulsion reactors at sea, see Mount, Sheaffer, and Abbott (1994). This chapter also does not cover end-of-operating-life activities for the handful of nuclear-powered satellites launched by the Soviet Union and the United States during the 1960s, 1970s, and 1980s. For a description of measures to place shutdown satellite reactor cores in safe orbit and a description of the environmental damage caused when one of these satellites, Kosmos 954 Radar Ocean Reconnaissance Satellite, came crashing down to Earth, see Harland and Lorenz (2006: 235–6).

with power reactor decommissioning. States with limited experience in decommissioning power reactors may benefit from information exchanges with more experienced states.

Table 10.1 Status of all nuclear power reactors in the world, 31 December 2015

Status	Number	Number of states
Operational or in temporary shutdown	443	31 (plus Taiwan)[1]
Permanently shut down (includes reactors that have entered decommissioning process)	157	19 (Europe, North America, Kazakhstan, and Japan)[2]
In decommissioning process or decommissioned	124	18[3]

[1] Does not include the 5 megawatt electric (MWe) Yongbyon reactor (North Korea) or the Bataan nuclear power plant (the Philippines).

[2] Does not include a handful of experimental power reactors (fewer than 10).

[3] Does not include the Santa Susana Sodium Reactor Experiment.

Source: IAEA (2016c: 47–58).

The end point of decommissioning

There is no universal standard for determining when a facility is fully decommissioned—the definition depends on the type of regulatory controls in question and their underlying purpose. For example, the purpose of nuclear safeguards is primarily to verify that nuclear materials, equipment, and technology are used for exclusively peaceful purposes, rather than contributing to a nuclear weapons program. From a nuclear safeguards standpoint, a facility is decommissioned once it becomes effectively impossible to utilise the remaining structures or equipment at the site to process or use nuclear material (IAEA 2002: paragraph 5.31). By contrast, from a nuclear safety standpoint, it makes sense to say that a site is only fully decommissioned once radiological and other risks at the site have been reduced to a pre-defined acceptable level. The IAEA asserts that, as of October 2014, 17 power reactors have been 'fully decommissioned', although it does not expressly define the term and it appears to have left out some small experimental power reactors (IAEA 2015b: paragraph 74).

Most national regulators declare that decommissioning is complete once the licensee has completed the tasks in its decommissioning plan, no further dismantling or decontamination operations are foreseen, the reactor licence has been terminated, and the site is suitable for a new purpose, either with or without restrictions imposed by the regulator. This study identifies about 30 power reactors that could be referred to as 'fully decommissioned' in this sense (see Table 10.2 and Table 10.3). National legislation or regulations typically contain specific safety and environmental criteria for the end-state of decommissioning and the removal of regulatory controls from a site.

Table 10.2 Fully decommissioned nuclear power reactors in the US and current site uses, December 2015[1]

Reactor	Location	Type	Reference unit power (MWe)	Operating life[2]	Current site use
Big Rock Point NPP	Charlevoix, MI	BWR	67	1962–97	Unrestricted + dry cask spent fuel storage[4]
Boiling Nuclear Superheater	Rincón, Puerto Rico	BWR	17	1964–68	Unrestricted[3] with reactor entombed on site
Carolinas–Virginia Tube Reactor	Parr, SC	PHWR	17	1963–67	Adjacent to new nuclear power plant
Connecticut Yankee (Haddam Neck) NPP	Haddam Neck, CT	PWR	560	1967–96	Unrestricted + dry cask spent fuel storage[4]
Elk River Station	Elk River, MN	BWR	22	1963–68	Fossil fuel power station
Enrico Fermi APP, Unit 1	Monroe, MI	FBR	61	1966–72	Newer nuclear power plant (most of the components of Fermi 1 have been removed but the site currently hosts Fermi 2)
Fort St Vrain	Platteville, CO	HTGR	330	1974–89	Fossil fuel power station + dry cask spent fuel storage
Hallam	Hallam, NE	SCGR	75	1963–64	Fossil fuel power station + low-level radioactive waste storage and reactor vessel entombed onsite
Humboldt Bay 3	Eureka, CA	BWR	63	1963–76	Fossil fuel power station

Reactor	Location	Type	Reference unit power (MWe)	Operating life[2]	Current site use
Maine Yankee NPP	Wiscasset, ME	PWR	860	1972–97	Unrestricted + dry cask spent fuel storage[4]
Pathfinder APP	Sioux Falls, SD	BWR	59	1966–67	Fossil fuel power station
Piqua NPP	Piqua, OH	Other	12	1963–66	Unrestricted[3] + reactor vessel entombed onsite
Rancho Seco	Herald, CA	PWR	873	1974–89	Cooling towers remain; low-level radioactive waste storage and spent fuel storage; plan for solar power array on part of site
Santa Susana Sodium Reactor Experiment	Bell Canyon, CA	SCGR	~6[5]	1957–64	Industrial research
Saxton Nuclear Experiment Station	Saxton, PA	PWR	3	1967–72	Unrestricted[3]
Shippingport APP	Shippingport, PA	PWR	60	1957–82	New nuclear power plant
Shoreham NPP	East Shoreham, NY	BWR	820	1986–89	Fossil fuel power station
Trojan NPP	Rainier, OR	BWR	1095	1975–92	Unrestricted + dry cask spent fuel storage[4]
Yankee Rowe NPP	Franklin, MA	PWR	167	1960–91	Unrestricted + dry cask spent fuel storage[4]

[1] This table does not include cases where one reactor has been decommissioned, while other reactors of the same type continued to operate on the same site (i.e. Dresden, OH; and San Onofre, CA).

[2] 'Operating life' is the period between the first grid connection and the last year in which the reactor supplied electricity to the grid.

[3] 'Unrestricted' means the site is now 'greenfield'. In some cases, part of the site is now parkland (for example, Trojan NPP) or a wildlife refuge (for example, Connecticut Yankee).

[4] 'Unrestricted + dry cask spent fuel storage' means most of the original site is now 'greenfield'. A small lot is licensed by the regulator for dry cask storage of spent fuel (independent spent fuel storage installation).

[5] Reactor not listed in IAEA (2015a, 2016c). Power approximated using Wald (2011).

Key

APP = atomic power plant
BWR = boiling water reactor
FBR = fast breeder reactor
HTGR = high-temperature gas-cooled reactor
MWe = megawatt electric
NPP = nuclear power plant
PHWR = pressurised heavy-water reactor
PWR = pressurised water reactor
SCGR = sodium-cooled, graphite-moderated reactor

Sources: IAEA (2015a, 2016c); US NRC (2015b).

Table 10.3 Fully decommissioned nuclear power reactors (and selected reactors at advanced stages of decommissioning) outside the US and current site uses, December 2015[1]

Reactor	Location	Type	Reference unit power (MWe)	Operating life[2]	Current site use
Chinon Units A-1, A-2, and A-3	Avoine, France	GCRs	70, 180, 360	1963–90	Newer nuclear power reactors; part of Chinon A-1 is now a museum; final dismantling to take place after shutdown of newer reactors
Saint Laurent Units A-1 and A-2	Saint-Laurent-Nouan, France	GCRs	390, 465	1969–92	Newer nuclear power reactors
HDR Großweltzheim (Kahl)	Karlstein a.Main, Germany	BWR	25	1969–71	Unrestricted;[3] current use is light manufacturing
Kahl VAK NPP	Seligenstadt, Germany	BWR	15	1961–85	Unrestricted[3] (commercial/manufacturing)
Niederaichbach NPP	Landshut, Germany	HWGCR	100	1973–74	Adjacent to new nuclear power plant
Stade	Bassenfleth, Germany	PWR	640	1972–2003	Awaiting final demolition of remaining (non-active) structures + storage of low-level radioactive waste

Reactor	Location	Type	Reference unit power (MWe)	Operating life[2]	Current site use
Wuergassen NPP	Beverungen, Germany	BWR	640	1971–94	Temporary storage of low- and intermediate-level radioactive waste from decommissioning
Japan Power Demonstration Reactor	Tokai-mura, Japan	BWR	12	1963–76	Nuclear research institute + very low-level radioactive concrete waste buried onsite
Tokai NPP, Unit 1	Tokai-mura, Japan	GCR	137	1965–98	Newer nuclear power reactor
Lucens reactor	Lucens, Switzerland	HWGCR	6	1968–69	Reactor was in underground cavern; appears to be sealed off with greenfield above
Windscale Advanced Gas Cooled Reactor	Sellafield, United Kingdom	GCR	24	1963–81	Newer nuclear facilities

[1] This table does not include cases where one reactor has been decommissioned, while other reactors of the same type continued to operate on the same site (i.e. Gundremmingen, Germany).

[2] 'Operating life' is the period between the first grid connection and the last year in which the reactor supplied electricity to the grid.

[3] 'Unrestricted' means the site is now 'greenfield'.

Key

BWR = boiling water reactor
GCR = gas-cooled, graphite-moderated reactor
HWGCR = heavy water–moderated, gas-cooled reactor
MWe = megawatt electric
NPP = nuclear power plant
PWR = pressurised water reactor

Sources: IAEA (2015a, 2016c); Schmittem (2016); Weigl (2008); *World Nuclear News* (2015).

The end-states of former power reactor sites tend to fall into three categories. First, decommissioned sites may host newer nuclear facilities, such as new reactors or new low-level waste disposal. Former power plant sites (and their residual electricity infrastructure) may also be released for re-use by fossil fuel power plants. Since the proposed new use for the site tends to determine remediation goals, including acceptable residual radiation levels, these examples are of limited value in studying site restoration (Laraia 2012).

Second, the majority of a site may be released from regulation, while the remaining part continues to host long-term dry cask spent fuel storage (an 'independent spent fuel storage installation'). This is an artefact of the current deficit of permanent disposal options for high-level radioactive waste and the persistent difficulties with using reprocessing as a source of fresh reactor fuel (Hiruo 2016). In some of these cases, despite being 'unrestricted use' from the standpoint of nuclear regulations, access to the site is limited by the owners (the utilities that completed the decommissioning) to activities in connection with spent fuel storage or groundwater monitoring. This is the case for Connecticut Yankee and Yankee Rowe in the United States, where there are currently no timetables for making a decision on disposition of former site property (Connecticut Yankee 2015). Compared with wet (pool) storage, dry casks require minimal maintenance for safe storage of spent fuel. However, protecting casks against sabotage is a necessary, ongoing expense (see US NRC 2016). In the United States, most independent spent fuel storage installations are far from major population centres and unlikely to be high-value targets for sabotage. However, terrorists who lack the strength, weaponry, or training to attack an operating reactor might choose to try to damage storage casks with the aim of dispersing radioactive material, causing economic damage, and producing panic among local residents.

Third, the entire site may be released without restrictions for general use in agriculture, park land, or commerce. To date, the only power reactors that fall into this category were either low-power or short-lived (see Tables 10.2 and 10.3). A key requirement for suitability for a new purpose without restrictions is verification by the regulator of reduction of the degree of radioactivity at all parts of the site to a limit set by legislation or regulations. In order to meet the limits on annual doses required for reactor licence termination and site release, the radioactive structures from the plant must usually be removed from the site. It may also be necessary to remove topsoil from beneath the former reactor.

It is often asserted that dose limits are set with a view to ensuring that radiological risks to humans and the environment are 'no longer present' (see, for example, Nuclear Energy Agency 2016: 51). However, there is no generally accepted definition of an effective dose (measured in sieverts, Sv) that constitutes a 'radiological risk', or a limit below which radiological risk is absent. There are no straightforward means of precisely measuring all stochastic health effects of radiation exposure at very low levels. Effects from doses of less than about 0.1 Sv per year are difficult to assess, particularly if exposure is spread over the course of the year rather than associated with an acute event.

The most obvious anticipated effect from low-dose exposure is an increased risk of cancer, but the effect on cancer rates is difficult to measure due to the time delay and the high naturally occurring background rate of cancer. Instead, effects at low doses are extrapolated from measurements of effects at higher doses. This is typically accomplished by assuming a linear relationship between dose and response with no threshold below which risk vanishes, known as linear no-threshold (LNT) theory (see Calabrese 2013; Morgan 2013). LNT is widely accepted as a prudent and conservative model for estimating radiological risk.

Applying LNT strictly, it is impossible to say that any location is, or ever was, completely devoid of radiological risk due to the presence of naturally occurring sources of radiation in all environments. The average annual radiation dose from natural sources for a human varies considerably with geographical location but is typically 1 to 5 milliSieverts (mSv) (UNSCEAR 2008).

For this reason, dose limits from artificial sources are usually set at arbitrary levels, based on an assessment that a particular level of risk is both acceptable and realistically achievable. For the return of a decommissioned or otherwise contaminated site to unrestricted use, national regulations and international standards tend to set annual dose limits attributable to the artificial source at levels that are lower-than-average annual effective doses for humans from either natural sources or medical procedures.[5] The dose attributable to a decommissioned reactor may be inferred by comparing activity levels after decommissioning with activity levels prior to the reactor's construction.[6] Applying LNT, any additional risk

---

5   For example, the NRC imposes a limit of 0.25 mSv annual dose equivalent from residual radioactivity associated with decommissioned nuclear sites as a requirement for release of the site for unrestricted use (US NRC Regulations 2015: Section 20.1402). Doses are typically calculated using measurements of external radiation and then adding on calculated values of internal radiation exposure for various organs based on assumptions about normal rates of inhalation and ingestion.
6   The IAEA's model provisions for national legislation on decommissioning recommend that the regulatory body require a baseline survey of the radiological conditions at the site prior to facility construction for comparison with the end-state after decommissioning (Stoiber et al. 2010: 72).

from such artificial sources of radiation tends to be comparable with or smaller than the risk from natural sources of radiation. The most notable contemporary exception to this trend is the reported use of a dose reference level of 20 mSv per year above natural background for evacuees from the region surrounding the Fukushima Daiichi nuclear power plant (Office of the Deputy Chief Cabinet Secretary 2011).

Financing decommissioning

Financing can be among the most contentious of issues in decommissioning. Consistent with the user/polluter pays principle, financing is the responsibility of the licensee in most states. However, states parties to the Joint Convention on the Safety of Spent Fuel Management and on the Safety of Radioactive Waste Management (1997: article 26) are obligated to ensure that 'qualified staff and adequate resources are available' to ensure the safety of decommissioning. Although a small amount of revenue may be generated during decommissioning by salvaging resaleable reactor components,[7] decommissioning is essentially a cost that is only realised once commercial activity has ceased. Financing decommissioning therefore involves the challenge of coping with a future financial liability. Ensuring sufficiency of funds is difficult because a variety of events during the reactor's lifetime may generate sudden, unexpected costs or losses of revenue. If a reactor needs to be shut down and decommissioned earlier than expected due to an accident or a change in national regulations, then ensuring the availability of funds at the right time may also be a challenge.

Funding strategies

There are two basic strategies for funding decommissioning: prepaid funding and accumulation. The former involves setting aside a set amount of money, proportional to the presumed cost of decommissioning, prior to construction as part of the upfront cost. The latter involves establishing a sinking fund and then gradually paying a small percentage of electricity

---

7   In order for scrap materials to be 'cleared' for unrestricted re-use or disposal, their radioactivity must fall below certain thresholds. These limits are often set very conservatively so that only materials exhibiting roughly natural background levels of radioactivity are cleared. Conservative activity limits and careful quality assurance are important because radioactive material may not be homogeneously distributed over the volume of scrap material.

revenue into the fund throughout the operating life of the reactor. Both strategies typically involve investing funds set aside with the aim of ensuring that some target value, presumed to be adequate capital to finance decommissioning, is available by the end of the projected operational life of the reactor.

Prepaid funding provides a degree of protection in the event that the reactor fails to generate projected revenue as a result of early shutdown, but it is difficult to finance. Accumulation has the advantage of allowing financing by electricity consumers as part of the cost of electricity but it is difficult to make arrangements for the possibility of loss of revenue following earlier than expected reactor shutdown. Neither prepaid funding nor accumulation can guarantee available and sufficient funds in every possible contingency. In principle, funds can include contingency estimates for incidents that increase decommissioning costs; however, due to the indeterminate nature of the timing and cost of decommissioning, there is not much guidance available for calculating contingency estimates (Nuclear Energy Agency 2016: 82). If adequate funds are not available when it comes time to decommission a reactor then it may be necessary to abrogate the polluter/user pays principle by passing the cost on to taxpayers or by charging a levy against future electricity consumers even though they are drawing electricity from other sources (Drozdiak and Busche 2015; Pagnamenta 2016). In April 2016, Germany's Commission on the Review of the Financing of the Nuclear Phaseout recommended that utility companies pay €23 billion of the cost associated with decommissioning their power reactors (*World Nuclear News* 2016a). There is an unresolved debate about whether or not the liability of these utility companies for decommissioning costs should be limited given that their projected revenue may have already been cut by the government's decision to phase out nuclear power early.

Calculating costs

The cost of decommissioning is heavily dependent on the specific conditions at the reactor site, the state's regulations, and the precise decommissioning strategy employed. However, a few general observations can be made about the influence of reactor size, age, and type on the cost of decommissioning. First, the extent of radioactive contamination/ activation at a reactor tends to increase gradually with the power of the reactor and the length of time over which it has operated. So far, several of the reactors that have reached an advanced stage of decommissioning (see especially Tables 10.2 and 10.3) are comparatively low-power reactors with short operating lives. By comparison, decommissioning modern 1 gigawatt electric reactors that have operated for over 40 years involves dealing with larger volumes of radioactive waste.

The US NRC Regulations (2015) require a licensee to have a set amount of funds for decommissioning that depends on the type of reactor, as well as the cost of labour, energy, and waste disposition (US Government Accountability Office 2012). The applicable values for the year 2016 range from a few hundred million to about US$1 billion (see US NRC 2013). These values provide some insights into the differences between the basic types of reactors.

The NRC has estimated pressurised water reactors (PWRs, a type of light-water reactor) to be among the least expensive. PWRs make up 64 per cent of operational reactors. PWRs use heat exchange between two cooling loops, one runs through the core and is highly radioactive, while the other runs through the turbine and is effectively non-radioactive. The turbine room tends to be, at most, mildly contaminated so it can be dismantled in a straightforward way, with components being recycled or disposed of by near-surface burial. Pressurised heavy-water reactors (i.e. Canada Deuterium Uranium CANDU reactors), tend to be similar except that there is an additional cost associated with storage or treatment of the heavy-water moderator, which tends to contain relatively high levels of radioactive tritium resulting from neutron absorption during reactor operation.

By contrast, boiling water reactors (BWRs), which make up the majority of reactors in Japan and Sweden, use a single cooling loop. This means that the 'radiation control area' of the facility is much larger, incorporating the turbine and condenser. Since more parts of the facility are contaminated, decommissioning tends to be more complicated and disposal of components tends to be more costly.

The NRC Regulations indicate that decommissioning gas-cooled, graphite-moderated reactors (GCRs, like the British MAGNOX reactors) will be several times more expensive than decommissioning PWRs or BWRs. The graphite moderator in GCRs accumulates radioactive carbon while the reactor is operating. This makes it difficult to safely access the reactor for the purposes of decommissioning and it creates an additional highly radioactive waste stream. Former Soviet graphite-moderated, light-water cooled reactors, like those at Chernobyl, are similar (*NucNet* 2015).

The main distinguishing feature of decommissioning liquid metal–cooled fast reactors is the cost of draining the coolant while preventing its oxidation and dealing with subsequent radioactive coolant residues (Goodman 2009). This represents an additional complexity compared with draining water from PWRs or BWRs, particularly if the coolant becomes highly contaminated with fission products as a result of breached fuel cladding. However, experience at six fast reactors (Santa Susanna Sodium Reactor Experiment, Hallam, Fermi-1, Phenix, EBR-1, and EBR-2) indicates that the overall cost of decommissioning is similar for fast breeder reactors (FBRs) and BWRs (see Michelbacher et al. 2009).

Overall, there is a lack of reliable and comparable data on decommissioning costs across states. Even for reactors of comparable type, power, and operating history, cost and time estimates for decommissioning vary considerably. Costs depend heavily on the desired final state of the reactor site—it will be more expensive to achieve release of the site for unrestricted use than it will be to make the site suitable for a new nuclear facility. There are large and complex differences in defining the scope of the 'decommissioning costs' of a reactor, as opposed to its 'operating costs' (Nuclear Energy Agency 2012). For most types of power reactors, nuclear fuel is repeatedly loaded and unloaded during the operational life of the reactor. The disposition (storage, disposal, or reprocessing) of spent fuel may therefore be considered an operational cost, rather than a decommissioning cost.[8] Some estimates include the cost of dry cask storage, either at the reactor site or off-site, while others do not.

Even where costs are broken down by dismantling activities, project management activities, and waste management activities, costs estimates for individual items are heavily dependent on the cost of labour and other services in the specific state (see Nuclear Energy Agency 2016: 67). Costs may also depend on the amount of prior experience that the regulators, licensees, and contractors have with decommissioning.

---

8   For a summary of the approaches taken by seven countries with nuclear facilities, see Nuclear Energy Agency (2016: 61).

Depending on the state, much of the cost of decommissioning may be disposal of components of the reactor that are either low-level or intermediate-level radioactive waste.[9] Some states, including France and the United States, have tended to possess sufficiently capacious storage or disposal sites for low-level waste to allow dismantling of facilities with large components intact. These states tend to have more flexibility in removing radioactive components from sites relatively soon after shutdown without the need for complex cutting procedures, significantly decreasing the cost of maintaining and ultimately dismantling the remainder of the components on the site.

The steps involved in decommissioning

This section explains some of the risks involved in each of the steps from shutting down the reactor to releasing the site, as well as choices and trade-offs in specific approaches to decommissioning.

Pre-decommissioning 1: Shutting down the reactor

The decision to cease operations and permanently shut down a power reactor is usually based on a mix of technical, financial, and political considerations. Six reactors have been permanently shut down as a direct result of 'major accidents', 'serious accidents', or 'accidents with wider consequences' during operation, as defined by the International Nuclear and Radiological Event Scale (INES): St Lucens, Three Mile Island Unit 2, Chernobyl Unit 4, and Fukushima Units 1 to 3.[10] By definition, these events involve release of radioactive material either producing severe damage to a reactor core, causing deaths from radiation, or requiring implementation of counter-measures (IAEA n.d.).[11]

---

9   For a conceptual overview of waste classification, including disposal of various levels of radioactive waste, see IAEA (2009).

10   The other units at Fukushima Daiichi were not in operation at the time of the Tohoku earthquake. Some nuclear reactors were restarted after experiencing accidents, including the Santa Susana Sodium Reactor Experiment and Saint Laurent Units A-1 and A-2. The Santa Susana accident pre-dates the scale and does not appear to have been given a definitive rating.

11   Note that events, such as generator failures at PWRs, are sometimes referred to in the media as 'accidents' even though they may not have direct radiological consequences for workers, the public, or the environment.

Most reactors are shut down around the time that they reach the end of their expected operating lives. Typically, the shutdown date is based on an assessment that it is no longer economical to run the reactor due to the increasing cost of maintenance or upgrades as components age or regulatory requirements change. Decreasing demand for electricity and decreasing cost of fossil fuels are also factors in some recent reactor shutdowns.

Once a decision is made to permanently shut down a reactor, it is common practice to leave the final irradiated fuel load inside the reactor core to cool. Once the temperature and pressure in a water-cooled reactor fall below a certain level, it is no longer essential to circulate cooling water to prevent it from boiling off ('cold shutdown'), significantly reducing the potential for accidents to cause a catastrophic loss of cooling. At the time that the Tohoku earthquake and tsunami hit, Units 5 and 6 at Fukushima Daiichi were in cold shutdown, so these reactors did not experience hydrogen explosions. Units 1 to 3 achieved cold shutdown in December 2011, but they continued to require injections of water due to leaking through cracks in the reactors (Brumfiel 2011).

The period following shutdown also corresponds to a significant cultural and organisational change for the reactor licensee. Worker morale is often adversely impacted by the end of electricity production (Laraia 2012). Some of the employees at the power plant will need to undertake additional training for new short-term tasks involved in the post-shutdown period (such as removing or deactivating equipment), only to then face the prospect of unemployment. For specialised decommissioning tasks, it may be necessary to hire new contractors, who must then be informed of the precise layout, history, and conditions of the plant.

Pre-decommissioning 2: Closing down the reactor

Once the radiation (and resulting heat) from the spent fuel drops to a certain level, the licensee may remove the spent fuel rods from the core and the wet (pool) storage, and then drain the coolant. On average, about 99 per cent of radioactivity at shutdown is associated with the spent fuel itself (Nuclear Energy Agency 2016: 50).

If the reactor has undergone a core melt, then, in addition to the fuel rods, it is also necessary to remove melted core debris. Removal of melted debris from Fukushima Daiichi Units 1 to 3 is not scheduled to be completed until at least the early 2020s, and a disposal method for retrieved debris has not been finalised (Schneider et al. 2015; International Energy Agency 2016: 78). Removal from these units will require the use of new remotely controlled, radiation-resistant equipment (Inter-Ministerial Council for Contaminated Water and Decommissioning Issues 2015: 18).

Once all spent fuel is removed from the reactor and on-site storage pools, the reactor is said to be 'closed-down' (IAEA 2002: paragraph 5.30) and the risk of a large-scale release of radioactivity at the reactor is greatly reduced. At this point, the regulator may grant the licensee an exemption from maintaining full emergency response mechanisms at the site (Cama 2014). According to US Senator David Vitter, there have not been any incidents in the United States during decommissioning (i.e. post-fuel removal) that resulted in harm to public safety (Cama 2014). The spent fuel itself tends to be safer in dry casks as well (US Senate 2014). Unlike the fuel in Units 1 to 4, spent fuel that had already been loaded into dry casks at the Fukushima Daiichi site withstood the earthquake and tsunami without suffering significant damage (Suzuki 2015: 597).

Around the time that spent fuel and coolant are being removed, the licensee is usually responsible for submitting an updated decommissioning plan, ensuring that radioactive contamination and activation throughout the site have been carefully measured and mapped out ('site characterisation').[12] Depending on the country, it may be a regulatory requirement to complete this process before shifting from closing down the reactor to actually dismantling and removing reactor components (IAEA 2014: 15).[13]

Strategies for decontamination, dismantlement, and disposal

Once the reactor has been closed down, the licensee has developed a decommissioning plan, and the regulator has provided the necessary approvals, decommissioning can begin. With the spent fuel gone, the main radiological risks are exposure of workers in residual radioactive structures during the hands-on processes involved in dismantling the

12 For examples of dose rate maps at Fukushima Daiichi reactors, see Kotoku (2016).
13 For example, in the United States, the First Circuit Court of Appeals has ruled that removal of reactor components is a form of decommissioning for the purposes of the National Environmental Policy Act. The NRC could not allow Yankee Atomic Energy Company to conduct an 'early component removal' prior to submitting a complete decommissioning plan. See *Citizens Awareness Network Inc. v United States Nuclear Regulatory Commission* (1995).

various systems that were designed to move, store, shield, and cool fuel assemblies. Unlike the routine operation of a reactor, the process of decommissioning is likely to involve personnel accessing moderately or highly contaminated parts of the facility for extended periods of time. However, licensees and regulators must also be mindful of the long-term, low-dose effects of residual radioactivity on the surrounding environment.

There are three basic options for decommissioning a nuclear reactor: immediate dismantling, safe storage, and entombment. Each approach has its own implications for activity-dependent costs (i.e. the 'hands-on' activities involved in decommissioning, including equipment and labour), period-dependent costs (i.e. management, licensing, security, electricity, insurance, property tax, and other site maintenance), and contingency costs (i.e. costs allocated for unforeseeable disruptions, adverse weather, changes to regulations, or loss of an essential service provider) (Atomic Industrial Forum 1986).

The strategies differ primarily in their approach to the relatively highly contaminated or neutron-activated parts of the facility, including the reactor pressure vessel, the reactor internal components, the coolant piping, radiation-shielding concrete, and spent fuel storage racks. Regardless of the strategy employed, building materials and auxiliary systems that are far enough from the reactor core and the coolant to avoid being contaminated or activated should exhibit only natural background radiation—these parts may be treated as normal demolition waste and disposed of whenever it is convenient.

Immediate dismantling involves commencing decommissioning as soon as possible after permanent shutdown, although past experience suggests that it still takes at least 10 years to move from permanent shutdown to site release (see Tables 10.2 and 10.3). Despite ongoing efforts at Fukushima Daiichi, decommissioning of its reactors is projected to take 30 to 40 years (Inter-Ministerial Council for Contaminated Water and Decommissioning Issues 2015: 8). That said, immediate dismantling tends to minimise period-dependent costs, making it an attractive option from a financial standpoint, provided that the funds are readily available. It also ensures that employees who are familiar with the reactor's operation and who have been responsible for maintaining records can be involved in its decommissioning.

Immediate dismantling appears to have political and practical appeal for several states in Europe, as well as Taiwan, that are either reducing their reliance on nuclear power or phasing it out (Adelman 2016). France, Germany, Italy, Lithuania, Slovakia, Spain, Sweden, and the United States are increasingly opting for immediate dismantling of shutdown reactors or requiring accelerated decommissioning of reactors previously in safe storage (Adelman 2016; Autorité de sûreté nucléaire 2009: 4; *Nuclear Energy Insider* 2016; Thomas 2016). Immediate dismantling permits earlier re-use of the site and reduces the burden on future generations. By decontaminating and dismantling containment structures in a timely manner, immediately decommissioning also reduces the risk of radiological release over time as a result of weathering, deterioration, or corrosion of the reactor building.

Safe storage is deferred dismantling of the radioactive parts of a facility. Currently, just over half of all reactors undergoing decommissioning are in safe storage. Depending on national regulations, reactor licences may be extended for up to about a century to accommodate projected storage times. Since many of the radioactive contaminants and activation products at a reactor have a half-life of years or less, radiation levels at the reactor decrease significantly during the storage period. In principle, this allows decommissioning workers to dismantle the reactor at a time when they will receive comparatively low radiation exposure. The volumes of higher level radioactive wastes requiring disposal are also reduced by deferring dismantlement.

If the stored reactor is co-located with another nuclear facility, such as an operational reactor, then keeping it in safe storage involves minimal additional site maintenance costs. Safe storage is also attractive where a licensee needs extra time to shore up decommissioning funds or to build up a decommissioning workforce. Finally, licensees and regulators that choose safe storage may stand to benefit from future technological developments, including improvements to long-term waste disposal options and robotics for remote decommissioning operations (Nagata 2016).

Entombment involves encasing a facility's radioactive structures (including the reactor pressure vessel) in a long-lived material like concrete. Since entombment does not require complete dismantlement of the plant, it minimises workers' contact with radioactive components and decreases volumes of waste that must be disposed of elsewhere. However, entombment may require very long-term surveillance and maintenance to ensure that the structure remains intact while radioactivity gradually declines. Currently, the IAEA cautions against the use of entombment for most types of power reactors (IAEA 2016a: 35). The Soviet Union chose entombment for Chernobyl Unit 4 since neither immediate dismantling nor safe storage could be achieved without posing considerable risks to workers or the surrounding environment. An additional 'New Safe Confinement' steel structure is being placed over the entombed reactor in 2016/2017 at a cost of over US$1 billion (*World Nuclear News* 2016b). Entombment has also been used for a few small, short-lived reactors with relatively low levels of contamination (see Table 10.2).

The choice of decommissioning strategy involves trade-offs between minimising the exposure of workers, the long-term impact on the local environment, the financial costs, and the volumes of radioactive waste requiring disposal. Similar trade-offs are inherent in dismantlement and decontamination of individual radioactive structures. Often, only a small volume is highly contaminated, while the majority of the structure is only slightly contaminated. For example, the inside of a metal coolant pipe may be highly radioactive, while the concrete around the pipe is effectively non-radioactive. From the standpoint of minimising the volume of material that must be sent for long-term storage or burial as intermediate-level waste, it is desirable to carefully cut the metal pipe out of the concrete, but during this process workers could be exposed to relatively high levels of radiation. Similarly, the surface of a wall could be highly contaminated, while the bulk concrete in the wall is only very slightly radioactive. Workers could use a technique like dry abrasive blasting to decontaminate the wall, but this would tend to increase their exposure to radiation. The International Commission on Radiological Protection (2005) recommends balancing these considerations by placing one set of limits on exposure to the public (for example, 1 mSv per year from artificial sources) and another set of limits on exposure to workers at nuclear facilities (for example, 20 mSv per year), in part because workers knowingly accept some degree of risk (see Clarke 2011: 31).

Current and future challenges for decommissioning

Regardless of the future direction of nuclear power, decommissioning is a high-priority area for human resources and technology development. Worldwide, experience with power reactor decommissioning and long-term site restoration is limited. Due to current trends in reactor shutdowns, demand for decommissioning services will increase in coming years. Attracting, motivating, and training the necessary experts, personnel, and contractors will be a particular challenge for states where nuclear energy is in decline.

It is important to start planning for decommissioning early in the life cycle of a reactor. For reactors that are currently operating, improvements made over the last 50 years to detailed record-keeping and site characterisation activities will help to ensure that events that could affect levels of radioactive contamination (and therefore decommissioning plans) at various parts of the site are well documented. For future reactors, the state should require a decommissioning plan to be drawn up during the design phase with an explanation of the proposed end-state for the site after decommissioning. Where decommissioning cost estimates are used as a basis for collecting funds, the estimates should make it clear when funds will be available, what contingencies are foreseen, and what assumptions are being made about end-use of the site. Furthermore, cost projections should take into consideration possible future scenarios for spent fuel management, particularly the possibility that a lack of high-level waste disposal options will make it necessary to continue to use interim independent spent fuel storage installations.

From the standpoint of transparency to regulatory and public scrutiny, it would be desirable to standardise cost estimate methodologies for decommissioning across states. At present, this appears to be almost impossible due to the large variation in the cost of decommissioning services among states. However, a competitive market for international decommissioning service contracts is emerging (see Schmittem 2016; Fell 1999). As this occurs, the question of standardisation should be revisited with a view to ensuring that cost projections are consistent across all states and accurately reflect the true cost of all activities associated with decommissioning.

More broadly, information exchange and promotion of best practice among practitioners will become increasingly important during the projected upsurge in decommissioning activities. Forums like the IAEA's International Decommissioning Network should focus on how lessons learned from decommissioning the first generation of reactors in North America and Europe can be used to assist countries in Asia that are currently planning, constructing, or operating reactors.

For nuclear experts, decommissioning involves the familiar challenge of communicating information about risks associated with stochastic and unobservable processes. Education, communication, and engagement with the public should be the shared responsibility of operators and governments. Licensees in the United States currently have the option to create citizen advisory panels to improve public participation in aspects of decommissioning, including final site-use and environmental remediation (US NRC 2015a). The use of these panels, and publication of their discussions, should be more widely encouraged. As the number of reactors undergoing decommissioning rises, transparency about the steps involved in the process and the projected end-state of former reactor sites will be increasingly important for maintaining public trust in nuclear technologies.

Acknowledgements

The author wrote this chapter while he was working as a Stanton Nuclear Security Postdoctoral Fellow for the Belfer Center for Science and International Affairs at the John F. Kennedy School of Government, 2015–16. This chapter represents the views of the author. It does not represent the views of any institution.

References

Adelman, Oliver, 2016. 'Prompt' decommissioning a potential trend, some in industry say. *Nucleonics Week* 57(10): 5.

Atomic Industrial Forum, 1986. Guidelines for producing nuclear power plant decommissioning cost estimates. National Environmental Studies Project AIF/NESP–036, Washington, DC.

Autorité de sûreté nucléaire, 2009. La politique de l'ASN en matière de démantèlement et de déclassement des installations nucléaires de base en France. Revision 0.v3, Paris, April. www.asn.fr/Media/Files/La-politique-de-l-ASN-en-matiere-de-demantelement-et-de-declassement-des-installations-nucleaires-de-base-en-France (accessed 24 January 2017).

Bråkenhielm, Carl Reinhold, 2005. Ethical guidance in connection with decommissioning of nuclear power plants. Nuclear Energy Agency Report No. NEA/RWM/WPDD(2005)4. Paper presented to Topical Session on Funding Issues in Connection with Decommissioning of Nuclear Power Plants, Paris, 9 November 2004. www.iaea.org/inis/collection/NCLCollectionStore/_Public/45/026/45026338.pdf (accessed 24 January 2017).

Brumfiel, Geoff, 2011. Fukushima reaches cold shutdown. *Nature*, 16 December. doi.org/10.1038/nature.2011.9674

Calabrese, Edward J., 2013. Origin of the linearity no threshold (LNT) dose-response concept. *Archives of Toxicology* 87(9): 1621–33. doi.org/10.1007/s00204-013-1104-7

Cama, Timothy, 2014. Senators: Nuclear decommissioning process is flawed. *The Hill*, 14 May. thehill.com/policy/energy-environment/206110-senators-nuclear-decommissioning-process-is-flawed (accessed 24 January 2017).

Citizens Awareness Network Inc. v United States Nuclear Regulatory Commission, 1995. *Federal Reporter*, third series, 59: 284.

Clarke, Roger H., 2011. Changes in underlying science and protection policy. In *Evolution of ICRP Recommendations 1977, 1990 and 2011-42.07: Changes in Underlying Science and Protection Policy and Their Impact on European and UK Domestic Legislation*, edited by Nuclear Energy Agency, 11–42. NEA No. 6920. Boulogne-Billancourt: Nuclear Energy Agency.

Connecticut Yankee, 2015. CY property. www.connyankee.com/html/future_use.asp (accessed 24 January 2017).

Drozdiak, Natalia, and Jenny Busche, 2015. Germany's nuclear costs trigger fears. *Wall Street Journal*, 22 March.

Fell, Nolan, 1999. Decommissioning: A rapidly maturing market. *Nuclear Engineering International*, 29 October.

Goodman, L., 2009. Fermi 1 sodium residue cleanup. In *Decommissioning of Fast Reactors after Sodium Draining*, 39–43. IAEA TecDoc 1633. Vienna: IAEA.

Harland, David M., and Ralph D. Lorenz, 2006. *Space Systems Failures: Disasters and Rescues of Satellites, Rocket and Space Probes*. Chichester: Praxis Publishing.

Hiruo, Elaine, 2016. DOE should take fuel from reactors in order they shut, former executive says. *Nuclear Fuel* 41(11): 8.

IAEA (International Atomic Energy Agency), 1976. Decommissioning of nuclear facilities. Report of a technical committee meeting held in Vienna, 20–24 October 1975. IAEA TecDoc 179.Vienna: IAEA.

IAEA (International Atomic Energy Agency), 2002. *IAEA Safeguards Glossary: 2001 Edition*. 3rd edn. Vienna: IAEA.

IAEA (International Atomic Energy Agency), 2009. Classification of radioactive waste: General safety guide. Safety Standards Series No. GSG-1, STI/PUB/1419. Vienna: IAEA.

IAEA (International Atomic Energy Agency), 2014. Decommissioning of facilities: General safety requirements part 6. IAEA Safety Standards for Protecting People and the Environment. No. GSR Part 6. Vienna: IAEA.

IAEA (International Atomic Energy Agency), 2015a. *Nuclear Power Reactors in the World*. Reference Data Series No. 2. 35th edn. Vienna: IAEA.

IAEA (International Atomic Energy Agency), 2015b. Nuclear technology review 2015: Report by the director general. General Conference Document GC/59/INF/2, 2 July.

IAEA (International Atomic Energy Agency), 2016a. *IAEA Safety Glossary: Terminology used in Nuclear Safety and Radiation Procedures, Draft*. Vienna: IAEA.

IAEA (International Atomic Energy Agency), 2016b. Research reactor database. nucleus.iaea.org/RRDB/RR/ReactorSearch.aspx?rf=1 (accessed 24 January 2017).

IAEA (International Atomic Energy Agency), 2016c. *Nuclear Power Reactors in the World*. Reference Data Series No. 2. 36th edn. Vienna: IAEA.

IAEA (International Atomic Energy Agency), n.d. INES: The international nuclear and radiological event scale. www.iaea.org/sites/default/files/ines.pdf (accessed 24 January 2017).

IAEA (International Atomic Energy Agency) Department of Nuclear Safety and Security, 2007. *IAEA Safety Glossary: Terminology Used in Nuclear, Radiation, Radioactive Waste and Transport Safety*. Vienna: IAEA.

Inter-Ministerial Council for Contaminated Water and Decommissioning Issues, 2015. Mid-and-long-term roadmap towards the decommissioning of TEPCO's Fukushima Daiichi nuclear power station. 12 June. www.meti.go.jp/english/earthquake/nuclear/decommissioning/pdf/20150725_01b.pdf (accessed 24 January 2017).

International Commission on Radiological Protection, 2005. Low-dose extrapolation of radiation-related cancer risk. *Annals of the ICRP* 35(4): publication 99.

International Energy Agency, 2016. *Energy Policies of IEA Countries: Belgium: 2016 Review*. Paris: International Energy Agency.

Joint Convention on the Safety of Spent Fuel Management and on the Safety of Radioactive Waste Management, 1997. Opened for signature 29 September, 2153 UNTS 37605 (entered into force 18 June 2001).

Knoll, Glenn F., 2010. *Radiation Detection and Measurement*. 4th edn. Brisbane: John Wiley & Sons.

Kotoku, Tetsuo, 2016. Robot challenges for nuclear decommissioning of Fukushima Daiichi nuclear power station. Presentation to the IAEA International Conference on Advancing the Global Implementations of Decommissioning and Environmental Remediation Programmes, Madrid, 25 June. irid.or.jp/_pdf/20160523.pdf (accessed 24 January 2017).

Laraia, Michele, ed., 2012. *Nuclear Decommissioning: Planning, Execution and International Experience.* Philadelphia, PA: Woodhead Publishing. doi.org/10.1533/9780857095336

Michelbacher, J. A., S. P. Henslee, C. J. Knight, and S. R. Sherman, 2009. Decommissioning of experimental breeder reactor-II complex, post sodium draining. In *Decommissioning of Fast Reactors after Sodium Draining,* 59–65. IAEA TecDoc 1633. Vienna: IAEA.

Morgan, William F., 2013. Issues in low dose radiation biology: The controversy continues: A perspective. *Radiation Research* 179: 501–10. doi.org/10.1667/RR3306.1

Mount, Mark E., Michael K. Sheaffer, and David T. Abbott, 1994. Kara Sea radionuclide inventory from naval reactor disposal. *Journal of Environmental Radioactivity* 25(1–2): 11–19. doi.org/10.1016/0265-931X(94)90004-3

Nagata, Kazuaki, 2016. Toshiba unveils remote-controlled device to remove reactor 3 fuel assemblies at Fukushima No. 1. *Japan Times,* 18 January.

Nuclear Energy Agency, 2012. *International Structure for Decommissioning Costing (ISDC) of Nuclear Installations.* Radioactive Waste Management Series, NEA No. 7088. Paris: NEA, www.oecd.org/publications/international-structure-for-decommissioning-costing-isdc-of-nuclear-installations-9789264991736-en.htm Accessed 13 February 2017).

Nuclear Energy Agency, 2016. *Costs of Decommissioning Nuclear Power Plants.* Nuclear Development Series, NEA No. 7201. Paris: NEA. www.oecd-nea.org/ndd/pubs/2016/7201-costs-decom-npp.pdf (accessed 24 January 2017).

Nuclear Energy Insider, 2016. European nuclear decommissioning activity to rise 8% per year. 2 June. analysis.nuclearenergyinsider.com/content/european-decommissioning-activity-rise-8-year (accessed 24 January 2017).

Nucleonics Week, 2016. China's nuclear waste market to get a boost. 57(13) (31 March): 5.

NucNet, 2015. Russia announces successful decommissioning of El-2 LWGR. 191(28 September). www.nucnet.org/all-the-news/2015/09/28/russia-announces-successful-decommissioning-of-el-2-lwgr (accessed 24 January 2017).

Office of the Deputy Chief Cabinet Secretary (Japan), 2011. Report: Working group on risk management of low-dose radiation exposure. 22 December. www.cas.go.jp/jp/genpatsujiko/info/twg/Working_Group_Report.pdf (accessed 24 January 2017).

Pagnamenta, Robin, 2016. Early closure of nuclear power station could cost £22bn. *The Times* (London), 19 March.

Samseth, Jon, Anthony Banford, Borislava Batandjieva-Metcalf, Marie Claire Cantone, Peter Lietava, Hooman Peimani, and Andrew Szilagyi, 2013. Closing and decommissioning nuclear power reactors. In *United Nations Environment Programme Year Book 2012*, 35–49. Nairobi: United Nations Environment Programme.

Schmittem, Marc, 2016. Nuclear decommissioning in Japan: Opportunities for European companies. Tokyo: EU–Japan Centre for Industrial Cooperation, March. www.eu-japan.eu/sites/default/files/publications/docs/2016-03-nuclear-decommissioning-japan-schmittem-min_0.pdf (accessed 24 January 2017).

Schneider, Mycle, and Antony Froggatt, with Julie Hazemann, Tadahiro Katsuta, M. V. Ramana, and Steve Thomas, 2015. *The World Nuclear Industry Status Report 2015*. Paris: Mycle Schneider Consulting Project.

Stoiber, Carlton, Abdelmadjid Cherf, Wolfram Tonhauser, and Maria de Lourdes Vez Carmona, 2010. *Handbook on Nuclear Law: Implementing Legislation*. Vienna: IAEA.

Suzuki, Tatsujiro, 2015. Nuclear energy policy issues in Japan after the Fukushima nuclear accident. *Asian Perspective* 39(4): 591–606.

Taebi, Benham, and Andrew C. Kadak, 2010. Intergenerational considerations affecting the future of nuclear power: Equity as a framework for assessing fuel cycles. *Risk Analysis* 30(9): 1341–62. doi.org/10.1111/j.1539-6924.2010.01434.x

Thomas, Karen, 2016. Sweden's plant closures to hike skills demand from 2017. *Nuclear Energy Insider*, 21 March. analysis. nuclearenergyinsider.com/swedens-plant-closures-hike-skills-demand-2017 (accessed 24 January 2017).

UNSCEAR (United Nations Scientific Committee on the Effects of Atomic Radiation), 2008. *Sources and Effects of Ionizing Radiation*. Vol. 1. Annex B. New York: United Nations.

US Government Accountability Office, 2012. NRC's oversight of nuclear power reactors decommissioning funds could be further strengthened. GAO-12-258, April.

US NRC (Nuclear Regulatory Commission), 2013. Report on waste burial charges: Changes in decommissioning waste disposal costs at low-level waste burial facilities: Final report. NUREG-1307, Rev. 15. Washington, DC: Nuclear Regulatory Commission, January. www.nrc.gov/docs/ML1302/ML13023A030.pdf (accessed 24 January 2017).

US NRC (Nuclear Regulatory Commission), 2015a. Communication strategy for the enhancement of public awareness regarding power reactors transitioning to decommissioning. February. www.nrc.gov/docs/ML1501/ML15013A068.pdf (accessed 24 January 2017).

US NRC (Nuclear Regulatory Commission), 2015b. Backgrounder on decommissioning nuclear power plants. May. www.nrc.gov/reading-rm/doc-collections/fact-sheets/decommissioning.html (accessed 24 January 2017).

US NRC (Nuclear Regulatory Commission), 2016. Background on the proposed security rulemaking for independent spent fuel storage installations. www.nrc.gov/about-nrc/radiation/related-info/isfsi-security/background.htm (accessed 4 September 2017).

US NRC (Nuclear Regulatory Commission) Regulations, 2015. § 50.75: Reporting and recordkeeping for decommissioning planning. www.nrc.gov/reading-rm/doc-collections/cfr/part050/part050-0075.html (accessed 24 January 2017).

US Senate, 2014. Nuclear reactor decommissioning: Stakeholder views. Hearing Before the Committee on Environment and Public Works, 130th Congress, 2nd Session, 14 May.

Wald, Matthew L., 2011. Keeping score on nuclear accidents. *New York Times*, 12 April.

Weigl, M., 2008. Decommissioning of German nuclear research facilities under the governance of the Federal Ministry of Education and Research. Paper presented to the Hazardous Wastes & Environmental Management Symposium, Phoenix, 25 February.

World Nuclear News, 2015. Vattenfall, EOn team up on German decommissioning. 29 May. www.world-nuclear-news.org/WR-Vattenfall-EOn-team-up-on-German-decommissioning-2905154.html (accessed 24 January 2017).

World Nuclear News, 2016a. Proposal for financing German nuclear phase-out. 28 April. www.world-nuclear-news.org/WR-Proposal-for-financing-German-nuclear-phase-out-2804164.html (accessed 24 January 2017).

World Nuclear News, 2016b. EU increases Chernobyl funding on eve of anniversary. 25 April. www.world-nuclear-news.org/WR-EU-increases-Chernobyl-funding-on-eve-of-anniversary-25041602.html (accessed 24 January 2017).

11

Sustainable energy options

Andrew Blakers

Abstract

Photovoltaics (PV) and wind are overwhelmingly dominant in terms of new, low emissions generation technology because they cost less than alternatives. PV and wind constitute half of the world's new generation capacity installed each year. Wind and PV are essentially unconstrained by resource, environmental, materials supply, water supply, or security issues. Their prices are now competitive with new fossil and nuclear power plants. Conventional hydro cannot keep pace with wind and PV due to lack of rivers to dam, and biomass availability is severely limited. Heroic growth rates are required for nuclear, carbon capture and storage (CCS), concentrating solar thermal, ocean, and geothermal to span the 20- to 200-fold difference in scale to catch wind and PV—which are themselves moving targets since both are growing rapidly and both access massive economies of scale. Pumped hydro energy storage (PHES) constitutes 99 per cent of all storage for the electrical supply industry. The combination of PV, wind, and PHES, each of which has more than 150 gigawatts (GW) deployed, allows high (80–100 per cent) renewable energy penetration of electricity markets. The conversation of land transport and urban heating to electrical supply may allow renewable electricity to supply more than three-quarters of end-use energy in the medium term.

Energy options

Energy and greenhouse gas emissions

About three-quarters of global greenhouse gas emissions arise from use of fossil fuels in the energy sector, as illustrated in Figure 11.1. In order to avoid dangerous climate change, it is necessary to replace this fossil fuel use with energy sources that do not emit greenhouse gases.

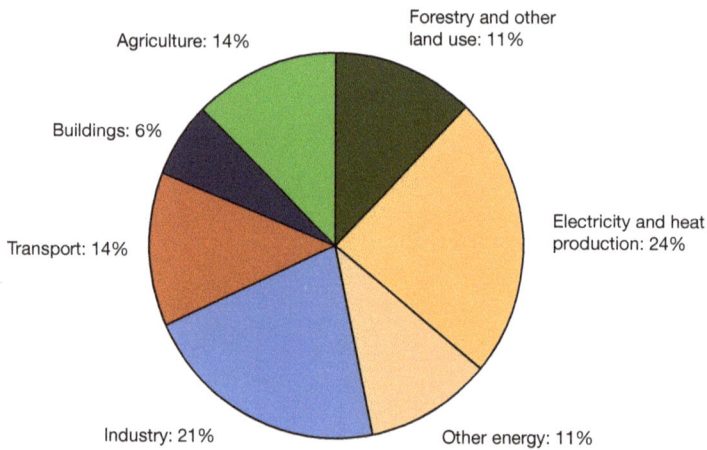

Figure 11.1 Global warming potential of greenhouse gases by economic sector over a 100-year time frame
Source: IPCC (2014: 88).

Energy technologies

The energy sources that are available and potentially have low greenhouse gas emissions comprise:

• solar—both directly from the Sun (photovoltaics (PV) and solar thermal) and indirectly (such as wind, hydro, biomass, and wave energy);

• fossil fuels (coal, oil, and gas) with carbon capture and storage (CCS);

• nuclear (fission and fusion);

• geothermal; and

• tidal.

The focus of this chapter is on pathways to deep reductions in energy-related greenhouse gas emissions over the next two decades. The energy technologies that can meet this goal must have large resource bases, should not introduce other serious problems, and must be at a point in their technology and economic development that does not require heroic assumptions in relation to future deployment rates and cost reductions. Wind and PV are likely to be the dominant low-emission energy technologies deployed over the next two decades. Some of the other energy technologies listed above will have significant supporting roles.

The worldwide PV and wind industries are being deployed on a large scale (>100 gigawatts (GW) of new plant constructed per year combined), and are likely to grow into enormous industries over the next decade. In 2015, renewable energy (primarily hydro, wind, and PV) provided 64 per cent of net new electricity generation capacity worldwide, with fossil fuel (primarily gas and coal) power stations providing most of the balance (see Figure 11.2). On current trends, wind and PV will both pass fossil and nuclear combined in 2018 in terms of annual new generation capacity. PV and wind presently constitute nearly all new generation capacity in Australia and several other countries.

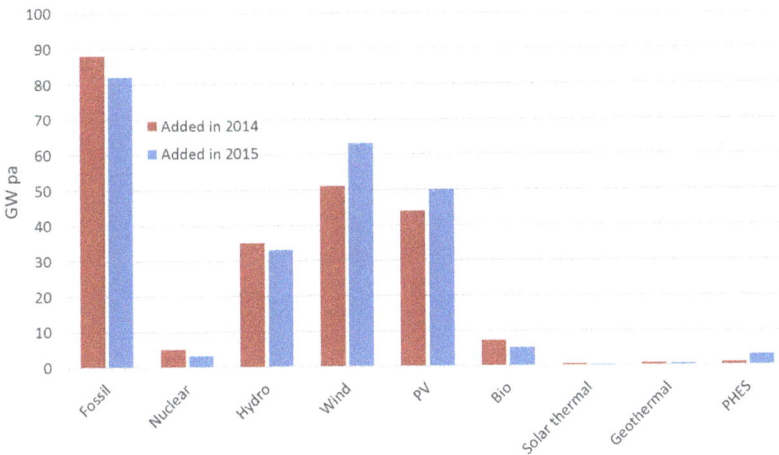

Figure 11.2 New generation capacity added in 2014 and 2015 by technology type

In 2015, 113 GW of net new wind and PV were deployed, which is nearly as much as everything else combined.

Source: REN21 (2016); Frankfurt School–UNEP Collaborating Centre (2014); IRENA (2016).

In 2015, new additions of PV and wind capacity combined grew 19 per cent over the previous year. Installation rates of the other generation technologies stayed steady or declined. Wind and PV are unconstrained by resource, environmental, material supply, or security issues.

Hydro cannot keep pace due to lack of rivers to dam, and biomass availability is severely limited. Heroic growth rates are required for the other potential low-emission technologies (nuclear, CCS, concentrating solar thermal, ocean, and geothermal) to span the 20- to 100-fold difference in scale to catch wind and PV—which are themselves moving targets since both are growing rapidly and both access massive economies of scale.

It appears that PV and wind have 'won' the race to dominate the low-emission generation sector. It will be difficult for another low-emission technology to catch PV and wind before they saturate the electricity market. Since their marginal cost of energy production is nearly zero (as with hydro), it will be difficult for another renewable electricity technology subsequently to undercut PV and wind in the market.

In the discussion below of the various energy technologies, the focus is upon those that have the potential for large-scale deployment (>100 GW per year by 2030). Current world electricity production is about 23 million gigawatt hours (GWh) per year. Of this, 22 per cent is derived from renewable energy (mostly hydro). To set this in context, deployment of 175 GW per year of each of wind and PV capacity would be sufficient to achieve a 23 million GWh per year renewable electricity target by 2035. This is obviously possible since it only requires a threefold increase in annual deployment rates of wind and PV compared with current practice (2015). Of course, demand for electricity is likely to grow because the world population is increasing, affluence is increasing, and it is probable that much of the fossil fuels used for motor transport, heating, and industry will be replaced by electricity from low-emission sources.

Fossil fuels

The carbon dioxide (CO_2) emitted from combustion of fossil fuels can be captured and stored underground. In practice, this has proved to be challenging and expensive on a large scale. It is necessary to separate CO_2 from other gases (nitrogen, argon, oxygen, and water vapour), transport it to a suitable site, pressurise it to form a liquid, and inject it into a secure location deep (kilometres) underground. Parasitic energy use by CCS reduces the overall efficiency of a coal- or gas-fired power station, and CCS adds substantially to the capital and operating costs. Because it is difficult and expensive to retrofit existing power stations with CCS equipment, widespread deployment of CCS would therefore require that future fossil fuel power stations be fitted with CCS.

The Boundary Dam project in Saskatchewan, Canada, is the first large CCS project in the power sector (Global CCS Institute 2014). About 1 million tonnes per annum of CO_2 will be captured and stored underground. For comparison, this amount of avoided CO_2 emissions could also be achieved by deploying about 0.3 GW of wind or 0.6 GW of PV. To provide context, in 2016, 60–70 GW each of wind and PV will be deployed worldwide.

Carbon dioxide is sometimes used for enhanced oil recovery by pressurising underground oil reserves in order to extract more oil. Whilst this is economically desirable, and does sequester significant amounts of CO_2, the resulting oil adds to greenhouse gas emissions.

Prospects for widespread deployment of new fossil fuel power stations fitted with CCS are poor due to immature technology, high cost, high risk, and strong competition from hydro, wind, and PV, which are being deployed at vastly larger scale and lower cost than the prototype CCS systems (Frankfurt School–UNEP Collaborating Centre 2014).

Nuclear energy

All nuclear reactors obtain energy from the fission of heavy elements, usually uranium. Nuclear energy is well-established and produces about 11 per cent of the world's electricity. Nuclear energy is associated with problems such as nuclear weapons technology proliferation, potential for fissile material production, reactor accidents, and waste disposal. Nuclear reactors are characterised by strong local opposition, long lead-times, substantial security requirements, and perceptions of high risk. This strongly constrains rapid deployment of nuclear energy. Current net deployment rates of nuclear energy (new reactors minus retiring reactors) are 15 times smaller than each of wind and PV.

Fusion energy drives the Sun. Magnetic and inertial confinement of deuterium and tritium (isotopes of hydrogen) can be used to achieve the temperatures and times required for net release of fusion energy. However, fusion reactors are a very challenging endeavour and are unlikely to be commercially available before 2050.

Geothermal and tidal

Geothermal and tidal energy are significant in some countries. However, economically harvestable global resources are too small to make a difference at a global level, although they could be important in some regions. Geothermal energy is derived from heat within the Earth. In volcanic regions, hot rocks are available near the Earth's surface—for example, in Iceland and Indonesia. Steam can be harvested for use directly or to generate electricity.

In some regions, masses of slightly radioactive rock buried kilometres below the surface become hot, allowing harvesting of heat at a temperature of around 300 degrees Celsius. Cold water is injected under pressure to fracture the rock, and allow steam to be extracted. This hot dry rock technology is challenging, and is only applicable in certain regions with the right geology. There has been no significant deployment to date.

Energy can be extracted from tidal flows using standard hydro technology. In a typical system, a weir is constructed across an estuary, and water flows through turbines as the tides rise and fall. Suitable sites with large tidal ranges and sufficiently low environmental impact are rare.

Solar energy supply

Solar energy is vast, ubiquitous, and indefinitely sustainable. Its utilisation generally has minimal environmental, social, and security impacts over unlimited timescales. Recent large-cost reductions now place solar-derived energy in the same cost range as fossil and nuclear energy. Renewable energy derived from the Sun, principally comprising solar, wind, and hydro energy, now constitutes most new electricity generation capacity constructed worldwide each year.

The Sun will continue to shine for billions of years. Each year the Earth receives about four orders of magnitude more solar energy than human commercial energy consumption. After accounting for conversion losses (only 15–50 per cent of solar energy incident on a solar collector is successfully collected and converted to a useable form) and inaccessible regions (oceans, the poles, mountains, and forests), there is hundreds of times more available solar energy than human commercial energy consumption.

The two major direct solar energy conversion technologies are PV and solar thermal. The former directly converts sunlight into electricity. The latter includes solar collection for heat in buildings and industrial processes (such as solar hot water and solar heating of buildings), solar thermal electricity (produced by concentrating sunlight onto a receiver to create high-temperature steam), and solar-assisted thermochemical production.

Solar energy drives world energy systems, which yield indirect energy in the form of wind energy, hydro energy, wave energy, ocean thermal energy, and biomass energy.

The solar resource

The Sun provides about 1.3 kilowatts (kW) per square metre (1.3 kW/m²) to the upper atmosphere of the illuminated half of the Earth. Most is transmitted though the atmosphere to the surface of the Earth, while some is absorbed and reflected by the atmosphere. The solar intensity at noon on a sunny day at the Earth's surface is about 1 kW/m². Each year, the Earth receives about 3.8×10^{24} joules of solar energy. This is about 50,000 times more than current worldwide electricity consumption. Much of this energy is in the form of direct beam radiation; that is, it comes directly from the visible disc of the Sun. At ground level, a sizeable fraction appears as indirect (diffuse) radiation due to scattering from clouds, aerosols, and other atmospheric constituents, plus reflected light from the ground. A relatively small fraction of the solar energy is converted into energy forms that can be harvested as wind energy, hydroelectricity, ocean energy, and biomass.

The sum of the direct and diffuse radiation received by a solar collector is termed global radiation. Some collector systems such as non-concentrating PV panels respond to both the direct and diffuse components of sunlight. Other collectors, such as concentrating PV (CPV) and solar thermal systems, respond primarily to the direct beam component—fundamental physical laws limit concentration of the scattered diffuse light. For this reason, concentrating systems are best suited to dry locations with low levels of cloud and air pollution. For example, the proportion of annual radiation in Australian cities that is diffuse is about one-third, meaning that one-third of the available solar power will be discarded in a solar concentrating system. Tropical cities and desert regions have annual diffuse radiation amounting to about half and one-quarter of the incoming solar radiation, respectively.

The available solar radiation depends upon latitude, weather patterns, and air pollution levels. The seasonal variation in solar radiation is important because it is expensive to store energy harvested in summer for use in winter. In general, low latitudes have much less seasonality in both solar energy availability and energy demand (for heating and cooling).

About two-thirds of the world's population live in the latitude range +/-35° where there is generally good solar availability and moderate seasonal variation of both solar energy supply and energy demand compared with higher latitudes. This latitude range is home to most of the populations of Africa, Central and South America, Australasia and Oceania, Southeast Asia, India and South Asia, and the Middle East. However, a group of highly influential countries with energy-intensive economies lie at higher latitudes, including European countries, South Korea, Russia, Canada and much of the United States, China, and Japan. Perceptions of the availability and suitability of solar energy are sometimes skewed by the current economic and political power of these countries.

Environmental and social aspects of solar and wind energy

Solar and wind energy collection utilises only very common materials, with few exceptions. For example, PV systems utilise silicon (for the solar cells); silicon, oxygen, and sodium (for the cover glass of the solar module); oxygen, carbon, and hydrogen (for the encapsulating plastic); aluminium (for the frame of the solar module); iron (for the steel support posts); plus some metals in small quantities such as phosphorus, boron, copper, and silver. These elements are ubiquitous in the Earth's crust and atmosphere, and it is difficult to envisage ever running out of them. The amount of rock that needs to be moved during mining, for a given level of solar energy production, is orders of magnitude smaller than the equivalent for fossil and nuclear energy systems, principally because of the absence of mined or extracted fuel.

Solar and wind energy are available nearly everywhere in vast quantities; it is unlikely that people will ever go to war over access to solar and wind energy, in contrast to the situation with fossil fuels. Utilisation of solar and wind energy entails minimal security and military risks. The highly dispersed locations of millions of solar and wind energy collectors entails a robust and resilient energy system with limited utility for warfare and terrorist activity.

Less than 1 per cent of the world's land area would be required to supply all of the world's commercial energy requirements from PV using current technology. A large segment of the world's energy can be supplied from roof-mounted solar collectors, which effectively alienate no land. Another large segment of the world's energy can be supplied from solar collectors in arid regions, in conjunction with long-distance high voltage direct current (HVDC) transmission of electricity. Wind generators alienate only a few square metres per megawatt (MW) of capacity (the site of the tower) and farming operations can continue around the base of the tower. Relatively little alienation of productive farmland, forests, and ecosystems is required to achieve a world economy where most of the commercially traded energy is derived from solar and wind energy.

Solar and wind energy systems do not emit greenhouse gases during operation. However, greenhouse gases, principally CO_2, are emitted during the manufacturing phase. The time required to generate enough electricity to displace the CO_2 emissions equivalent to that invested in construction of a solar or wind energy system is currently in the range 0.5–2 years, compared with typical system lifetimes of 20 to 30 years. CO_2 manufacturing intensity and price are directly linked (via material consumption and efficiency), and so CO_2 payback times continue to fall as prices fall. CO_2 payback times are also falling as the proportion of low-emission generators in electricity systems increase. CO_2 payback times will eventually fall to a small fraction of one year.

Solar and wind energy system manufacturing and operation entails minimal pollution and noise. Social acceptance is generally high, although there is opposition to wind generators by some people, primarily based on aesthetic considerations. Both the risk and consequences of accidents are very low compared with fossil fuel and nuclear energy systems.

The future of solar and wind energy

Renewable energy technologies can eliminate fossil fuel use within a few decades at low cost, allowing a fully sustainable and zero carbon energy future. Roof-mounted solar energy systems can provide PV electricity, hot water for domestic and industrial use, and thermal energy to heat and cool buildings. Grid parity for PV at a retail level has already been achieved for much of the world's population. This is leading to rapid growth in sales in the residential and commercial sectors without the need for subsidies.

Large PV and solar thermal concentrator power stations, in conjunction with wind and hydro energy, can provide most of the world's industrial energy. In addition to direct solar energy collection, indirect forms of solar energy such as wind, biomass, wave, and hydro can make important contributions.

Solar- and wind-generated electricity, coupled with a shift to electrically powered cars and public transport, can provide most of the world's transport energy.

Photovoltaics

PV is likely to eventually dominate energy production worldwide because the solar resource utilised by PV is much larger and more ubiquitous than wind energy. PV is an elegant technology for the direct production of electricity from sunlight without moving parts. Most of the world's PV market is serviced by crystalline silicon solar cells (Reinders at al. 2015). Sunlight is absorbed by the solar cell and the solar power is converted to electrical power with a conversion efficiency of 15–25 per cent. The remaining solar power (75–85 per cent) becomes heat. This process of conversion is called photovoltaics (photo = light, voltaics = voltage).

In a silicon solar cell, sunlight causes electrons to become detached from their host silicon atoms. Near the upper surface is a 'one-way membrane' called a pn-junction. When an electron crosses this junction it cannot easily return, causing a negative voltage to appear on the sunward surface (and a positive voltage on the rear surface). The sunward and rear surfaces are connected via an external circuit containing a battery or a load in order to extract current, voltage, and power from the solar cell (see Figure 11.3).

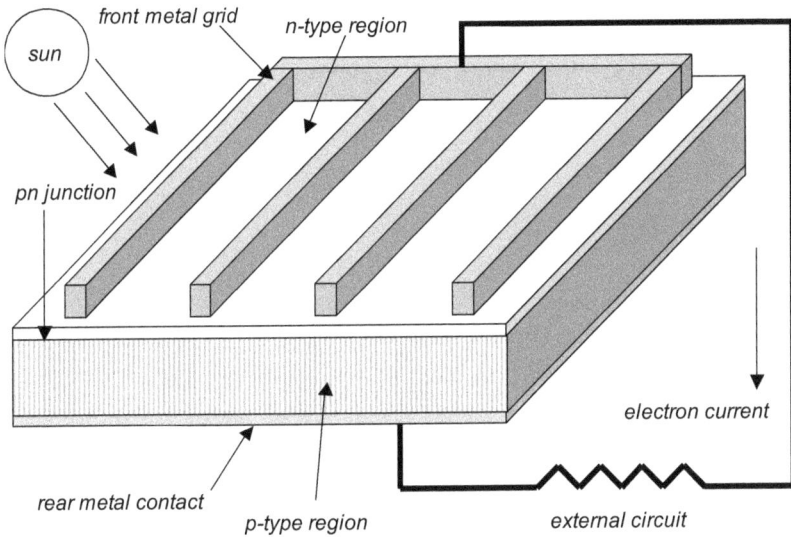

Figure 11.3 Schematic of a typical solar cell
Source: Author.

More than 90 per cent of the world's photovoltaic market is serviced by mono- and multi-crystalline silicon solar cells, and this will continue for the foreseeable future. Silicon has important advantages including elemental abundance (number two in the Earth's crust), moderate cost, non-toxicity, high efficiency, device performance stability, simplicity (it is a mono-elemental semiconductor), physical toughness, a highly advanced state of knowledge of silicon material and technology, and the advantages of incumbency. The latter comprises extensive and sophisticated supply chains, large-scale investment in mass production facilities, deep understanding of silicon PV technology and markets, and the presence of thousands of highly trained silicon specialists—scientists, engineers, and technicians.

In order to produce solar cells, silicon crystalline ingots are grown from a silicon melt at 1400 degrees Celsius. Many thin (0.15–0.2 mm) wafers with diameters of 156 mm are cut from the ingots using a saw, followed by wafers saw damage removed in a silicon etch. The next step is to diffuse a tiny amount of phosphorus into the sunward surface of the wafer to a depth of about 0.001 mm to create the pn-junction. Then follows deposition of a thin sheet of metal on the rear surface and a grid of metal on the sunward surface to allow extraction of electricity.

Groups of 60–80 solar cells are electrically connected and encapsulated in thin layers of plastic and laminated behind a tough, 3 mm thick glass cover to form solar modules, each with a power of about 300 watts (W). A junction box is added to house the electrical terminals. Dozens to millions of solar modules are mounted together and electrically interconnected to form a solar power system.

Some PV systems are mounted on fixed support structures that are tilted up to face the equator, with a tilt equal to the angle of latitude. This maximises annual electricity production. Large PV systems are usually mounted on sun-tracking systems to maximise annual output. Electricity produced by PV modules is conducted to a power conditioning unit that optimises voltages, converts the direct current produced by solar cells to the alternating current used in electrical grids, transforms the voltage to match that of the local grid, and manages interfacing with the local grid.

Photovoltaic systems have unmatched reliability and low maintenance cost due to the lack of moving parts. Manufacturers typically warrant PV modules for 25 years, and in dry locations they may continue to operate for 50 or more years. Specimen modules are exposed to severe accelerated failure testing in order to elucidate and prevent failure mechanisms. Degradation modes of PV modules include physical destruction caused by human action or violent hailstorms; slow chemical changes leading to yellowing of transparent encapsulation materials; and slow ingress of moisture causing corrosion of metallic components.

Photovoltaic technologies

Crystalline silicon solar cells constitute more than 90 per cent of the world PV market. The leading commercial crystalline silicon PV solar cell technology is based upon screen printing of the metallic contacts. Commercial solar cell efficiencies of 14–20 per cent are achieved with this technology. Interdigitated back contact and Heterostructure with Intrinsic Thin Layer silicon solar cells have commercial efficiencies of 22–24 per cent, albeit at a premium price, and are typically deployed where space is limited. The best laboratory cells have efficiency of 25–26 per cent, compared with the theoretical maximum efficiency of 29 per cent. The Passivated Emitter and Rear Cell design (Blakers et al. 1989) is likely to achieve dominance in world markets by 2020 due to improved efficiency (Reinders et al. 2015).

Currently, the leading non-silicon PV technology utilises cadmium telluride, with about 4 per cent market share primarily due to the commercial success of the company First Solar. A material called CIGS (comprising copper, indium, gallium, and selenium), and another called amorphous silicon, have market shares of 1–2 per cent each. Solar cells based on many other materials are under development but not yet in significant commercial production, notably perovskite materials. The latter has created substantial interest due to rapid improvement of efficiencies to above 20 per cent for small area laboratory cells, and the possibility of creating tandem solar cells in conjunction with crystalline silicon.

A potentially important branch of photovoltaics is CPV. Tracking of the Sun is required for concentrator systems. Mirrors or Fresnel lenses are used to concentrate light by 100 to 1,000 times onto a small number of highly efficient (and expensive) solar cells. Typically, active cooling of the solar cells is required to remove excessive heat. The best concentrator solar cells have conversion efficiencies approaching 50 per cent, and comprise three or more layers of different semiconductor materials drawn from the group 3 and group 5 columns of the periodic table. Such cells are very expensive per square centimetre compared with conventional silicon solar cells. However, concentration greatly reduces the effective cost per square metre of collector area—essentially most of the solar cells are replaced by cheaper lenses and mirrors. CPV may become important in the future but has only small market share at present, and will be restricted to areas with plenty of sunshine and low pollution because only direct beam light can be concentrated. CPV technology has much in common with concentrating solar thermal power, since much of the system infrastructure (such as trackers, controllers, lenses, and mirrors) could be used for either.

Photovoltaic markets

PV is unusual in that the unit cost of energy is similar for large (MW) and small (kW) systems—large systems have lower capital costs but higher financing costs and vice versa. Virtually all other energy sources have strong economies of scale. This confers a major advantage on PV, since it has markets at every scale from W to GW for the same basic product—the silicon solar cell.

In earlier decades, PV found widespread use in niche markets such as consumer electronics, remote-area power supplies, and satellites. Throughout the world, remote-area energy solutions are based upon various combinations of PV, wind, diesel, and batteries. Active load management is an important additional feature to minimise the requirement for diesel and batteries. In recent years, the industry has expanded (see Figure 11.4) and costs have declined very rapidly. PV systems are now installed on tens of millions of house roofs in cities, and also in large ground-mounted power stations. Mass production is causing rapid reductions in cost.

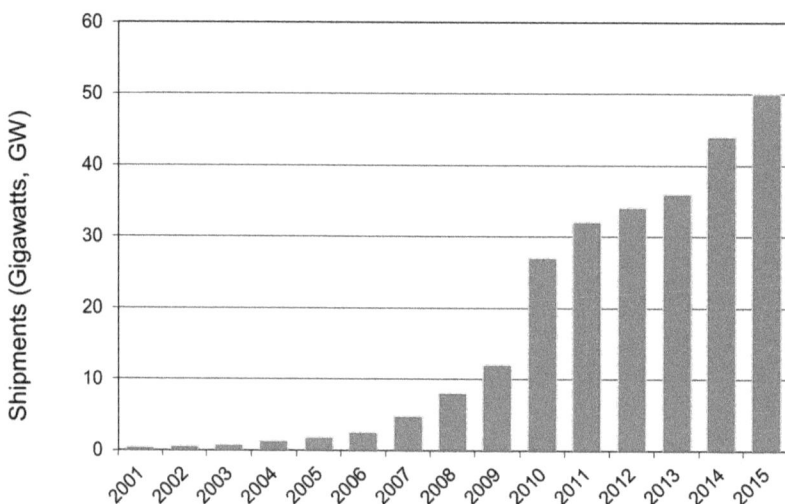

Figure 11.4 Shipments of photovoltaics since 2001
Source: Author data.

PV electricity is now less expensive than domestic and commercial retail electricity from the grid throughout much of the world, and is approaching cost-competitiveness with wholesale conventional electricity (IRENA 2015; Breyer and Gerlach 2013). This fact is the reason for the rapid growth in deployment of PV worldwide. Direct competitiveness with fossil fuels for wholesale energy supply is assisted by carbon pricing and the removal or equalisation of hidden support for fossil fuels. The cost of PV systems can be confidently expected to continue to decline for decades.

Solar energy has much to offer developing countries since nearly all are in low latitudes with good solar resources and low seasonal variability of both energy demand and solar availability. Small amounts of PV electricity can make dramatic differences to living standards, through the enabling of electric services such as lighting, computing, telecommunications, refrigeration, grain grinding, and water pumping.

Developing countries generally lack a widespread and robust electricity distribution grid. There are good prospects for organic development of thousands and millions of small solar- and wind-powered systems, which gradually merge to become a national grid. These countries may bypass the centralised electricity distribution systems of high-income countries. An analogy is telephony, for which low- and middle-income countries will rely heavily upon distributed mobile telecommunications rather than fixed lines.

Wind energy

Modern MW-scale wind generators located at good sites are amongst the lowest cost electricity generation technologies available. It is likely that wind and PV will have the largest deployment rates of electricity generation technologies in many countries for the next several decades. Wind and PV are often a good combination in that they counter-produce; it is often windy when not sunny, and vice versa. A modern wind generator comprises a tower, a rotating nacelle atop the tower housing a generator and control electronics, and three blades facing into the wind. Fields of hundreds of wind generators spaced apart from each other constitutes a wind farm. Typically, wind generators are located in farmland, and farming continues around the towers. The generators are located 5–10 rotor diameters apart. Wind farms located in shallow offshore waters are likely to become widespread in the future, both because it expands the available space for wind farms and because wind speeds are generally higher over water. However, there are significant technical and economic impediments to be overcome.

Commercial wind generators have power ratings of 1–8 MW. The largest currently available wind turbine is the Vestas V164, which is designed for offshore use and has a rated capacity of 8 MW (Wikipedia 2015), a maximum blade-tip height of 220 metres, and a rotor diameter of 164 metres. Even larger turbines are under development. Although costs increase rapidly with turbine scale due to the engineering required for such massive machines, the greater height of the rotors means that energy production also increases rapidly.

The power available from a wind generator is approximately proportional to the cube of the average wind speed at the top of the tower. This means that there is a strong incentive to find windy sites and to utilise tall towers. Accurate methods are available for assessing wind speeds as a function of location and height above ground. This is supported by sophisticated modelling to allow selection of the best sites for each wind generator in a wind farm.

The capacity factor refers to the annual output of a wind generator compared with the output that would be achieved were a generator to operate at its maximum power for the entire year. Higher average wind speed correlates with higher capacity factor. As larger machines become available, and as more offshore windfarms are constructed, they capture the stronger and more consistent winds blowing at greater heights from the ground and at sea, and the capacity factor rises. The cost of wind energy is inversely proportional to the capacity factor. Larger and more modern machines are gradually replacing older and smaller machines in good sites on land, and this is also leading to increased capacity factor.

Increases in machine size, and capacity factor, are leading to a decreasing cost of wind energy. It is unlikely that the bottom of the cost curve will be reached in the near future. Wind electricity is now fully competitive with fossil and nuclear electricity in many places throughout the world (IRENA 2015). This fact is the reason for the rapid growth in deployment of wind energy worldwide (see Figure 11.5).

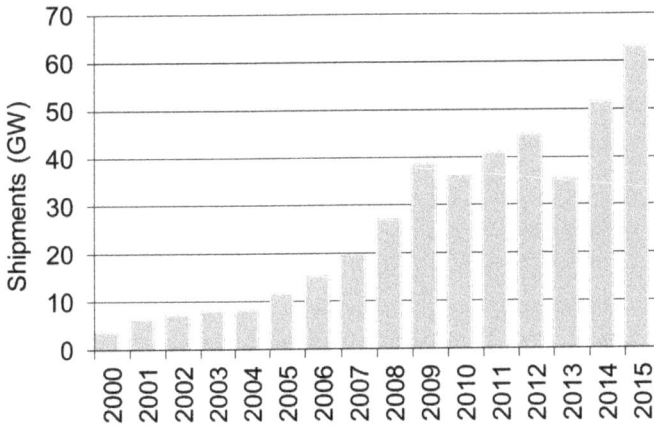

Figure 11.5 Wind energy shipments since 2000
Source: Global Wind Energy Council (2015).

Other solar energy technologies

Hydroelectric energy

Hydroelectric energy is a highly developed technology and accounts for about 16 per cent of worldwide electricity production. Hydro, wind, and PV constitute almost all current renewable electricity production. Generally, hydro involves construction of a dam on a river impounding a lake; construction of pipes or tunnels; and installation of an electrical turbine and power lines. Some hydro systems are 'run-of-river', which means that only small offtake weirs are needed. Most developed countries have utilised most potential hydro sites. Developing countries have many opportunities. Strong environmental and social opposition to hydro is often encountered, due to extensive flooding of sensitive river valleys, farmlands, and even cities.

Solar thermal

Good building design, which allows the use of natural solar heat and light, together with good insulation, minimises the requirement for space heating. Solar water heaters are directly competitive with electricity or gas in many parts of the world. Solar thermal electricity technologies use fields of sun-tracking mirrors to concentrate sunlight onto a receiver. The resulting heat is ultimately used to generate steam, which passes through a turbine to produce electricity. Concentrator methods are equally applicable to CPV systems. The commercially established methods of concentrating sunlight are line focus concentrators (troughs, both reflective and refractive) and central receivers (heliostats and power towers).

A potential application of concentrated sunlight is the generation of thermochemicals and the storage of heat at high temperature in molten salt to allow for 24-hour power production. Concentrated solar energy can achieve the same temperatures as fossil and nuclear fuels, either directly (using mirrors) or through the use of chemicals created using concentrated solar energy.

Solar thermal concentrators only utilise direct beam sunlight, and must be sited in dry locations with low levels of diffuse radiation in order to achieve economical application. Solar thermal electricity addresses a similar market to PV, but must be used at large scale in order to obtain competitive costs. Thus, there is a substantial cost barrier and financial risk factor compared with PV, for which energy costs from small and large systems are similar. Furthermore, the city market for which PV is well suited is unsuitable for solar thermal concentrators. Presently, new PV systems are being constructed 100 times faster than solar thermal. It is unclear whether solar thermal with thermal storage can become competitive with PV in conjunction with load management and storage (batteries and pumped hydro energy storage (PHES)).

Bio and ocean energy

Bio energy is obtained when sunlight allows growth and accumulation of biomass. Conversion of sunlight into chemical energy (i.e. biomass) is a very inefficient process compared with the PV and solar thermal energy collection. Bio energy conversion efficiency is generally much less than 1 per cent, while solar thermal and PV is 15–50 per cent efficient. Furthermore, growth of biomass requires large amounts of land, water, fertilisers, and pesticides, and competes with food and timber production.

Burning of waste biomass is a significant contributor to commercial energy production in many countries. However, the inherent limitations of bio energy production mean that it can never be more than a small fraction of commercial energy in an advanced economy. Bio energy is very widely used for heating and cooking in developing countries. It is observed in many countries that bio energy is gradually replaced by electricity or gas as incomes rise. The great flexibility of PV systems in terms of scale of deployment is likely to make a large impact in this respect.

Ocean energy comprises energy from waves, ocean currents, temperature gradients within the deep ocean, and gradients of salt concentration. In some countries with exposure to high levels of wave action and suitable seafloor conditions, wave energy could become significant when the technology is further developed. However, it is likely to be a small source of energy at a global scale.

Large-scale deployment of renewable energy

Large-scale energy storage

Wind and solar are likely to take an ever-expanding share of electricity markets. This raises questions about how to manage the variability of wind and solar power. It is observed that high penetration of electricity grids does not necessarily lead to unstable grids and the consequent need for substantial storage until quite high penetration is reached (Fraunhofer Institute for Solar Energy Systems 2015). For example, the state of South Australia obtains about half of its annual electrical energy from wind and PV, has no significant hydro or other storage, and is weakly connected electrically to other parts of Australia.

A key point in relation to storage is that a few hours of storage is sufficient to facilitate high penetration of wind and solar. Both power supply and demand are constantly fluctuating, but must remain in balance. Typically, this requirement is met through traditional hydro and low-duty cycle gas plants due to their rapid response time.

Short-term storage (4–24 hours) covers the day/night cycle, extreme demand events such as hot summer afternoons and cold winter mornings and evenings, offsets periods of low supply such as wind lulls and cloud, offsets plant and transmission line failure, and covers the time required to bring a low-duty cycle biomass, coal, or gas-fired power station online if the demand shortfall is likely to be extended. Additionally, short-term storage improves the capacity factor of constrained power lines; for example, those connecting wind and solar farms in remote windy and sunny regions to national grids. Owners of storage facilities can engage in arbitrage (buying energy when prices are low and selling when prices are higher).

PHES constitutes 99 per cent of all energy storage around the world (155 GW; see IRENA 2016: 17) because it is cheap compared with alternatives such as batteries. It is likely to continue its dominance in the wholesale storage market. PHES can provide excellent inertial energy, spinning reserve, and rapid start and black start capability. PHES involves pumping water to an upper reservoir when there is excess electricity, and later recovering that energy by allowing the water to flow back to a lower reservoir through a turbine. Response time (from off to fully on) is less than one minute. The operational lifetime is more than 50 years, with low operational costs. Round-trip energy storage efficiencies of 80 per cent are possible for well-designed systems, accounting for losses in the pumps, pipes, and turbines; i.e. 20 per cent of the stored energy is lost. The amount of energy stored in a PHES system is proportional to both the elevation difference ('head') between upper and lower reservoirs (typically 100–1,000 metres) and to the volume of water stored in the upper reservoir.

Most existing PHES is integrated with hydroelectric generation systems on rivers. In contrast, in a closed PHES system, the same water circulates indefinitely between upper and lower reservoirs, thus eliminating the need for a river ('off-river PHES'). There is limited capacity to build new hydroelectric dams in many countries. Constraints include limited, undeveloped economical dam sites, the environmental impact of the added power generation infrastructure, and the need to create additional power line easements, frequently in remote mountainous national parks. However, there is large scope to construct off-river PHES as the lowest cost mass storage option. Low costs for off-river PHES are facilitated by large heads (400–1,000 metres), absence of the need for flood control, short and steep pipelines, small reservoirs (1–20 hectares), and co-location with loads, power lines, and wind/solar farms—which are available at thousands of sites around the world.

Off-river PHES takes advantage of the vastly larger area of land that is off-river compared with on-river, which provides the opportunity to find numerous good sites close to loads and transmission infrastructure. Importantly, the upper reservoir can be on top of a hill rather than in a river valley, allowing three to five times larger head. This is a major advantage since energy storage capacity and power capacity both scale with head, and therefore cost scales inversely with head.

Off-river PHES sites comprise pairs of small, hectare-scale 'turkey nest' reservoirs, for which the walls are made from spoil scooped from the centre. There is no net generation. They can be located in hilly country near loads and power transmission networks, and connected by pipes or tunnels incorporating a pump/turbine. An off-river PHES system can deliver 1 GW for five hours utilising twin 15-hectare reservoirs with an average depth of 20 metres and an altitude difference of 750 metres. In contrast, typical conventional hydroelectric systems have lake areas of thousands of hectares, expensive flood control, and much smaller heads. Indicative costs of off-river PHES are US$0.8 million per MW for four hours of storage.

Several other large-scale energy storage technologies may be available by 2025. These include compressed air in caverns, advanced batteries, and solar thermal storage by means of molten salt. Should the cost of these technologies decline rapidly then they may become competitive with pumped hydro. Thermal storage would be a key element of any large-scale solar thermal electric generation at significant scale. However, none of these storage technologies have been deployed at significant scale compared with PHES, and future costs and technical feasibility are more speculative than for PHES.

Long-distance transmission of electricity

Transmission grids within industrialised countries are based around a relatively small number of large fossil, nuclear, and hydro power stations. Load management is practised by offering reduced prices at times of low demand. Increasing scale of interconnection confers robustness of supply, allows smoothing of total demand by increasing the variety and timing of loads, and allows the incorporation of more varied power sources including pumped hydroelectric storage. Additionally, wide geographical spread of generators connected with high voltage cables reduces the chance of simultaneous absence of sufficient sun and wind. Continent-wide transmission grids are emerging and are being strengthened. Long-distance transmission increases competition within markets, as well as allowing 'time shifting' through several time zones from one side of a continent to the other.

Greater penetration of variable renewable electricity sources is occurring. Although PV allows distributed generation of electricity including in cities, renewable energy generation is often far removed from cities. Examples include offshore wind generation and solar generation in desert regions. Transmission of electricity over long distances generally utilises HVDC technology. Since the power transmission capacity scales with the square of the transmission voltage, voltages in the range of hundreds of kilovolts (kV) up to a megavolt are used. High power alternating current transmission is technically infeasible over long distances. HVDC has additional advantages relating to reduced transmission easements and reduced induced current flowing through people living and working near direct current overhead transmission lines (Andersen 2006; Hammons 2008; Hammons et al. 2011; Kutuzova 2011).

HVDC technology was first used to link Gotland with mainland Sweden in 1954. This link was capable of transporting 20 MW, at 100 kV, over a 98 km underwater cable (Peake 2010). Since this installation, distances spanned, voltages attained, and power transmitted have all experienced massive increases. The longest HVDC line to date is in China, connecting the Xiangjiaba Dam to Shanghai, spanning a distance of 2,071 km and transmitting up to 6.4 GW at ±800 kV (Hammons et al. 2011).

The right of way needed for HVDC is substantial. On land, corridors of around 60 metres wide are necessary for the installation of a 5 GW cable (Kutuzova 2011). Obtaining access rights is a significant impediment to HVDC transmission. Underground cables require a much narrower corridor, but are more expensive and less capable of transmitting large amounts of power.

Transmission losses associated with an 800 kV DC power line with 5 GW capacity is quoted by Siemens as 3 per cent per 1,000 km (Siemens 2012). In addition, there is a few per cent conversion loss at the two ends of the cable. Costs of large HVDC cables can be expected to decline substantially as many more are constructed in coming decades. Costs below US$300 per MW kilometre are likely (Blakers, Luther, and Nadolny 2012).

Solar energy systems in cities

Tens of millions of roof-mounted PV systems have been deployed around the world. These systems have a power capacity of 0.1–10 kW for an individual dwelling, and tens to thousands of kW for commercial buildings. The roof area required amounts to 7–10 m^2 per kW. Low-density suburban housing generally has ample roof area to yield enough energy over the course of the year to equal the annual end-use energy of the dwelling. Low-rise commercial and light-industrial building roofs can yield substantial excess quantities of PV electricity for export to the electricity grid. However, in high-density regions of cities there is insufficient unshaded roof space for high penetration of PV electricity into the building energy requirement.

Roof-mounted PV systems generally compete with the retail price of electricity, which is typically two to four times larger than the wholesale price. The levelised cost of obtaining PV electricity from building roofs in moderate latitudes (<35 degrees Celsius) is well below the retail electricity tariff in many cities (Breyer and Gerlach 2013). This is driving strong growth in deployment of roof-mounted PV systems; for example, about one in six Australian houses had a roof-mounted PV system in 2015.

The deployment of millions of roof-mounted PV systems is causing a large shift in demand profiles of the distribution networks. Large increases in deployment of thermal and electrical storage will allow high penetration of PV electricity in urban areas, which will cause a dramatic shift in the economics of the electricity distribution industry. Storage is required because energy demand is often out of phase with solar energy availability. When teamed with effective energy management and control strategies, combined electrical and thermal storage is attractive both for the building owner and the electricity grid operator.

Battery storage can be used to increase self-consumption of roof-mounted PV generation, and effectively manage the electrical network and power system, by taking advantage of the fast response, power-on-demand nature of batteries. However, the high (but declining) costs of batteries means that using battery storage alone to match PV generation to building demand is costly compared with other methods of energy storage.

Storage of energy in the form of hot water in an insulated tank is very widely deployed. Conventional solar, gas, and electric hot water systems are under commercial pressure from the rapidly declining price of electricity from PV systems on building roofs. PV, combined with advances in highly efficient air-to-water heat pumps, allows PV-driven heat-pump hot water to become a cost-effective hot water supply option. Heat pumps use electrical energy to move heat from one place (outside the building) to another at a higher temperature (hot water tank). Several units of thermal energy can be delivered per unit of electricity. Heat-pump hot water storage can be easily controlled alongside other storage elements in conjunction with roof-mounted PV generation rates and household energy loads.

Thermal storage to provide space heating and cooling in buildings can be accomplished by raising or lowering the temperature of a building when low-cost energy is available, relying on thermal mass to store heat, and good insulation to reduce thermal losses. Reverse-cycle air conditioners (which are also heat pumps) are increasingly cost-effective in this application, and can be powered using roof-mounted PV. Heat banks rely on ceramic bricks to store at a high temperature and fans to circulate heat as required. They can be charged using PV electricity during the day for use at night. Cold storage can be accomplished using PV-driven heat pumps to produce cold water and ice during the day. Thermal energy storage via building space heating and cooling can be controlled, along with hot water and battery storage, to optimise the use of PV generation and to manage net household demand.

Reduction in the retail use of gas, and its replacement with PV-driven delivery of thermal energy in conjunction with heat pumps, is an emerging trend in urban buildings. Gas burning in domestic dwellings for delivery of hot water and space heating is relatively expensive and inefficient compared with PV. Additionally, gas appliances for water and space heating are relatively expensive. Hotplate cooking, a minor but highly valued use of gas, can be replaced by induction cooktops.

Generation of medium-temperature (>100 degrees Celsius) heat in cities for industrial purposes is usually achieved with gas for substantial applications. This temperature range is beyond the supply capacity of conventional thermosiphon and evacuated solar collectors. However, provision of 100–150 degrees Celsius heat is within reach of PV-driven heat pumps and resistive elements and is expected to rapidly increase.

Transport systems and solar fuels

Fossil fuels used for transport typically account for about 20 per cent of a developed country's greenhouse gas emissions. Cars, buses, and commercial vehicles comprise most of this, which can be avoided by moving to electric vehicles and electrically powered public transport, provided that electricity comes predominantly from renewable energy. Electric vehicle sales are rising rapidly, due to cost reductions in vehicles and improvements in automotive systems and batteries. For a driving distance of 8,000 km per year, a 1 kW photovoltaic panel is sufficient to provide the annual electricity requirements. The panel will occupy space of about 7 m², and would preferably be mounted in a sunny place such as a rooftop. The fully installed cost of the PV panels will be US$1,500–$2,000, and typically they will last 25 years—twice the typical lifetime of the car. Thus an outlay of a few thousand dollars provides the electricity requirements of an electric car for its whole lifetime, at a cost of about one cent per km.

Conversion of most land transport (vehicles and trains) to electricity derived from renewable energy sources appears to be feasible, which will greatly reduce the requirement for fossil transport fuels. However, some transport functions, such as ships, aircraft, and heavy machinery, cannot be met from electrical sources because of the impracticable size, weight, and cost of the required battery storage. Some industrial processes may also be difficult to service with (renewable) electricity. Cost-effective renewable energy-driven fuel synthesis, in competition with fossil fuels, is still some time away from large-scale utilisation, not least because there are many energy losses along the way.

Synthesis of chemical fuels, utilising solar energy to drive the chemical reactions, allows solar and other renewable electricity sources to substitute for fossil fuels for both transport and as an industrial fuel. There are a limited number of suitable chemical fuels, taking account of material abundance, toxicity, storability, and other factors. Notable candidates are carbon-based compounds (methane CH_4, diesel $C_{12}H_{23}$, kerosene $C_{12}H_{26}$), hydrogen (H_2), and ammonia (NH_3). The likelihood is that most synthetic fuels will be 'drop-in' replacements for existing fuels to avoid the need to redesign existing engines, i.e. based upon carbon.

Synthesis of carbon fuels will require an energy source derived from a low-emission technology. A renewable source of carbon is also required, and the two candidates are direct extraction of CO_2 from air or seawater, and biomass. In the latter case, provided that the biomass is merely the source of carbon, rather than additionally the source of energy, then the amount of biomass required is reduced. This is important because biomass has very low efficiency of solar energy collection and conversion (less than 1 per cent) and requires large amounts of land, water, pesticides, and fertilisers.

Chemical fuel synthesis utilising solar energy can be driven either by heat or electricity. It is not possible to transfer heat over long distances— it must be generated and used locally. Additionally, land for solar collectors is expensive in industrial areas. Thus direct (local) utilisation of high-temperature solar heat in industry requires locations that have high direct beam irradiation and low land cost. This is problematical for nearly all current manufacturing localities around the world, including much of China and India (severe air pollution), much of Southeast Asia (tropical cloudiness), and much of Europe and North America (substantial cloudiness and substantial seasonality of solar insolation). Loss of 50 per cent or more of the global radiation because it is diffuse makes solar concentrators considerably less economic.

Electric-driven fuel synthesis takes advantage of rapid reductions in the price of electricity from renewable wind and PV. PV and wind collectors can be located remotely in windy/sunny locations to transmit electricity to an industrial centre. Thus, renewable electricity has a significant advantage over solar concentrator heat for heavy industry. A key requirement in fuel synthesis and other industrial chemical processes such as ammonia production is to obtain hydrogen. Currently, most hydrogen used in industry comes from natural gas (CH_4). Renewable fuel synthesis can, however, obtain hydrogen from electrolysis of water. If renewable carbon fuel synthesis relies upon extraction of CO_2 from the air, then by far the largest energy requirement is electrolysis of water to obtain hydrogen, which is an electrical process.

100 per cent renewable energy

Recent work at The Australian National University (Blakers, Lu, and Stocks 2016) shows that 100 per cent renewable electricity is feasible at low cost in Australia, and by extension in similar countries and regions. The analysis avoids heroic assumptions about future technology development, and only includes technology that has already been deployed in large quantities (>150 GW), namely PV, wind, HVDC, and PHES.

In the modelling, wind and PV contribute 90 per cent of annual electricity, while existing hydroelectricity and biomass contributes about 10 per cent. PV and wind are overwhelmingly dominant in terms of new, low emissions generation technology because of their low cost: they constitute half of the world's new generation capacity installed each year, and all new generation capacity installed in Australia. The modelling uses historical data for wind, sun, and demand for every hour of the years 2006–10, and maintains energy balance between supply and demand by adding sufficient PHES, HVDC, and excess wind and PV capacity.

The key outcome of the modelling is that the cost of balancing energy on an hourly basis for 100 per cent renewable penetration is relatively small, in the range US$15 per megawatt hour (MWh). This covers the cost of PHES, HVDC, and spillage of wind and PV electricity when supply exceeds demand and the storages are full. The total cost of a fully renewable electricity supply in Australia is estimated at US$50 per MWh (in the post-2020 time frame), including the cost of wind and PV as well as the cost of energy balancing mentioned above. A large fraction of the cost of electricity balancing relates to periods of several days of overcast and windless weather that occur once every few years. Substantial reductions in electricity cost are possible through contractual load shedding, the occasional use of coal and gas generators to charge the PHES reservoirs, and management of the charging times of batteries in electric cars.

Wholesale movement of transport and low temperature heat to electric vehicles and electric-driven heat pumps, respectively, will add to electricity demand but can sharply reduce greenhouse gas emissions at low net cost. In the longer term, complete electrification of the entire energy system in a developed country results in an approximate tripling of electricity demand after taking account of the greater efficiency of electric devices in most applications (in terms of joules of energy required to deliver an energy service). There is far more than enough availability of solar and wind resources to achieve this outcome, and the overall cost is likely to be little different from the cost of the current fossil fuel–dominated energy system.

Acknowledgements

This work has been supported by the Australian Government through the Australian Renewable Energy Agency (ARENA). Responsibility for the views, information, or advice expressed herein is not accepted by the Australian Government.

References

Andersen, Bjarne, 2006. HVDC transmission – Opportunities and challenges. In *The 8th IEE International Conference on AC and DC Power Transmission, 2006 (ACDC 2006)*, 24–9. London: Institution of Electrical Engineers, 28–31 March.

Blakers, Andrew, Bin Lu, and Matthew Stocks, 2016. 100% renewable electricity in Australia. Unpublished.

Blakers, Andrew, Joachim Luther, and Anna Nadolny, 2012. Asia Pacific super grid – Solar electricity generation, storage and distribution. *GREEN – The International Journal of Sustainable Energy Conversion and Storage* 2(4): 189–202.

Blakers, Andrew, Aihua Wang, Adele Milne, Jianhua Zhao, and Martin Green, 1989. 22.8% efficient silicon solar cell. *Applied Physics Letters* 55: 1363–65. doi.org/10.1063/1.101596

Breyer, Christian, and Alexander Gerlach, 2013. Global overview on grid-parity. *Progress in Photovoltaics: Research and Applications* 21(1): 121–36. doi.org/10.1002/pip.1254

Frankfurt School–UNEP Collaborating Centre, 2014. Global trends in renewable energy investment 2015. Frankfurt: Frankfurt School–UNEP Centre.

Fraunhofer Institute for Solar Energy Systems, 2015. *Current and Future Cost of Photovoltaics: Long-term Scenarios for Market Development, System Prices and LCOE of Utility-Scale PV Systems*. Study on behalf of Agora Energiewende. Berlin: Agora Energiewende.

Global CCS Institute, 2014. *The Global Status of CCS: 2014*. Melbourne: Global Carbon Capture and Storage Institute.

Global Wind Energy Council, 2015. Global annual installed wind capacity 2000–2015. www.gwec.net/wp-content/uploads/2012/06/Global-Annual-Installed-Wind-Capacity-2000-2015.jpg (accessed 22 November 2016).

Hammons, Thomas James, 2008. Integrating renewable energy sources into European grids. *International Journal of Electrical Power & Energy Systems* 30(8): 462–75. doi.org/10.1016/j.ijepes.2008.04.010

Hammons, Thomas James, Victor F. Lescale, Karl Uecker, Marcus Haeusler, Dietmar Retzmann, Konstantin Staschus, and Sébastien Lepy, 2011. State of the art in ultrahigh-voltage transmission. *Proceedings of the IEEE* 100(2): 360–90. doi.org/10.1109/JPROC. 2011.2152310

IPCC (Intergovernmental Panel on Climate Change), 2014. *Climate Change 2014: Synthesis Report. Contribution of Working Groups I, II and III to the Fifth Assessment Report of the Intergovernmental Panel on Climate Change* (Core Writing Team, Rajendra K. Pachauri and Leo A. Meyer, eds). Geneva: IPCC.

IRENA (International Renewable Energy Agency), 2015. Renewable power generation costs in 2014. Abu Dhabi: IRENA. www.irena. org/menu/index.aspx?mnu=Subcat&PriMenuID=36&CatID=141& SubcatID=494 (accessed 21 November 2016).

IRENA (International Renewable Energy Agency), 2016. Renewable capacity statistics 2016. Abu Dhabi: IRENA. www.irena.org/ DocumentDownloads/Publications/IRENA_RE_Capacity_ Statistics_2016.pdf (accessed 21 November 2016).

Kutuzova, N. B., 2011. Ecological benefits of DC power transmission. *Power Technology and Engineering* 45(1): 62–8. doi.org/10.1007/ s10749-011-0225-5

Peake, Owen, 2010. The history of high voltage direct current transmission. *Australian Journal of Multi-disciplinary Engineering* 8(1): 47–55.

Reinders, Angèle, Pierre Verlinden, Wilfried van Sark, and Alexandre Freundlich, eds, 2015. *Photovoltaic Solar Energy: From Fundamentals to Applications.* Chichester, West Sussex: Wiley & Sons.

REN21, 2016. *Renewables 2016: Global Status Report.* Paris: REN21 Secretariat.

Siemens, 2012. Factsheet energy sector. Abu Dhabi: Siemens.

Wikipedia, 2015. Vestas V164. en.wikipedia.org/wiki/Vestas_V164 (accessed 1 April 2015).

12

Lessons of Fukushima: Nine reasons why

Peter Van Ness

Abstract

Following the disaster in Fukushima, we brought together a group of specialists to address the issue of nuclear power in East Asia, and held two international workshops to investigate the topic. This chapter is not a consensus report on our work, but rather a personal statement—one participant's point of view—on our collective deliberations about nuclear power. In my view, there are nine reasons why nuclear power is a bad choice for any country, unless they are already a nuclear weapons power or they aspire to become one. Even then, there are serious problems in a potential commitment. My nine reasons follow the nine-item agenda for the second workshop that we held at The Australian National University in 2014.

Introduction

The Global Nuclear Power Database: World Nuclear Power Reactor Construction, 1951–2017 (Schneider et al. 2017, also published by the *Bulletin of the Atomic Scientists* 2017*)*, analyses the construction to date of the world's 754 nuclear power reactors in 41 countries, of which 90 have been abandoned. This is the most comprehensive global analysis of nuclear power published to date.

The Database provides a global understanding of the history of nuclear power for all countries interested in building new reactors and power plants. Our investigation of the experience in East Asia, especially looking into the vastly different situations in China and Japan, complements the work of the database, and helps to clarify questions that should be asked by any country, such as Australia or the 10 members of the Association of Southeast Asian Nations (ASEAN), when each considers whether it should, or should not, opt for nuclear power.

As of 1 January 2017, the Database reported that 55 reactors around the world were under construction, of which at least 35 were behind schedule. Forty of the 55 reactors under construction were in nuclear weapons states, and 20 of these were in China. Westinghouse had built the most nuclear power reactors (90, and a further 12 had been abandoned), and it currently had four reactors under construction in the US, and four in China. However, in February 2017, Toshiba (which owns Westinghouse) announced that it was taking a US$6.3 billion loss, and that it would not seek new opportunities to export nuclear power reactors. General Electric, France, and Russia are other major builders in different parts of the world.

When we held an international workshop at The Australian National University in August 2014 on the topic 'Nuclear Power in East Asia: The Costs and Benefits', participants from the United States, Japan, Singapore, Taiwan, and Australia attended. The objective of the meeting was to provide an empirical assessment of a commitment to build nuclear power plants, and to get beyond the typical debates between proponents and opponents of nuclear power in which they talk past each other and often make unsupported claims.

We identified nine aspects of a potential nuclear power project, and invited qualified specialists to address those specifics, no matter whether in the past they had publicly opposed or supported nuclear power. The key specifics were the initial cost of construction; the requirements of professional staff to operate and to maintain the nuclear reactors; the establishment of an independent and transparent regulatory authority; liability in the event of accident; the cost and procedure of decommissioning, under normal circumstances and under crisis circumstances (for example, Chernobyl and Fukushima); the relationship of nuclear power generation to nuclear weapons; problems of nuclear waste disposal; the health implications of exposure to radiation; and nuclear power and climate change.

Twenty-five scholars participated in three days of meetings: natural scientists, social scientists, physicists, biologists, and historians. As convener of the workshop, I was delighted by the quality of the insights that emerged and the dispassionate tone of our discussions. By the end of the third day, I was surprised by what I had learned.

1. The initial cost of construction

Building a new nuclear power plant is expensive, and a proper costing of the construction of a nuclear power reactor is a complicated task. Comparisons with the costs of alternative energy producers, like coal, gas, or renewables (see Andrew Blakers' chapter) are difficult to make because of the unique features of nuclear power. For example, should estimates of the cost of decommissioning the plant be included, or the expected cost of the disposal of its high-level nuclear waste? And, if so, how should it be valued in light of the long timescale involved in the latter expenditure?

Concerns about safety have become paramount, especially after the disasters of Chernobyl and Fukushima. But how much safety is enough? Which safety features might be required to deal with possible dangers to the integrity of the plant, and how much might the additional cost add to construction expenditure?

Critics also note the extraordinarily long time horizon involved: the length of time taken to build a nuclear power plant, the duration of its operation, and the lifetime of potentially dangerous radiation emanating from the plant. These issues further complicate the assessment. Cost overruns and failure to complete on schedule have markedly increased the comparative cost of nuclear power. Recall the Database report that of the 55 nuclear power reactors currently under construction around the world, 35 are behind schedule.

Comparisons of the costs of alternative energy sources are often made in terms of the so-called levelised cost of energy (LCOE), an economic procedure that attempts to take into account all of the relevant costs in the production of electricity, but in these calculations one should note what is included and excluded, and which values are used. Government subsidies, as described in Doug Koplow's chapter, and taxes also affect the comparisons.

2. Professional staff needed to operate and maintain the nuclear reactors

Almost never included adequately in a comparative costing of energy sources is the requirement for specially trained staff to build the reactors, to maintain them appropriately, and to assure the public that they can safely deliver the needed electricity. The nuclear physicists and engineers required are personnel with sustained academic and vocational training, which, in turn, means that a country making a commitment to nuclear power will have to make a substantial and continuing investment in its educational institutions to provide the technical specialists that are needed. Failure to continue to train and to support the technical staff required to run a nuclear power plant over the years can lead to serious management problems in the plant's operation. In an emergency, the need for appropriately trained staff is even greater.

3. Establishment of an independent and transparent regulatory authority

Among the problems here, the regulatory authority must have the required technical skills and understanding of the industry necessary in order to carry out its duties. However, given the security concerns about nuclear weapons, it is immensely difficult for the authority to operate transparently.

The problem of independence has also proven to be serious. In Japan, they speak of the 'nuclear village', which Jeff Kingston (2012: 1) has defined as: 'the institutional and individual pro-nuclear advocates who comprise the utilities, nuclear vendors, bureaucracy, Diet (Japan's parliament), financial sector, media and academia'. It became a closed shop of vested interests among the companies involved, the government, the press, and even some of the academic community in promoting the production of nuclear power, and when the Fukushima disaster occurred, those in charge were completely unprepared to respond or even to provide the public with reliable information. Japanese families affected by the disaster did not believe the information that the Tokyo Electric Power Company (TEPCO) or the government provided, and came to distrust what they were told about potential dangers to their health and their safety.

Mely Caballero-Anthony and Julius Cesar I. Trajano, in their chapter, address many of these issues from the perspective of ASEAN. One possibility in Southeast Asia is a regional regulatory authority, as Sulfikar Amir (2014) points out, because '[t]he logic is simple. If only one country in the region has a nuclear power plant, only that country will enjoy the benefits. But the geography of the region dictates that many nations will share the risks'. He notes that since ASEAN members are geographically located so close to one another, a nuclear accident in one country will mean that neighbouring countries may also suffer.

Typically, regulation in the nuclear power industry amounts to the industry regulating itself. Moreover, the accountability of the authority is further compromised by the degree of secrecy and security required because of the relationship between nuclear power production and the possibility of building nuclear weapons. In short, it has proven thus far almost impossible to construct an independent, transparent, and accountable nuclear power regulatory authority.

4. Liability in the event of accident

Liability has been a major concern for private investors in nuclear power because of the immense potential costs in the event of an accident. The final bill for the meltdowns in Chernobyl and Fukushima has not yet been calculated. Governments determined to build nuclear power plants with private capital have had to devise strategies to limit the liability of investors. As one workshop participant commented, '[t]he issue is that private vendors want to sell reactors and be indemnified in the event of an accident'.

In the United States, the Price–Anderson Nuclear Industries Indemnity Act, first passed in 1957, and renewed and extended by Congress through to December 2025, is designed to cap the liability of private investors in nuclear power by means of a combination of insurance held by a company in addition to a pooled amount of US$12 billion to compensate claimants in the event of a nuclear accident. The Act is reported to have worked effectively in the case of the Three Mile Island accident in 1979 in the United States (NAIC and Center for Insurance Policy and Research 2016), but such government insurance arrangements have been seriously tested by the more traumatic nuclear accidents in Chernobyl and Fukushima.

5. Decommissioning costs and procedures under normal circumstances, and under crisis circumstances (for example, Chernobyl and Fukushima)

There is much debate about what should be required for successful decommissioning of a nuclear power plant. Unlike a coal-fueled or petroleum-fueled facility, you cannot simply turn a switch and walk away when the plant is no longer needed. Kalman Robertson analyses in detail the problems involved.

In short, the objective after decommissioning should be to return the plant site to a so-called 'greenfield' condition. Decommissioning is a lengthy and extremely technical procedure for dismantling a nuclear power plant, dealing with existing radiation, and preparing the site for alternative use. Nuclear power plants are typically under strict governmental regulation. Successful decommissioning would result in a condition that would no longer require regulatory oversight. But what happens if before decommissioning is completed, the company involved goes bankrupt, or the government with regulatory responsibilities changes?

Robertson notes that in the next 15 to 20 years, an unprecedentedly large number of reactors will have to be decommissioned as they reach their designed expiration date.

6. The relationship between nuclear power generation and nuclear weapons

Only one of this volume's case-countries, China (see the chapter by M. V. Ramana and Amy King), is a nuclear weapons power, while three others (Japan, South Korea, and Taiwan) have nuclear power plants but no nuclear weapons. For all three, however, the weapons option has been part of their nuclear history. Gloria Kuang-Jung Hsu describes this history in her chapter on Taiwan. Japan is a classic case of proliferation in a country that has nuclear power but no nuclear weapons. In the 1960s, Japan proclaimed its three non-nuclear principles—not to produce, possess, or permit the introduction of nuclear weapons—and later joined the Nuclear Non-Proliferation Treaty. But from its nuclear power industry, Japan has all that would be required to build nuclear weapons, along with enough plutonium for 1,000 nuclear warheads. Some analysts estimate that given a decision to build a nuclear weapons capability, Japan could achieve this within one year—a capacity that has been labelled as a de facto nuclear state, or 'having a bomb in the basement'.

Because of this possibility of employing nuclear physics to produce nuclear weapons, a country that builds nuclear power plants will inevitably require a degree of security that is not needed in the construction of other sources of energy (such as coal, gas, or renewables). In turn, this need for security has contributed to a serious problem of lack of transparency, and explains in part how accountability in the nuclear power industry has been so very difficult to establish and to maintain. Tatsujiro Suzuki, in his chapter, describes the extent of public distrust that developed in Japan, illustrated sharply in the Fukushima disaster.

From a broader national security perspective, the establishment of a nuclear power plant provides adversaries with a strategic opportunity to create a nuclear crisis in the country. The building of a nuclear power plant gives a country's adversaries a potential target for attack, either with an explosive device from a terrorist or an enemy country, or a cyber intervention to distort the management of a nuclear reactor in order to cause a nuclear meltdown. Such a cyber-attack, for example, might be designed to disguise the source of the attack, but nonetheless cause immense damage and disruption in the country, especially in heavily populated areas. Both Lauren Richardson, in her chapter on South Korea, and Tilman Ruff, in his analysis of health implications of radiation exposure, draw attention to the possibility of such attacks.

Another aspect of the relationship between nuclear power and nuclear weapons emerged in the British debate about whether or not to go ahead with construction of the Hinkley Point C nuclear power plant, potentially the biggest building project underway in the world. Following years of discussion and debate, British Prime Minister Theresa May confirmed in September 2016 that the UK would build the first new nuclear power plant in Britain in 20 years at Hinkley Point. Hinkley Point C nuclear power plant is to be built by Électricité de France S.A. (EDF), with Chinese investment and 33 per cent ownership of the proposed US$30 billion project, based on the European pressurised reactor (EPR) design. The UK has promised to pay twice the current market price for a unit of electricity produced by the project for 35 years. A key consideration for the Chinese was a promise that it might in the future build a second nuclear power plant, employing its own reactor design, in Essex.

The decision by May to go ahead with the Hinkley Point C project baffled critics because it made no sense from an economic, technological, environmental, or even national security point of view. The French company in charge, EDF, was in serious financial difficulties; the EPR nuclear technology was untested; the promised purchase price for the electricity to be produced was too high; the renewables alternative would be much better for the environment; and there was a national security concern about the deep Chinese investment involved, i.e. potentially having a major influence on the British national power grid.

Why did May decide to go ahead? A University of Sussex study tested a variety of possible explanations, and concluded that, despite the serious reservations, the British prime minister approved the project principally to sustain Britain's nuclear deterrent capability by keeping the nuclear power industry in the UK up to such a level as to support the construction of new nuclear-powered ballistic missile submarines intended to replace the existing Vanguard-class submarines (Cox, Johnstone, and Stirling 2016).

Although critics have raised concerns about Chinese investment in Hinkley Point C because of its potential influence over the UK power grid, a more serious national security concern is that China's participation in the development of the British nuclear power industry, which will be focused on producing the UK's new nuclear-powered ballistic missile submarines, could compromise Britain's Trident nuclear deterrent.

7. Nuclear waste disposal

One of the most serious problems confronting the production of nuclear power is how to dispose safely of high-level radioactive waste (HLW). As Ramana (2017: 415) has argued:

> Some of the radioactive elements produced during the operation of nuclear reactors have extremely long half-lives, and have to be isolated from human contact for hundreds of thousands of years … This requirement for stewardship is unprecedented in human history.

Currently, there is no operational site for the permanent storage of high-level nuclear waste anywhere in the world. Deep geological disposal has been proposed as the answer to this problem, and sites have been suggested in different countries, including Australia. Meanwhile, all countries that have working nuclear power plants have made temporary arrangements to store nuclear waste, some with tragic consequences, as outlined in Hsu's description of the situation on Orchid Island in Taiwan.

Yucca Mountain in the United States, a remote Nevada desert site in which the government has invested US$15 billion, was officially designated as a site in 2002, but its current status is uncertain, and there was a serious accident in 2014 at the Waste Isolation Pilot Plant (WIPP) in New Mexico (Alvarez 2014). Globally, the most promising project is a US$3.2 billion facility on Olkiluoto island in Finland (Gibney 2015), which is not yet operational.

8. Health implications of exposure to radiation

The chapters by Tim Mousseau and Anders Pape Møller, and Ruff, on the biological and human health implications of exposure to radiation, were written by some of our most experienced researchers. Mousseau had worked for 10 years on Chernobyl before beginning research about Fukushima after the disaster, and Ruff, a public health physician, has combined a lifetime of research, teaching, and activism focused on investigating the public health dimensions of nuclear technology. In my view, there are no better statements of the serious risk to human health and ecological viability involved in a decision to build a nuclear power plant, and their work should provide warnings to countries considering whether or not to invest in nuclear power.

9. Nuclear power and climate change

Christina Stuart, in her chapter on the exceptional situation in France where more than 70 per cent of the country's electricity is provided by nuclear power, notes that when the United Nations Framework Convention on Climate Change Conference of the Parties (COP) was held in Paris in December 2015 (COP21), and universal agreement on climate change among the 195 participating countries was achieved, a role for nuclear power was not directly addressed. Having observed the meetings, she reports that, surprisingly, nuclear was only indirectly discussed.

Yet some prominent climate scientists, like James Hansen et al. (2015), have made forceful arguments for nuclear power, and the Nuclear Energy Institute insists:

> There is widespread agreement that nuclear energy is part of the climate change solution. Mainstream analyses conducted by independent organizations have shown that reducing carbon emissions will require a diverse energy portfolio and that nuclear energy is the only low-carbon option to help meet forecasted global electricity demand (Nuclear Energy Institute n.d.).

Compared with the production of carbon dioxide by fossil fuels, there is no question that nuclear power is vastly superior with respect to concerns about climate change. The principal argument made by proponents is that nuclear power is the only low-carbon option that can provide so-called 'baseload' electric power. Blakers' chapter describes the current capacity of renewables to meet this demand, and Mark Diesendorf (2016) argues that not only are baseload power stations not needed, but also that renewable storage, such as the newly designed batteries and pumped hydro, as well as other technological innovations, can provide the flexibility needed to produce reliable electric power.

The victims of Chernobyl and Fukushima might add that the risks associated with nuclear power are simply too great to justify the nuclear option when other energy resources are available to respond to climate change concerns and to meet the requirements for the production of reliable electric power.

Conclusion

The lessons of Fukushima, learned in our years of investigation since the disaster in Tohoku in March 2011, are based first and foremost on the conclusion that the fundamental costs of nuclear power are incalculable. They cannot at present be measured either in financial terms or in terms of compromised public health and lives lost. One reason is that estimates of the cost of a truly successful decommissioning of a nuclear power reactor still vary widely, while the expected cost of decommissioning of reactors in crisis, such as at Chernobyl and Fukushima Daiichi, continue to escalate.

The problem of estimating the cost of processing high-level nuclear waste is even more difficult because, at present, there is no available permanent site for HLW anywhere in the world. And, in terms of public health, Ruff has spelled out in meticulous detail the implications of exposure to radiation, and Mousseau and Møller have demonstrated the adverse effects of radiation on the broader ecology.

As we studied each of the nine aspects listed above, questions were raised that could not be adequately answered by the proponents of nuclear power. However, if a country is already a nuclear weapons power, like China or the United Kingdom, or aspires to become one, then the calculations change fundamentally. It is not, then, a matter of finding the best way to produce reliable electricity for the cheapest price, but rather involves questions of national security. Countries that decide to build nuclear weapons must maintain a nuclear industry, no matter what the cost. For them, the problems raised in our study become secondary. They must decide what price they are prepared to pay for 'national security'.

References

Alvarez, Robert, 2014. The WIPP problem, and what it means for defense nuclear waste disposal. *Bulletin of the Atomic Scientists,* 23 March. thebulletin.org/wipp-problem-and-what-it-means-defense-nuclear-waste-disposal7002 (accessed 14 March 2017).

Amir, Sulfikar, 2014. The transnational dimensions of nuclear risk. *Bulletin of the Atomic Scientists*, 25 April. thebulletin.org/needed-ability-manage-nuclear-power/transnational-dimensions-nuclear-risk (accessed 14 March 2017).

Bulletin of the Atomic Scientists, 2017. Global nuclear power database: World nuclear power reactor construction, 1951–2017. thebulletin.org/global-nuclear-power-database (accessed 14 March 2017).

Cox, Emily, Phil Johnstone, and Andy Stirling, 2016. Understanding the intensity of UK policy commitments to nuclear power. Science Policy Research Unit Working Paper Series SWP 2016-l6. Brighton: University of Sussex, September.

Diesendorf, Mark, 2016. Dispelling the nuclear 'baseload' myth: Nothing renewables can't do better! *Ecologist,* 18 March. reneweconomy.com. au/dispelling-the-nuclear-baseload-myth-nothing-renewables-cant-do-better-94486/ (accessed 14 March 2017).

Gibney, Elizabeth, 2015. Why Finland now leads the world in nuclear waste storage. *Nature*, 2 December. doi.org/10.1038/nature.2015.18903

Hansen, James, Kerry Emanuel, Ken Caldeira, and Tom Wigley, 2015. Nuclear power paves the only viable path forward on climate change. *Guardian*, 4 December.

Kingston, Jeff, 2012. Japan's nuclear village. *Asia-Pacific Journal: Japan Focus* 10(37)(1) 9 September: 1–22.

NAIC (National Association of Insurance Commissioners) and Center for Insurance Policy and Research, 2016. Nuclear Liability Insurance (Price–Anderson Act). 8 December. www.naic.org/cipr_topics/topic_nuclear_liability_insurance.htm (accessed 14 March 2017).

Nuclear Energy Institute, n.d. Climate change. www.nei.org/Why-Nuclear-Energy/Clean-Air-Energy/Climate-Change (accessed 14 March 2017).

Ramana, M. V., 2017. An enduring problem: Radioactive waste from nuclear energy. *Proceedings of the IEEE* 105(3): 415–18. doi. org/10.1109/JPROC.2017.2661518

Schneider, Mycle, and Antony Froggatt, with Julie Hazemann, Tadahiro Katsuta, M. V. Ramana, Juan C. Rodriguez, and Andreas Rüdinger, 2017. *The World Nuclear Industry Status Report 2017*. Paris: Mycle Schneider Consulting Project.

www.ingramcontent.com/pod-product-compliance
Lightning Source LLC
Chambersburg PA
CBHW050806270326
41926CB00026B/4576